# ATLAS OF PACIFIC SALMON

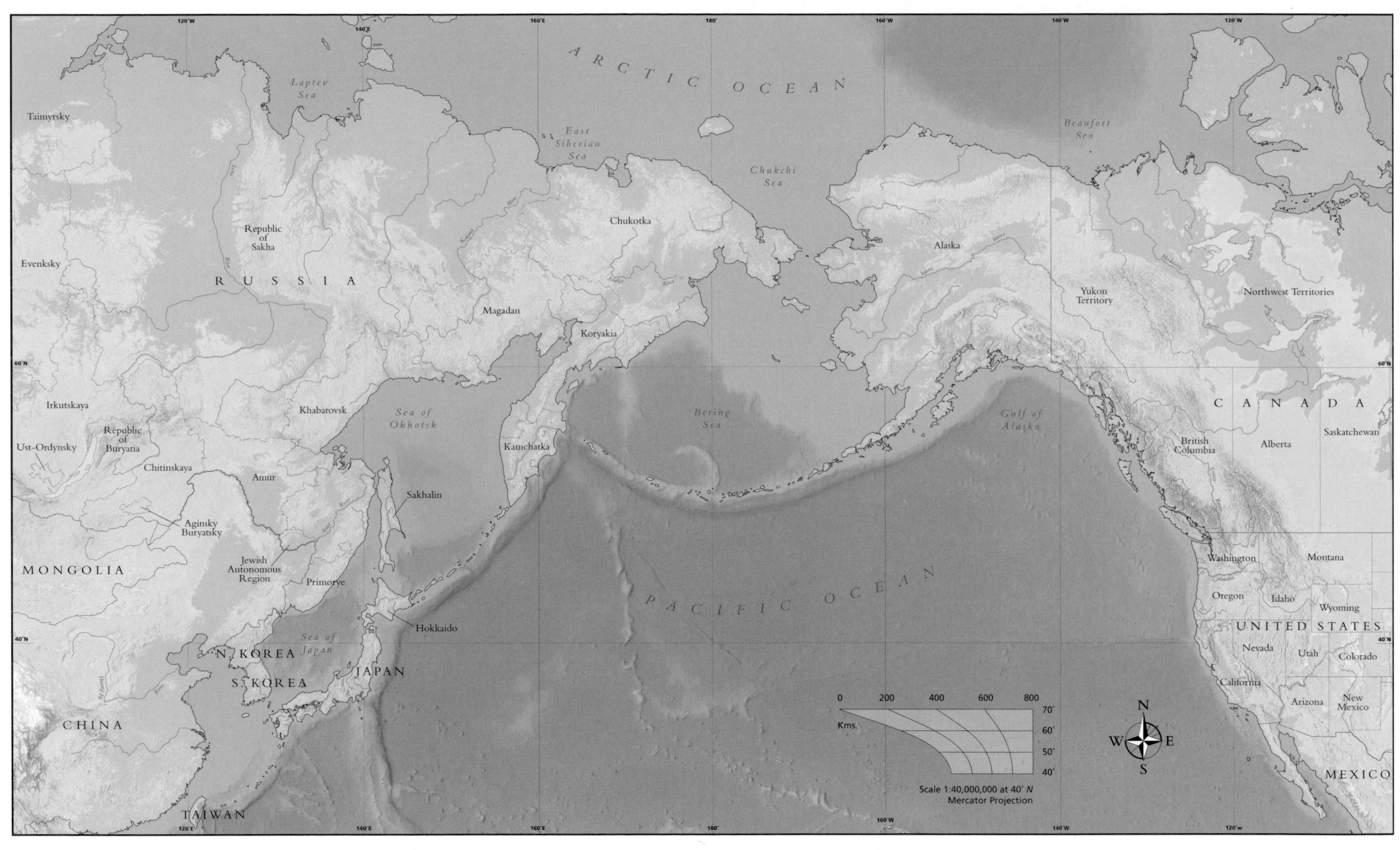


THE NORTH PACIFIC

# ATLAS OF PACIFIC SALMON

## THE FIRST MAP-BASED STATUS ASSESSMENT OF SALMON IN THE NORTH PACIFIC

Xanthippe Augerot
with
Dana Nadel Foley

CARTOGRAPHY
Charles Steinback

DESIGN
Andrew Fuller

WITH PHOTOGRAPHY BY
Natalie Fobes

SALMON ILLUSTRATIONS BY
Kate Spencer

University of California Press
BERKELEY • LOS ANGELES • LONDON

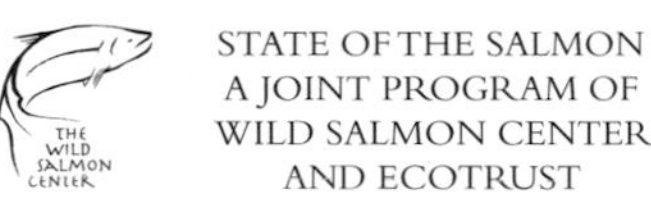


STATE OF THE SALMON
A JOINT PROGRAM OF
WILD SALMON CENTER
AND ECOTRUST

STATE OF THE SALMON
*Knowledge across borders.*
721 NW Ninth Avenue, Suite 280
Portland, Oregon, USA 97209
info@stateofthesalmon.org
www.StateoftheSalmon.org

University of California Press
Berkeley and Los Angeles, California

University of California Press, Ltd.
London, England

Copublished by University of California Press, and State of the Salmon—a joint program of Wild Salmon Center and Ecotrust.



Cataloging-in-Publication Data is on file with the Library of Congress
ISBN 0-520-24504-0



*Atlas of Pacific Salmon* was published with the assistance of grants from the Carolyn Foundation, Charles Engelhard Foundation, Gordon and Betty Moore Foundation, Lazar Foundation, The David and Lucile Packard Foundation, Rockefeller Brothers Fund, Trust for Mutual Understanding, United States Environmental Protection Agency, University of California Press, Weeden Foundation, and WEM Foundation. Research for *Atlas of Pacific Salmon* was supported in part by Oregon Sea Grant, the National Sea Grant College Program of the U.S. Department of Commerce's National Oceanic and Atmospheric Administration under NOAA grant number NA36RG0451, and also by appropriations made by the Oregon State legislature.

The inside front and back covers depict detail of the major stream networks (see page vi) within the salmon ecoregions (see pages 6–7): front endpaper, the Russian Far East from Chukotka to Primorye and Sakhalin Island; back endpaper, North America from Alaska to Washington.

Printed by Dynagraphics, Portland, Oregon, USA
13 12 11 10 09 08 07 06 05
10 9 8 7 6 5 4 3 2 1

PHOTOS AND DIAGRAMS
Natalie Fobes: all photos, except those listed below.
Andrew Fuller: all charts and diagrams.
Kate Spencer: all illustrations, except those listed below.
Spencer B. Beebe 49 (Owyhee R.); Steve Blackburn 31 (gear illustration); Gary Braasch 110 (geophysicist); Igor Chereshnev 23 (Markovo tent); Brandon D. Cole/Corbis 106 (oil platform); Dave G. Houser/Corbis 107 (oil tanker); Barrie Kovish 2 (kokanee), 114 (Kennedy R. sockeye), 64 (B.C. chinook); Kirill Kuzishchin 13 (anadromous *mykiss*); Gunter Marx/Corbis 60 (Chilliwack R.); NASA 48 (Kamchatka volcano); National Library of Australia 8 (Ortelius map); Pat O'Hara/Corbis 56 (Seven Lakes Basin); Oregon Historical Society 30 (OrHi23141_BumbleBee_Cannery); John Francis Pratt/University of Washington Digital Collections 23 (Chilkat fish trap); Guido Rahr 44 (Kol R.), 46 & cover (Opala R.), 106 (gas development); Jim Richardson 14 (drying salmon), 112 (fisherman cradling catch); Dave Sinson/NOAA 49 (Icy Bay); National Archives of Canada 18 (Inuit spearfishing); Mikhail Skopets 13 (resident *mykiss*), 99 (poaching), 102 (Etergen Lake); USDA Forest Service 52 (Hubbard Glacier); Douglas P. Wilson/Corbis 58 (copepods).

GIS ANALYSIS AND CARTOGRAPHY
Charles Steinback

SECONDARY MAP PRODUCTION; MAP, BOOK, AND COVER DESIGN
Andrew Fuller

**NEW LEAF PAPER**
ENVIRONMENTAL BENEFITS STATEMENT

This project is printed on New Leaf Reincarnation Matte, made with 100% recycled fiber, 50% post-consumer waste, processed chlorine free. By using this environmentally friendly paper, *Atlas of Pacific Salmon* saved the following resources:

| trees | water | energy | solid waste | greenhouse gases |
|---|---|---|---|---|
| 108 fully grown | 23,543 gallons | 49 million BTUs | 5,150 pounds | 8,703 pounds |

Calculated based on research done by Environmental Defense and other members of the Paper Task Force.

© New Leaf Paper Visit us in cyberspace at www.newleafpaper.com or call 1-888-989-5323

# C O N T E N T S



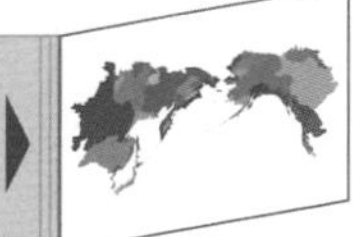

Ecoregion foldout map in back

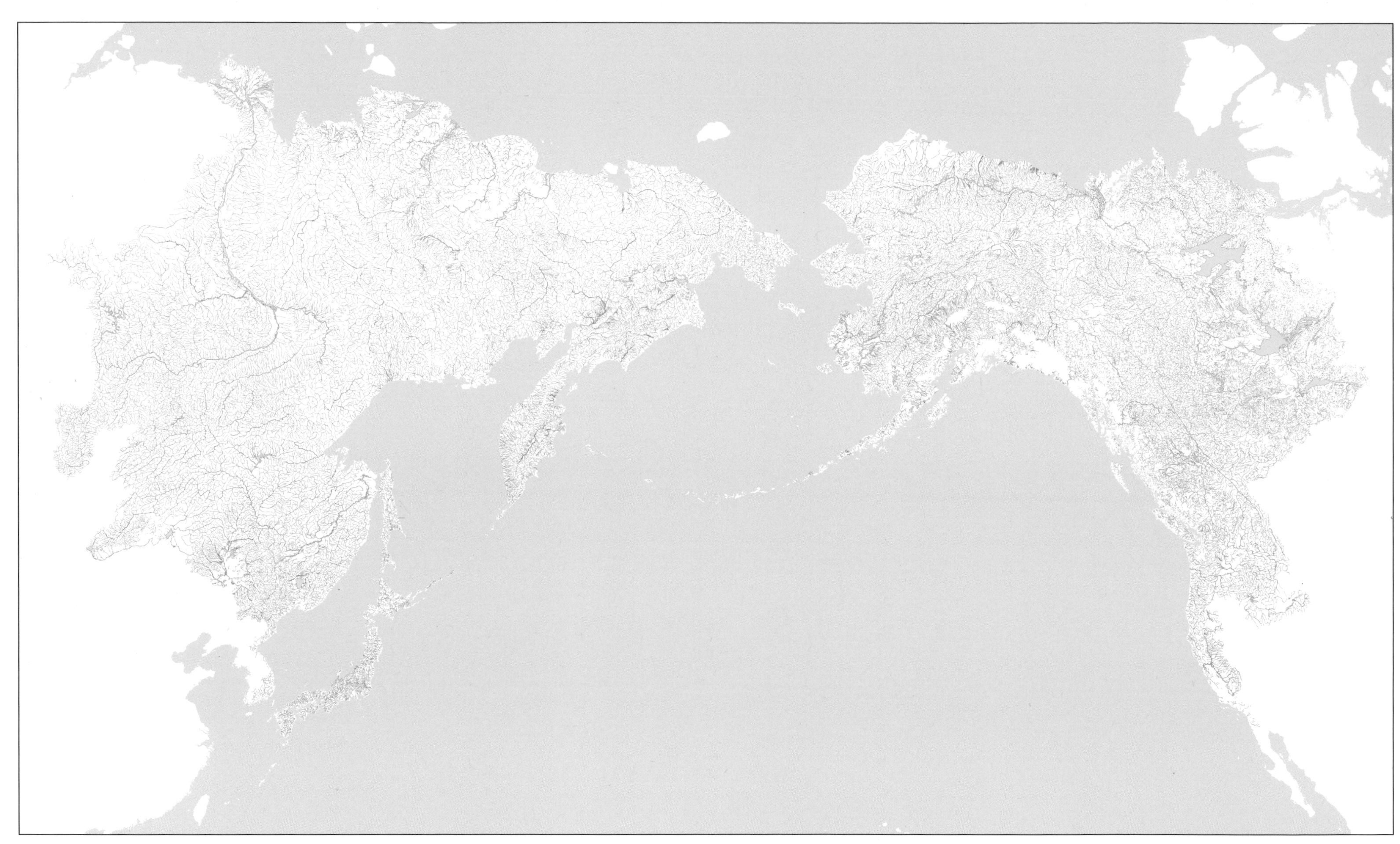

MAJOR STREAM NETWORKS
WITHIN THE SALMON ECOREGION

*This book is dedicated to the impassioned and embattled salmon biologists, commercial and recreational fishers, and native peoples across the North Pacific who, each in their own way, are fighting to protect wild salmon and the way of life they represent.*

# Forewords

**The crash of the wild Atlantic salmon** fishery took place within my lifetime; from a management perspective, it happened in the blink of an eye. Between 1970 and the end of the 20th century, wild Atlantic salmon catch fell by 80 percent across the North Atlantic. Today, more than half of Atlantic stocks are either at risk or extinct, and the declines continue, unabated.

Pacific salmon are swimming in the same direction. In this atlas, we demonstrate that 23 percent of Pacific salmon populations we assessed are at risk. Although this may be just half the risk demonstrated in Atlantic salmon stocks, we know how quickly these numbers can change. The window of recovery is quite narrow: once a population is at moderate or high risk, declines may be difficult or impossible to reverse. As we work to save stocks at risk, we must get ahead of the "extinction curve" to protect remaining healthy stocks and their ecosystems *before* they suffer declines.

The first step in doing so is establishing an information baseline, a common yardstick with which to measure salmon status and trends across the North Pacific, and we've done that in the pages that follow. Our findings are the first assessment of distribution and risk of extinction for Pacific salmon at one consistent scale across their entire range. They demonstrate the first clear evidence of a north-south cline in salmon risk of extinction, not just regionally but across the North Pacific. This is critical information, and what we do with it may determine whether wild Pacific salmon thrive or go the way of their Atlantic counterparts.

As we confront salmon declines that are marching up both sides of the North Pacific, in our last chapter we propose a North Pacific Ecosystem Approach to salmon conservation.

First, we need to establish baseline datasets against which to measure status and trends. To do so, we need to support a comprehensive strategy to complete river-by-river inventories of salmon stocks and work with government agencies to implement monitoring activities.

Second, we need to make a fundamental shift in fisheries management, from managing for biomass or yield to managing for biodiversity and ecosystem health.

Third, we need to identify and protect the most productive and species-rich salmon ecosystems in watersheds throughout the North Pacific while supporting the restoration of the most at-risk populations.

Fourth, we need to reform and strengthen the human institutions—from citizen watershed councils to international commissions and development agencies—charged with wild salmon management.

We do not have to look beyond the North Atlantic to see what will happen if we fail to act decisively. Let us build upon our thousand-year history with salmon, learn from our mistakes and successes, and achieve a better coexistence with wild salmon and the dozens of species they support—while we still have the chance.

Guido R. Rahr
President, CEO
Wild Salmon Center

Although this book has been a decade in the making, we consider its release a beginning. We're introducing a new ecosystem-wide perspective on Pacific salmon that can change the way we study, interact with, and celebrate these keystone species.

In the pages that follow, our colleagues present the current state of Pacific salmon. It's a picture that we are able to put together thanks to the descriptive capabilities of geographic information systems (GIS), the leadership of Xanthippe Augerot, and the persistence of the State of the Salmon team, codirected by Xan and Edward Backus, and generously supported by the Gordon and Betty Moore Foundation. This is a picture we need to share, study, and talk about—and repaint often, as our knowledge develops and our partnerships expand.

Our aim in this book is to enrich our understanding of these miraculous migratory fish, discuss the complexities of natural processes that affect salmon, and address the implications of the human-made threats heaped upon them by the industrial economy during the past 150 years.

But overall, our message is not just about the past 150 years, and it's not just about salmon. It's about places salmon use in the North Pacific and the people who share salmon country. It's about a thousand years of data, science, and policy that, however well-intentioned, have failed us. It's about the ghosts of cod and Atlantic salmon off Georges Bank and in the rivers of Europe, and the lessons that continue to elude us.

The salmon story is our story. We depend on salmon for food, jobs, spiritual renewal, recreation, and an indication of the condition of our rivers and oceans. Until we can reach a shared understanding of the role of salmon in our lives, unless we clarify the lines connecting salmon and ourselves, prospects for recovery for these irreplaceable species are poor.

Somehow salmon have endured in spite of our attempts to manage them. But as numbers and diversity within salmon populations diminish, we wonder how long the natural resilience of this remarkable genus can hold on. The problem—how do we reshape our human institutions to allow salmon populations to recover to healthy levels?—is far greater than the sum of our solutions. Now is the time to step back, assess, question, collaborate, and take in the big picture that is the North Pacific ecosystem.

Through full public disclosure on the state of Pacific salmon, we hope to galvanize broader interest, shape public attitudes, create a learning network, improve policies, and bridge gaps between scientists and fisheries managers. Through sweeping change, perhaps we can help salmon save themselves.

Spencer B. Beebe
President
Ecotrust

# Overview

**Even while the total catch of Pacific** salmon has increased within recent decades, many wild salmon populations have suffered precipitous declines, both in number and in biological diversity, particularly at the southern edges of their North Pacific range. The problem is not local or regional or national; the problem is pan-Pacific, affecting salmon populations from South Korea to Southern California. The waters salmon use—from steep, cold snowmelt streams to tributaries, through estuaries to the ocean, and back again—are continuous ecosystems that reach across jurisdictional boundaries.

In this atlas, we present the first map-based status assessment of seven species of Pacific salmon at one consistent scale, under one authorship. From this starting point, we now have a benchmark from which to study trends across time at an ecosystem-wide North Pacific scale that we may have missed at a finer resolution.

## SALMON ECOREGIONS (PAGE 6)

This atlas arose out of a need for a broad North Pacific perspective on the status of salmon. Before we could assess status, however, we first needed access to comparable datasets—but they did not exist. We had no common yardstick with which to measure changes in salmon abundance and biodiversity throughout the range of Pacific salmon. The seven Pacific Rim nations—the United States, Canada, Russia, Japan, China, and North and South Korea—enjoy vastly different relationships with salmon, economically, ecologically, and culturally. Data availability, resolution, and type were as varied as the scientists who collect them.

Therefore, to launch our study, we needed to create a spatial unit that would allow us to aggregate and measure data consistently across the North Pacific. We parsed the Pacific Rim into a series of catchment or basin ecosystems that salmon use, from rivers to coastal areas, to semi-enclosed seas, to straits, upwelling areas, and more. We called these territories (or units) "salmon ecoregions" and identified 66 discrete units along the Pacific and Arctic oceans.

## DATA (PAGE 64)

To create a full picture of Pacific salmon within their ecological, economic, and cultural context, we analyzed distribution and risk of extinction data for seven anadromous *Oncorhynchus* species: chinook (*O. tshawytscha)*, chum (*O. keta)*, coho (*O. kisutch)*, masu (*O. masou)*, pink (*O. gorbuscha)*, sockeye (*O. nerka)*, and steelhead (*O. mykiss)*.

Our distribution and risk maps make an important contribution by presenting for the first time georeferenced data for the western Pacific at a consistent measurable scale. To achieve this, we used best expert judgment data from field biologists in Russia, Japan, China, and South Korea. To evaluate eastern Pacific distribution, we relied primarily on datasets from provincial, state, and federal fisheries agencies, as well as previously published literature and best expert judgment.

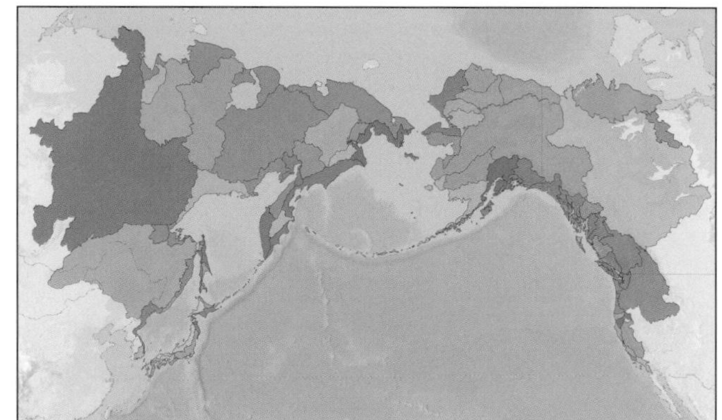

SALMON ECOREGIONS Using a consistent scale, we aggregated data within ecological, watershed-based units that classify salmon habitat.

Our risk of extinction maps were also built using best expert judgment from biologists in Russia and North America. In addition, we depended on previously published literature: for Washington, Oregon, California, Idaho (WOCI) (Nehlsen et al. 1991; Huntington et al. 1996), for British Columbia (Slaney et al. 1996), and for southeast Alaska (Baker et al. 1996).

For the rest of Alaska, we were unable to identify state agency resources conducting similar status assessments. We did not attempt to collect data for Japan because its wild populations have not yet been adequately assessed. Lastly, we omitted Arctic ecoregions due to lack of data.

We had sufficient information to assess the status of 7,519 discrete stocks—at most an estimated 10 percent of *Oncorhynchus spp.*—from 41 of the 52 Pacific Ocean ecoregions.

### DISTRIBUTION FINDINGS

The natural range of Pacific salmon sweeps around the coastline of the North Pacific Rim and reaches inland to headwaters. We found that:

- the distribution of Pacific salmon is shrinking at the southern edge of its range;
- the northern edge of the range is represented primarily by pink, chum, and chinook;

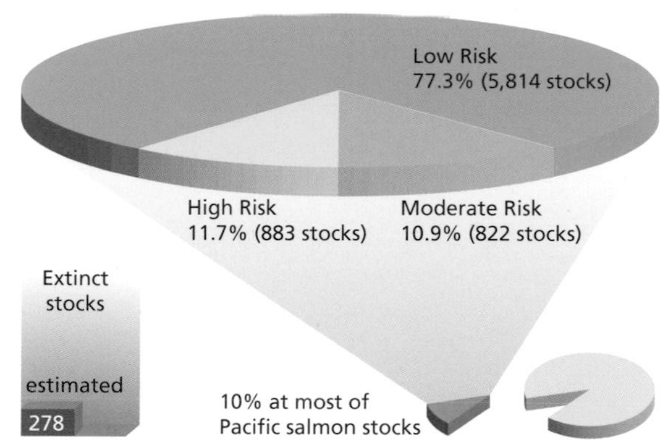


RISK OF EXTINCTION Representing at most an estimated 10 percent of extant North Pacific salmon stocks, our assessments revealed that 23 percent are at risk. We also identify an estimated 278 extinct stocks.

- chum exhibit the most extensive distribution;
- masu exhibit the least extensive distribution.

## RISK OF EXTINCTION FINDINGS

Pacific salmon have survived natural threats for millennia, but human-made threats present new and ever-increasing challenges. We found that:

- 23 percent of all salmon stocks identified are at moderate or high risk
- by species, the following salmon stocks are at moderate or high risk:

  7 percent of pink
  24 percent of masu
  29 percent of chum
  30 percent of coho and sockeye
  36 percent of chinook
  39 percent of steelhead

- ecoregions in WOCI have the highest concentrations of high-risk stocks (risk not assessed in Japan and parts of Alaska)
- Sea of Okhotsk and western Bering Sea ecoregions have a high proportion of moderate-risk stocks
- the western Pacific has more ecoregions classified entirely in the low-risk category than does the eastern Pacific

## THREATS (PAGE 97)

Anthropogenic threats to Pacific salmon include natural resources extraction, increased population pressures, and industrial influences. Generally,

- as human population density and land-use pressures increase at the southern edge of the distributional range, so does risk of extinction;
- dams, water diversions, and river channelization are most intense at the southern extent of the range, where many wild salmon populations have been extirpated;
- natural resources extraction often occurs in remote locations where regulatory oversight is less rigorous;
- poaching poses the biggest threat to salmon in the Russian Far East;
- global warming may significantly alter salmon populations through major ecological and hydrological changes in salmon habitat.

## RECOMMENDATIONS (PAGE 114)

We make four recommendations toward a North Pacific Ecosystem Approach for salmon conservation. First, we urge the creation and implementation of an international monitoring strategy that will give us an early warning system to detect risk and an analytical framework in which scientists may assess viability. Second, we must use these data to improve our ecological understanding of salmon and their habitats and inform management decisions. Third, we propose two conservation-based initiatives to be carried out at the basin scale: identifying and protecting the most species-rich ecosystems, and restoring at-risk populations. Finally, we encourage the reform of human institutions and the creation of partnerships that will enrich our understanding of the continuous North Pacific ecosystem that we all share.

Through these efforts across jurisdictions, cultures, professions, interests, and economies, we can create a context in which wild salmon populations thrive. Pacific salmon have triumphed over natural threats for millennia. More recently, anthropogenic threats present increased complexities for salmon fighting for survival. But if we can advance policy and understanding so that we respect biological diversity and natural resiliency, we can create opportunities for wild salmon to save themselves.

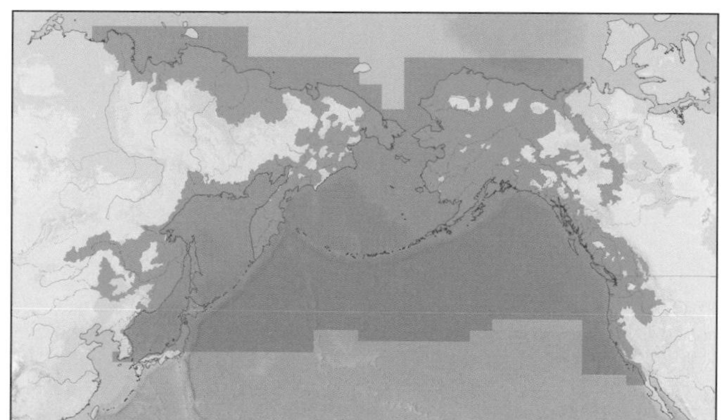

ORIGINAL *ONCORHYNCHUS* DISTRIBUTION The current range of Pacific salmon is shrinking in a human-changed environment.

# ATLAS OF PACIFIC SALMON

# 1

# The Fish

*Biodiversity and life history strategies in seven species of North Pacific salmon*

■ Salmon Ecoregions ■

■ Original Distribution of Genus *Oncorhynchus* ■

■ Salmon Diversity ■

■ Kamchatka Rainbow-Steelhead Life Histories ■

Sockeye spawners *(Oncorhynchus nerka)* find nesting gravels in an Alaskan river. The heads of the two males have deformed as they approach their final days.

**Pacific salmon (*Oncorhynchus spp.*) are** relatively new to this 4.5-billion-year-old planet. Their oldest vertebrate ancestors appeared 500 million years ago; these evolved into teleosts, bony fishes, 400 million years ago, which eventually gave rise to the family Salmonidae, the progenitor of modern salmon and trout, whitefish, and grayling, some 100 million years ago. Ancestral members of the genus *Oncorhynchus* emerged a mere 18 million years ago and continued to advance until around six million years ago, when the species took the distinct forms that we recognize today.

A coho salmon leaps up Washington's Soleduc River, in an ecoregion where more than half the coho populations are at risk of extinction.

However recently they evolved, salmon are ancient in the context of human presence on the earth. The earliest known salmonid fossil, *Eosalmo driftwoodensis,* a member of the subfamily Salmoninae, is believed to be 50 million years old; and although we date the oldest *Homo sapiens* fossil at around 100,000 years, we have no record of human presence in what is now North America prior to the last glacial maximum some 18,000 years ago. Once humans arrived in North America and until the mid-19TH century, North Pacific people and salmon formed an intimate relationship based on interdependence, as indigenous cultures relied on salmon for subsistence and spiritual and ecological renewal.

Yet within the past 150 years, the industrial economy has threatened the viability of the 18-million-year history of *Oncorhynchus* and the integrity of the relationship between North Pacific salmon and people. Today many salmon populations are declining in abundance and biological diversity. Given the importance of the resource within North Pacific nations, economic repercussions are grave. Anthropogenic threats, such as overfishing and natural resource extraction, have upset the balance. In Japan around 95 percent of the salmon catch is hatchery derived, and scientists are concerned about diminished biological diversity. In Russia officials estimate that poaching of salmon and other natural resources siphons the equivalent of US$1.5 billion from the economy each year. In 2002, for the fifth time in six years, Alaska's governor declared a state of economic disaster in the state's western region as catch prices

Landlocked sockeye, or kokanee, migrate homeward in the Russian River, one of the major rivers feeding agricultural interests in California.

dropped by as much as 80 percent, largely due to diminished returns as well as global increase in the supply of farmed salmon. In British Columbia and Washington, Oregon, California, and Idaho (WOCI), logging, overfishing, and mining have affected salmon viability: for example, in the 19TH century Columbia River salmon runs were estimated to be 10–16 million in a given year; today escapements are as low as 200,000.

What has happened? How severe are the losses? What can we do now? What can we expect in the future? In order to answer these questions, we must first establish a baseline status assessment of Pacific salmon.

## THE NORTH PACIFIC ECOSYSTEM

Pacific salmon exist on a grand scale. Migratory and anadromous, they use the rivers within and the oceans between seven countries that border the North Pacific Ocean. Inland, the domain of Pacific salmon stretches from the Tachia River in Taiwan to Albazino Station up along the Trans-

Siberian Railway and northward to the Chukchi Sea; and across western Alaska, down through the Yukon Territory and British Columbia, and well into WOCI, and as far south as Rio Santo Domingo in Baja California. Marine distribution reaches across the vast expanse of the Pacific Ocean between Japan and California, up into the Bering Sea—one of the most productive natural systems in the world—and into the Arctic waters of the Laptev, East Siberian, Chukchi, and Beaufort seas.

Data are as diverse as the scientists who collect them. What, when, and how we measure differs from country to country, basin to basin, river to river, and even tributary to tributary. Most of the data we have gathered were collected to support commercial harvest management rather than to assess biological diversity. We have only several decades of historic abundance data. And we are just beginning to confirm losses in biological diversity through studies of genetic and life history diversity.

This atlas represents the first compilation of Pacific salmon data aggregated at the same scale throughout the distributional range of *Oncorhynchus*. Using observational and best expert judgment data, we present salmon status and risk of extinction for seven species of North Pacific salmon in order to build the knowledge base that will help us answer the questions we posed above: In short, how do we stem losses in Pacific salmon biodiversity and abundance?

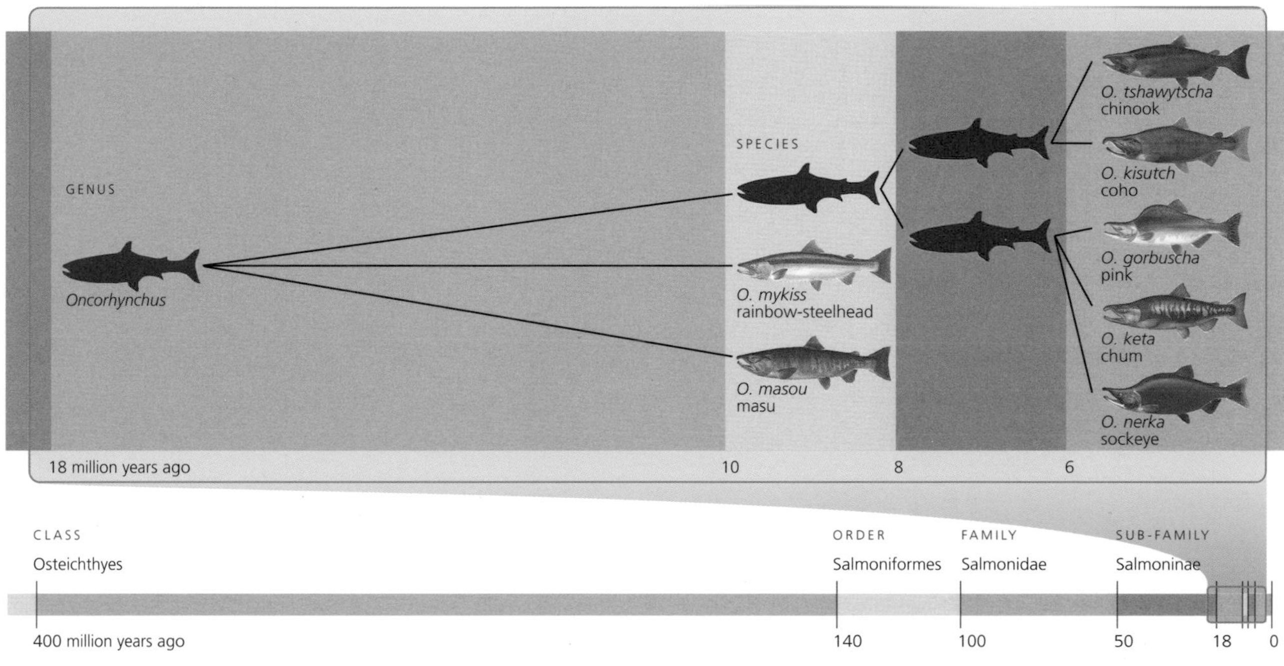


***ONCORHYNCHUS* FAMILY TREE** We study the anadromous forms of the seven commercially important salmonids that yield the most data. Atlantic salmon fall within the *Salmo* genus in the subfamily Salmoninae, where Russian scientists continue to classify the rainbow-steelhead complex.

Our subjects are siblings within the genus *Oncorhynchus*: chum (*O. keta*), pink (*O. gorbuscha*), sockeye (*O. nerka*), chinook (*O. tshawytscha*), coho (*O. kisutch*), and masu (*O. masou*). We focused on these species because they have been the most frequently studied, due to their commercial importance, and therefore offer the most comprehensive data. We also study rainbow-steelhead trout (*O. mykiss*), which were commercially fished until the early 20TH century. Of these *Oncorhynchus*, we study only the anadromous forms. Other members of the *Oncorhynchus* genus include Mexican golden trout (*O. chrysogaster*), gila (*O. gilae*), apache trout (*O. apache*), and cutthroat trout (*O. clarki*), but we did

## HOW WE TALK ABOUT SALMON: STOCK V. POPULATION

"How do we identify interbreeding individuals within a salmon species?" is a fundamental question in fisheries management and biology. In 1954 Dr. William E. Ricker of the Fisheries Research Board of Canada published the seminal paper "Stock and Recruitment," which provided fisheries science with a new vernacular for describing individual families of fish. The term "Ricker's stock" is now the universally accepted definition of a discrete group within a salmon species that spawns in a particular river and season and does not interbreed with other spawning groups.

In context, the term was used as a management unit for building biomass, not as a biological unit constituting genetic individuation, and we are sensitive to this connotation. Therefore we use the term "population" to describe interbreeding salmon from a biological perspective and "stock" to describe interbreeding salmon in terms of catch management. ■

As these pink salmon prepare for spawning, a territorial male hovers near a female as she digs her redd, in Washington's Skagit River.

not include them in our study because data are insufficient for proper analysis.

Each species has its own list of superlatives. Chum are a staple for the Japanese, who produce billions of fry in hatcheries each year. Pink, the most abundant across the North Pacific, largely adhere to a strict two-year life cycle. Sockeye, the most valuable to the commercial salmon fishery in North America, don spectacular colors when spawning. Chinook are the largest and the hardiest migrators. Silvery coho spawn in the narrowest and shallowest of coastal streams. Masu, found only in the western Pacific, have the most restricted distribution. A favored target for sport fishers, steelhead exhibit unparalleled life history and morphometric variability (see page 13).

**THE SALMON LIFE CYCLE** At the most basic level, anadromous Pacific salmon have similar life cycles: they begin life as fertilized eggs in a nest (redd) in freshwater; migrate to sea; return to their natal river to spawn; and, with the exception of *mykiss* and perhaps some *masou*, which are repeat spawners, die soon after spawning. The journey is arduous and the survival rate is quite low; on average a female may deposit approximately 5,000 eggs in a redd, yet only two need to survive to maintain population stability.

We can learn a great deal by following one representative egg through its life cycle. Assuming the egg is fertilized by the milt (sperm) of a male—a challenge in swiftly moving water—the resulting alevin (embryo) will be among a small percentage of its siblings that survive to this stage. This alevin may confront a host of environmental factors affecting its survival, including excessive current leading to scouring, elevated temperature, depleted dissolved oxygen, and siltation. The next challenge will be emerging from the redd through the gravel. If successful, the alevin becomes a fry that will, depending on the species, immediately begin downstream migration to the ocean or reside near the natal stream reach for months or even years before beginning migration to the sea. There the juvenile will face a host of natural and anthropogenic threats, such as predators (including humans), pollutants, storms, climate variability, and lack of prey, to name a few.

A newly hatched sockeye alevin emerges from its redd in Idaho's Snake River, home to one of the most endangered sockeye populations.

After some time (e.g., a year or two or five), during a migration that may extend hundreds or thousands of kilometers, the salmon relies on magnetite (small particles of iron) in its brain tissue as well as polarized light from the sun to navigate the open ocean. Once it makes landfall close to the river's mouth, the salmon increasingly relies on its sense of smell to guide itself back to its tributary and finally back to its natal stream.

For most species of Pacific salmon, reproduction takes place during the last several weeks of life, when the salmon cease feeding and direct all their energy toward migration,

## NUANCES IN RAINBOW-STEELHEAD AND MASU LIFE HISTORIES

Among the species we study, *O. mykiss* (rainbow-steelhead) and *O. masou* display nuances in life history type and subspeciation that require clarification. *O. masou* contains two subspecies that do not interbreed: *O. masou masou*, which is represented by both an anadromous life history type (*sakura* or cherry), and a resident life history type (*O. rhodurus,* also called *amago or yamame*). We present data only on *O. masou masou,* or cherry salmon.

*O. mykiss* contains two different subspecies: *O. mykiss irideus* and *O. mykiss gairdneri*, which each contain resident (rainbow) and anadromous (steelhead) populations. Recent studies demonstrate that rainbow produce steelhead offspring and, conversely, steelhead may produce rainbow offspring. However, this view is not yet universally accepted: rainbow and steelhead populations are managed separately across the North Pacific. In this book, with some noted exceptions, we discuss only the anadromous forms of *mykiss.* ■

courtship, and spawning. The male salmon typically arrives first to stake out territory, often defending it from rivals; the female arrives soon after to begin the process of digging her redd, which can take up to a week. During this time, the female will lie on her side and flap her caudal fin to excavate a depression in the gravel, from several centimeters to several meters across. Then the female steadies herself in the current, using her open mouth as a rudder to stabilize her body while she lays her eggs. One or many male salmon simultaneously deposit milt on the redd once the eggs have been released. In her final act of athleticism, after the eggs have been fertilized, the female covers the nest, undulating her body and fanning her fin to move the gravel gently onto the top of the redd.

When spawning is complete, the salmon (except for *mykiss* and some *masou*) die, depositing in their home streams an abundance of marine-derived nutrients such as nitrogen, phosphorus, carbon, trace minerals, and vital organic matter. Those carcasses will directly feed many river basin residents, including insects and bacteria, minks and otters, and jays and eagles.

For salmon populations to prosper, the whole sequence of habitats must be intact and functioning. But if several important habitats are impaired, populations may be at risk.

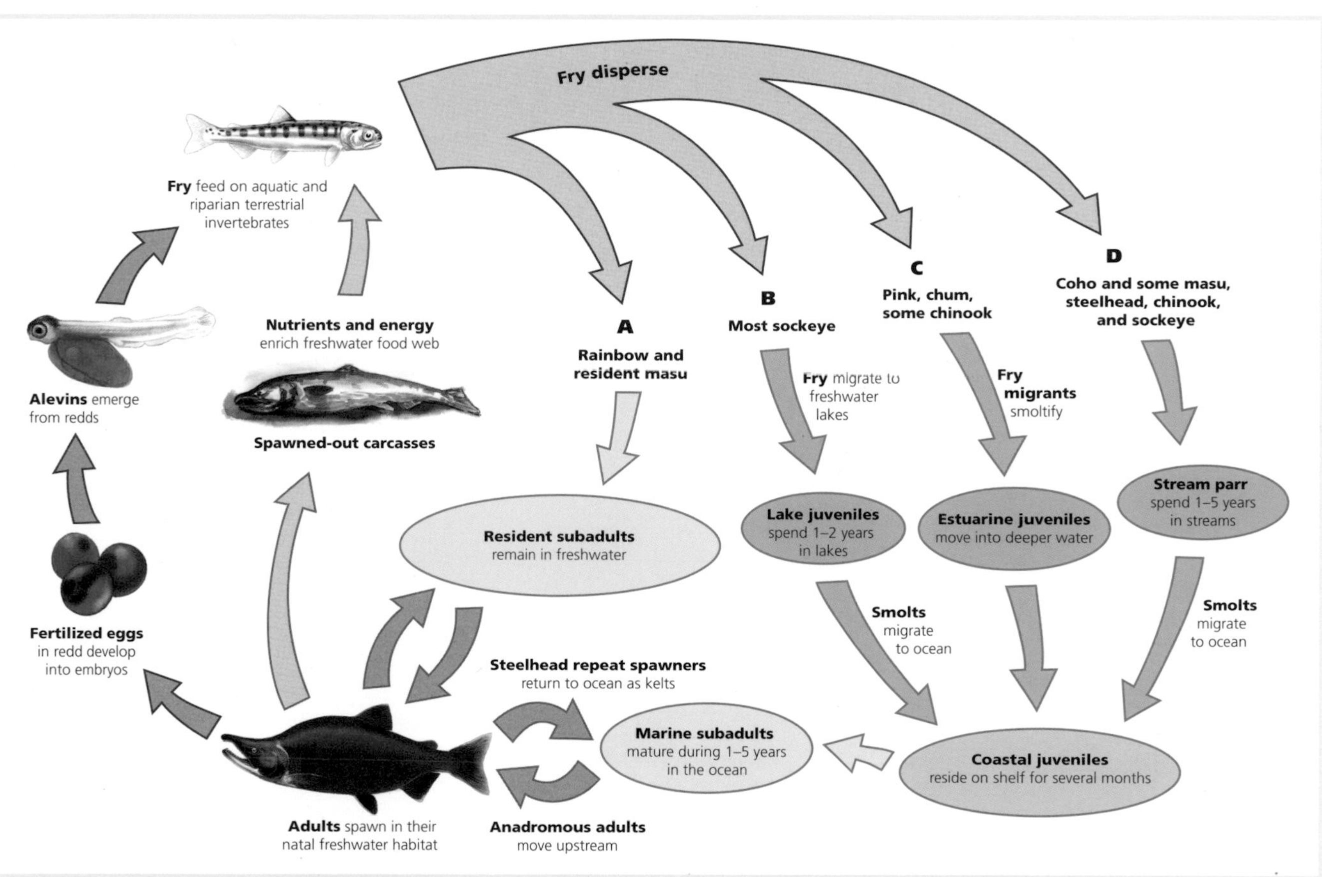


SALMON LIFE CYCLE From snowmelt to the high seas, water (salt and fresh) conveys salmon throughout the ecosystem. The odds for salmon survival are slim even if all aspects of the habitat are healthy. Only several of the thousands of eggs fertilized in a redd will grow into adults that return to spawn successfully. The cycle—hatch, migrate, spawn, die—repeats with every generation and has done so for millions of years.

## THE ROLE OF BIOLOGICAL DIVERSITY

At the macro level, there is impressive consistency in the 18 million-year history of *Oncorhynchus,* which has maintained an enduring presence in freshwater and estuaries throughout the North Pacific. But at the micro level, salmon life histories are variegated and diverse as they adapt to local conditions. Natural selection rewards resilience: as populations evolve with changing conditions, the suitability of habitat changes across the North Pacific; run timing shifts; spawning locations move; and migration distances and duration adjust so that adversity within one population will not affect the viability of the species. Those qualities that prove hardiest are written into the genetic road map that will be used and amended by generations to come.

Salmon have recolonized streams following ice ages, volcanic eruptions, climatic shifts, and ecological challenges. Those selfsame life history strategies, which spread risk over individuals and populations, give fishery managers and biologists cause for optimism; the extraordinary resilience of this genus may be its most effective tool on the road to recovery.

# Salmon Ecoregions

**In order to present data—risk of** extinction data in particular—at a uniform scale across the North Pacific, we needed to build a geographic framework with a consistent spatial unit for classifying information from widely divergent sources. Recognizing that water is what matters most to a migratory fish, we crafted a lens through which to view habitats occupied by salmon based primarily on natural hydrological boundaries. We parsed the North Pacific Rim into regions defined not by political boundaries but as parts of a full sequence of catchments and nearshore and ocean ecosystems that salmon use, each with distinct physical characteristics. From this we created our ecoregion template, which, in simple terms, maps at a coarse grain the neighborhoods that salmon populations call home—a set of 66 distinct ecosystems inhabited by salmon around the North Pacific.

The final criteria and boundaries for the ecoregions were developed in 1999 during a workshop in Corvallis, Oregon, attended by Japanese, Russian, Canadian, and U.S. scientists (see Acknowledgments, page 144). Workshop participants endorsed a four-level hierarchical classification that defined salmon ecoregions.

Once the ecoregions were determined, we used the Environmental Systems Research Institute's (ESRI) Digital Chart of the World to provide a digital geographic representation of the North Pacific. These datasets enabled us to identify stream networks and spatial boundaries of each ecoregion and to establish a geographically explicit system for data management. To define ecoregion boundaries, we digitized polygons around the stream networks associated with each of the four levels of the hierarchy.

Our ecoregional framework enabled us to present and compare data at a consistent scale that otherwise could not be aggregated. Ecoregions proved useful templates for our risk of extinction maps (chapter 4); and our maps of comparative species diversity (page 11), protected areas (page 45), terrestrial ecoregions (page 55), mineral deposit density (page 105), and stream runoff changes in a warming climate (page 111).

This framework does have its limitations. Data reported on an ecoregional basis run the risk of overrepresenting the geographical extent of salmon habitat. For example, data from a small portion of an ecoregion—such as risk in the Kamchatka portion of the Shelikhov Gulf ecoregion (31)—are applied to the entire ecoregion, which also includes Koryakia. Nonetheless we found the ecoregional framework useful for data aggregation and interpretation at a North Pacific scale.

Level 1 Ecoregions
Two ecoregions: the Arctic and the Pacific oceans, and associated freshwater drainages.

Level 2 Ecoregions
Eighteen ecoregions, including semi-enclosed seas and primary ocean circulation systems with distinct processes or bathymetric characteristics in the North Pacific and associated freshwater drainages. There are two Arctic Ocean and 16 Pacific Ocean regions defined at this level.

Level 3 Ecoregions
Thirty-nine ecoregions, including finer-scale coastal discontinuities within each semi-enclosed sea or major circulation system, including fjords, straits, and areas with distinct production processes (e.g., upwelling and downwelling areas). There are three Arctic Ocean and 36 Pacific Ocean regions defined at this level.

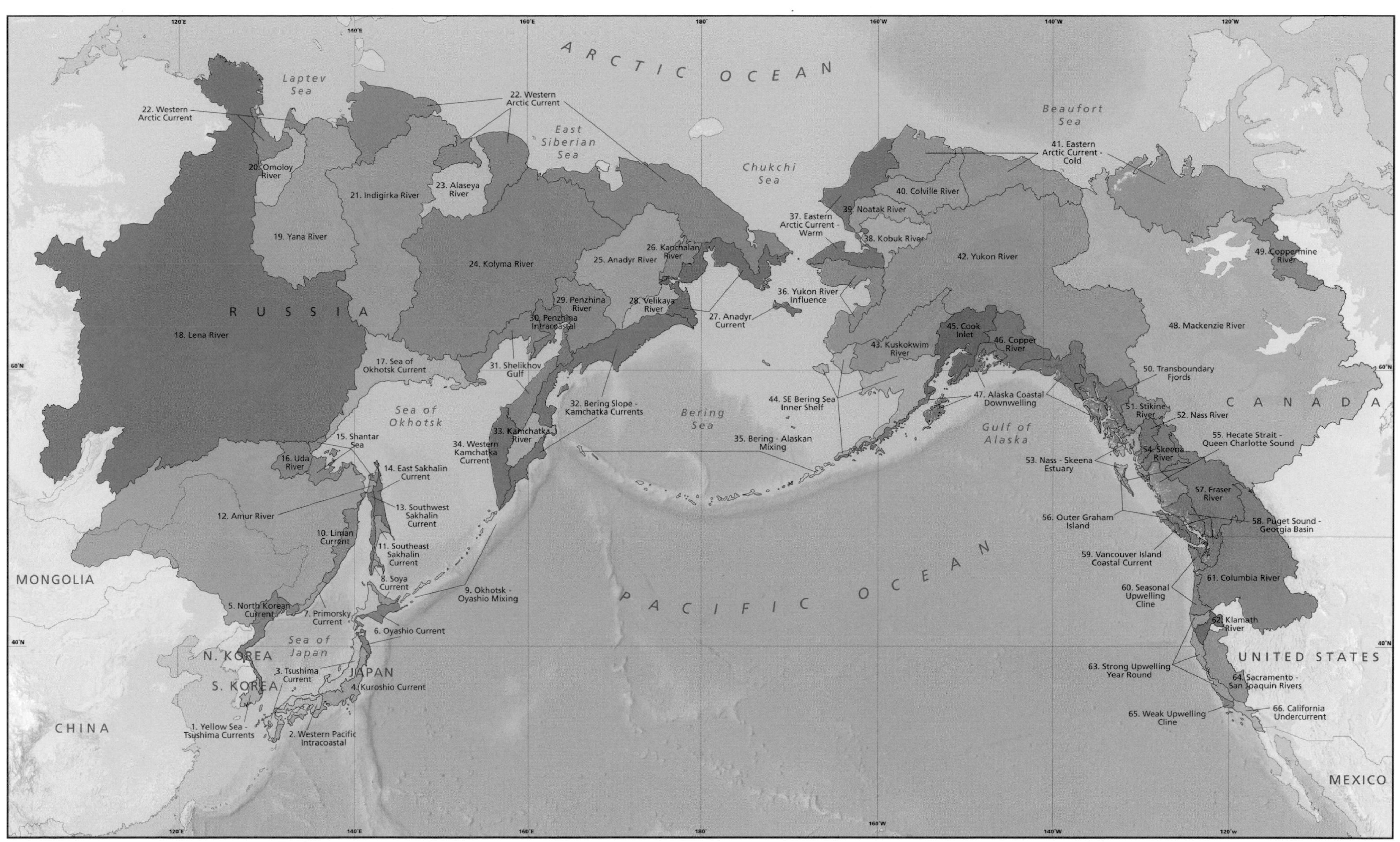


**LEVEL 4 ECOREGIONS** We have identified 66 distinct ecoregions that describe the coastal ecosystems used by salmon around the North Pacific. Through this common geographic framework, we are able to depict data at one consistent scale throughout the North Pacific. Although salmon are not ubiquitous within these ecoregions, they rely on the full expanse of habitats from headwaters to coastline and the entire ecosystem that supports them. The size of the ecoregion has no bearing on the extent of salmon presence: for example, the two largest ecoregions, Lena River (18) and Mackenzie River (48), support only limited salmonid distribution.

NOTE: The colors in this map were chosen for contrast to distinguish ecoregions. They do not represent ranges of comparable data or severity of condition, as do colors elsewhere in this book.

# Original Distribution of Genus *Oncorhynchus*

**The places salmon reach determine the** spatial extent of our study, and the domain is vast. Just their inland realm covers approximately 6 million square kilometers—about 4 percent of the world's entire landmass. (See page 66 for how we determined distribution.) That said, the area of salmon distribution is a fraction of the ecoregional footprint described in the pages prior. Salmon depend on the entire sequence of habitats, from headwaters to estuaries, to provide cold, clean waters, habitat structure, and nutrients that keep populations healthy.

Although *Oncorhynchus spp.* date back around 18 million years, only a few core populations survived the last glacial maximum some 18,000–22,000 years ago, spawning in the thawing refugia within British Columbia's and Alaska's glaciers (see page 52). Descendents of these creatures have since radiated from these ice-age refugia, adapting to habitats from Kamchatka to Ketchikan, from the Skeena River to the Sacramento River, from the Kuril Islands to Kodiak Island, from the Anadyr River to the Amur River. Salmon link these lands across the Pacific, connecting the oceans, nearshores, estuaries, rivers, creeks, rivulets, and, finally, the headwaters where the streams originate. Salmon have used these pathways before they had names, and as far back as the post-Pleistocene era.

Anadromous salmon share basic habits: juveniles make their way from freshwater to the ocean, mature at sea, and return to their natal streams to spawn and die. Within this framework, life history may vary dramatically. Some chinook and chum venture thousands of kilometers inland; pink and chum explore the Arctic Ocean; sockeye proliferate in the eastern Pacific; masu, a more ancestral form of *Oncorhynchus*, use waters largely along western Kamchatka, the Japanese archipelago, and Sakhalin and the Sea of Japan; steelhead form the most southerly populations.

Through published data and best expert judgment, we know where salmon have been and currently are. But beyond historic and present distribution, our understanding of trends regarding other measurements of salmon viability is less definitive. For example, abundance data were first collected in WOCI around the 1930s in conjunction with widespread dam construction. Because these data emerged in a multiplicity of forms from numerous sources at various scales of resolution, baseline abundance datasets have been impossible to create. Similarly trend analyses of biodiversity and productivity have also remained elusive, and therefore we have no way of ascertaining viability measurements for Pacific salmon populations over time.

## REDISCOVERING THE NORTH PACIFIC PERSPECTIVE

The vast majority of the maps throughout this book emphasize our panoramic North Pacific perspective. However familiar the image may become in the pages ahead, we want to remind readers of its relative novelty: the Pacific-centric map is a reinvention of the past decade. In 1994, for emerging work on Pacific salmon, Oregon Department of Fish and Wildlife biologist Jeff Rodgers needed to create a North Pacific map, so he stitched together digital images of the east and west Pacific from the "edges" of other maps and wrapped them around a consistent Mercator projection. The perspective was, at the time, quite unfamiliar.

## A NOTE ON MAP PROJECTION THROUGHOUT THIS BOOK

As anyone who has made a paper hat knows, distortion occurs when you apply a spherical shape onto a two-dimensional surface. Projection attempts to correct distortion, but solutions are never perfect, and some feature of the earth's surface will always be compromised. Weighing the options, our cartographers believed that Mercator was a fair tradeoff, even though it exaggerates areas at higher latitudes, particularly Alaska and northern Russia. For example, in the map on the facing page, the Mercator projection makes the state of Alaska appear massive, looming large over most other North Pacific jurisdictions. In fact, this American state is merely four times the size of California or Kamchatka. ■

Likely the first map of the Pacific Ocean, "Maris Pacifici quod vulgo Mar del Zur" was included in the first "modern" geographic atlas, *Theatrum Orbis Terrarvm* (Theater of the World), which was compiled and edited by Abraham Ortelius and published in Antwerp in 1570.

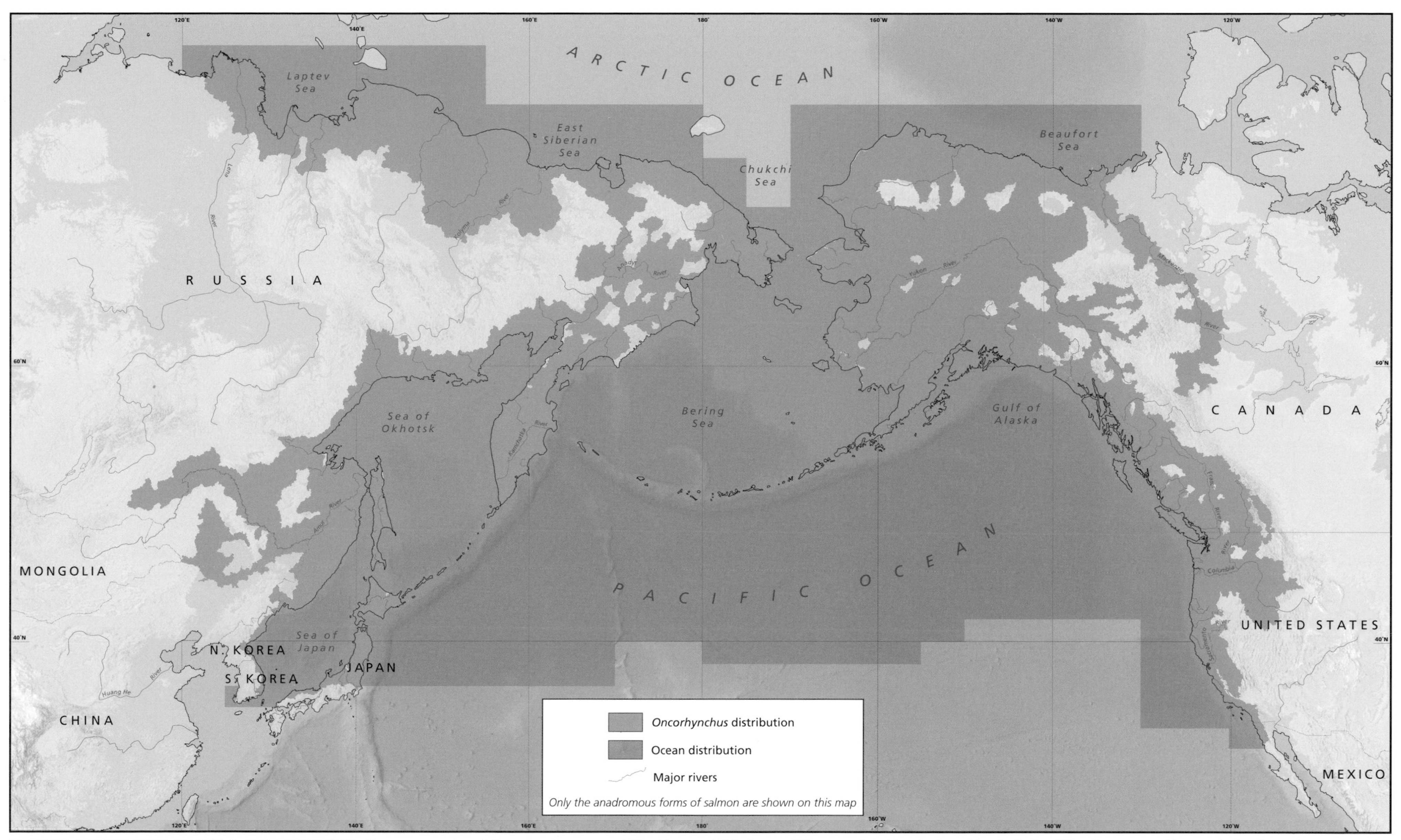


**ORIGINAL DISTRIBUTION OF GENUS *ONCORHYNCHUS*** The seven species of *Oncorhynchus* presented in this atlas use the entire Pacific Rim coastline and can venture hundreds of kilometers inland in every direction. Salmon use the tributaries, rivers, and estuaries without regard to jurisdiction, from South Korea to Southern California. Note that this map—a composite of the distribution maps for seven species of *Oncorhynchus* that appear in chapter 4—shows current and limited distribution as well as historic presence, effectively outlining the footprint of salmon throughout the North Pacific.

NOTE: See page 66 for our metholodology for determining distribution. Although we were unable to acquire data for North Korea, original distribution in the western Pacific has been extended between known locations in Primorye and South Korea.

# Salmon Diversity

**Scientists generally place the center-**of-range for North Pacific salmon in the Gulf of Alaska, the epicenter of the three major refugia that endured during the last glacial maximum (in what are now Kodiak, Queen Charlotte, and Vancouver islands; see page 52). In the area radiating outward from the Gulf of Alaska—as far west as Kamchatka, across the Bering Strait, in southeast Alaska and British Columbia, and down to the Columbia River basin—*Oncorhynchus*, with the exception of masu, demonstrate greater speciation than anywhere else in the Pacific.

At the edges of the range—in China, Sakhalin, and Primorye and across the Rim in California—biodiversity in *Oncorhynchus* diminishes. Off the coast of Southern California and along the southern coast of the Sea of Japan, the waters are too warm to accommodate anadromous salmon, with the exception of steelhead, whose southernmost populations remain south of California's Point Conception. Japan's salmonid diversity is further limited by the short rivers prone to flash flooding. Low species richness also manifests in inland climates, which can be dry and where lengthy migratory distances can be prohibitive, such as in the Lena River (18) and the Mackenzie River (48) ecoregions.

This map, however, does not demonstrate the remarkably complex population structure resulting from reproductive isolation. Even a single species may contain hundreds or even thousands of discrete populations, each with its own life history variation tailored to its habitat and environment.

It is important to note that this map represents a snapshot in time, as salmon diversity is measured using current distribution data. Salmon abundance and distribution across the landscape are in fact a moving picture, changing continually in response to dynamic influences.

In chapter 4 we demonstrate an increased risk of extinction for salmon populations at lower latitudes across our study region. This is a major feature of the North Pacific ecosystem, defined in part by its salmonid biogeography, oceanic and climatic processes, and human interactions with salmon. Trends over time will reveal more: Low-productivity salmon stocks at the southern edges of the range (e.g., coho, masu, and steelhead) may serve as bellwethers for the effects of climate change throughout the North Pacific.

When seen through a narrower perspective, trends may seem isolated and unrelated. Stepping back from local details to examine salmon at a panoramic scale enables us to recognize widespread patterns associated with salmon and their ecosystems.

## NAMING CONVENTIONS

As Robert J. Behnke explains in *Trout and Salmon of North America*, the lineage of salmon nomenclature is largely rooted in the language of the indigenous Koryak peoples of Kamchatka. Around 1740, while conducting research on the natural history of Kamchatka, naturalist George Wilhelm Steller phonetically transcribed the Koryak names for salmon into his German manuscript. (There were exceptions: *keta* is a Nanai word from the Amur region; *gorbuscha* means "hunchbacked" in Russian.) For the next 50 years, Steller's work was translated into Russian, English, and French and finally Latinized in 1792 by Dr. Johann Julius Walbaum in his *Genera Piscium* to conform to scientific naming conventions.

The Russian names for salmon, therefore, are the Latin names, but nicknames are common. The Japanese call masu salmon "cherry" because they run when the cherry blossoms emerge every spring. Chum salmon may be referred to as "calico" because as they change into spawning condition, their skin has streaks in hues of red, black, and blue or green. Chum may also be called "dog" salmon, perhaps because the male salmon have prominent canine teeth or because chum were fed to the sled dogs in northern climes. The name "steelhead" describes the silvery hue these fish take in ocean water. Chinook are called "king" because they are the largest of the *Oncorhynchus.* The name for "coho" likely derived from Native American appellations "cohose" and "kwahult." "Sockeye" may also have a Native American origin, where the fish was dubbed "sukkai"; and "sake" is the Japanese word for salmon.

Note that Russians classify the rainbow-steelhead complex in the genus *Parasalmo*. ■

| LATIN NAME | ENGLISH | RUSSIAN | JAPANESE |
|---|---|---|---|
| *O. keta* | chum, dog, calico | keta | shirozake, sake |
| *O. gorbuscha* | pink, humpie, humpback | gorbusha | karafutomasu |
| *O. nerka* | sockeye, red, blueback, silver trout, kokanee (resident) | nerka | benizake |
| *O. tshawytscha* | chinook, tyee, king | chavycha | |
| *O. kisutch* | coho, silver | kizhuch | ginzake |
| *O. masou* | masu, cherry | sima | sakuramasu yamame or yamabe (landlocked) |
| *O. mykiss* (*Parasalmo mykiss* in Russia) | steelhead (anadromous) rainbow (resident) | mikizha, syomga (steel-head) | niji masu |

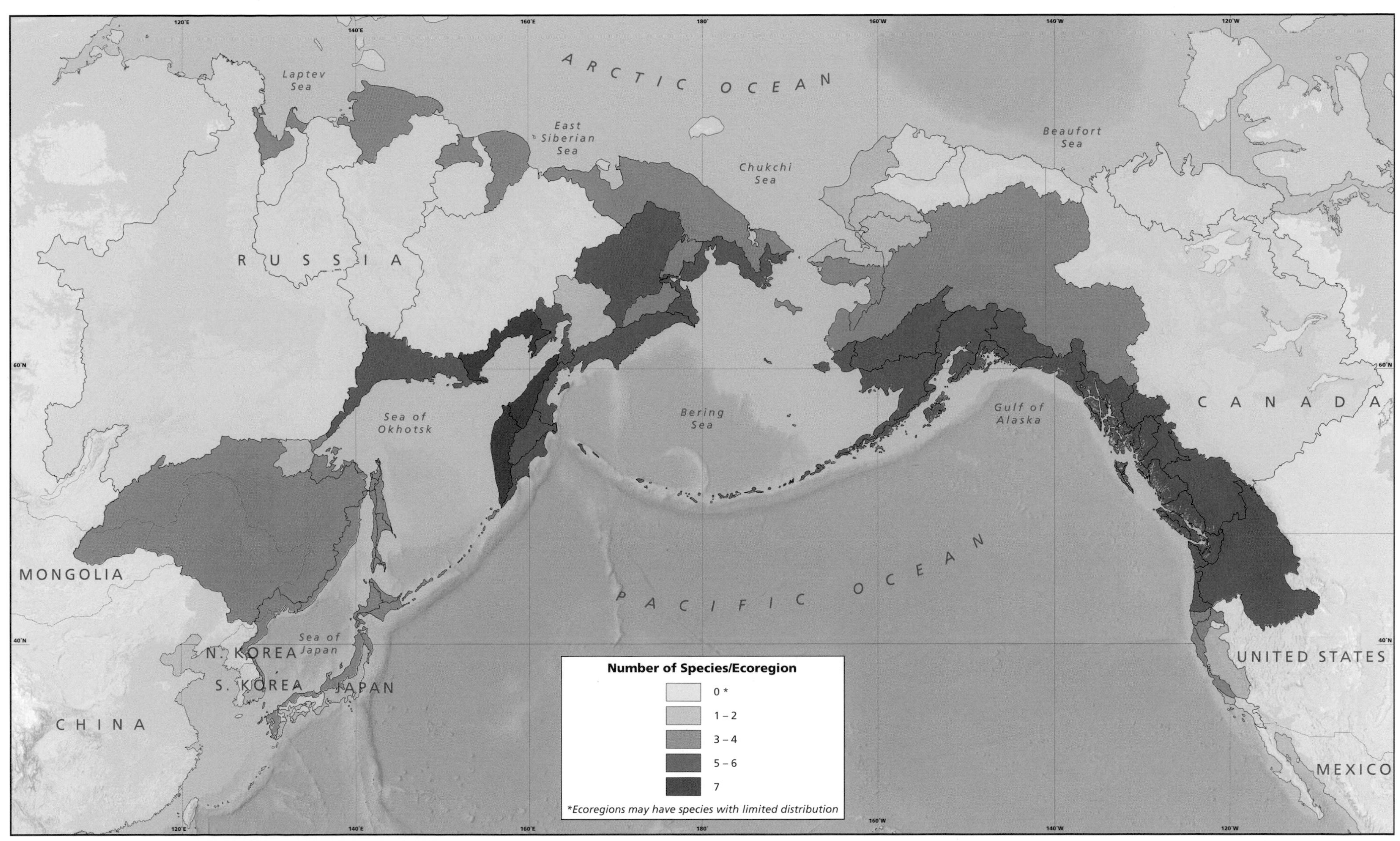


SALMON DIVERSITY BY ECOREGION (BASED ON CURRENT DISTRIBUTION) The center of range for *Oncorhynchus spp.* is generally the Gulf of Alaska. From the Russian coasts of Magadan, Kamchatka, and Koryakia, across the Aleutians, and down the North American coastline to Washington, speciation within the genus is high. Note that Kamchatka is the only region to host all seven species. Conversely, we can also see how the edges of the range contain limited speciation. Because this map is ecoregionally based, it may overrepresent diversity within a given ecoregion in cases where one or a few river basin host exceptional diversity.

NOTE: Degree of speciation may appear overreported due to the nature of our ecoregion contruct. For example, only the Kamchatka portion of the Shelikov Gulf ecoregion (31) contains seven species of salmon.

# Kamchatka Rainbow-Steelhead Life Histories

**In a given year, Kamchatka may** produce as much as one-quarter of the world's wild Pacific salmon. It is a remarkable setting in which to study diversity not just within the genus *Oncorhynchus* but also within particular species. Heterogeneity is especially notable in *O. mykiss*, the rainbow-steelhead complex.

Roughly the land area of California but with just about 3 percent of the population, Kamchatka, the product of volcanoes and tundra, has long cold winters and dewy summers. The Sredinnyy mountain range runs up the middle of the peninsula, separating the Vostochnyy range in the east from the tundra and plains in the west, which provide rich habitat in the marshes and rivers feeding into the Sea of Okhotsk.

The map to the right depicts data on six distinct phenotypes, or life history types, of *O. mykiss* within ten major rivers along Kamchatka's west coast. Depending on the river and the conditions, one redd may produce any number of life-history types; determination likely occurs by the fall. Estuarine *mykiss* are silvery and may grow to 90 centimeters, while resident rainbow might have fine spots and grow only to 35 centimeters. Anadromous A and B deviate further: generally the largest, Anadromous A spends several years in the ocean; B spends one summer at sea, overwinters in freshwater, then migrates back to the ocean, where it spends several years.

The middle of the Kamchatka peninsula features the richest life history diversity. The fast and winding Krutogorova River is home to six different steelhead life history types in almost even proportions. Similarly, the long, braided Sopochnaya River harbors varying food sources that support many life history types.

In the cooler waters of Kamchatka's northern rivers, like the volcano-based Sedanka, resident rainbow prevail; yet its tributary, the Snatolveyem, is home to anadromous steelhead. Similarly, to the south the Kol contains largely resident *mykiss*, but the neighboring tundra-based Kekhta nurtures anadromous *mykiss*.

Scientists are examining the interaction between the physical complexities of river systems and the nutrient subsidies that salmon provide, but the drivers remain elusive (see below). Most agree that the breadth of life history diversity documented in Kamchatka is more typical of the natural inherent diversity of the species than what is represented on the eastern side of the Pacific, but this remains undocumented. Research on *mykiss* diversity has yet to be conducted in North America at such a resolution; nonetheless, Russian science can tell us much about historic diversity elsewhere.

## *MYKISS* RESEARCH ON THE KAMCHATKA PENINSULA

Scientists study otoliths (ear bones) and scales under the microscope to assess emergent patterns demarcating life history transitions, such as timing and duration of migrations. Scale and otolith research is conducted worldwide, largely for the purposes of determining age, but Russian scientists use the data for broader study. Decades of morphometric data have led Russian scientists to classify *mykiss* in the genus *Salmo*—not as *Oncorhynchus* as other Pacific Rim scientists do.

In research on Kamchatka, Russian scientists have sampled scales from thousands of juvenile and adult *Parasalmo mykiss* salmon in major river systems. By analyzing scale patterns and relating data to geography, Russian scientists have documented the existence of six different life history types on the Kamchatka peninsula. Such a wide spectrum of life history diversity may exist elsewhere but has not yet been documented.

Teams of Russian and American scientists are also observing river morphology on Kamchatka and the extent to which physical aspects of rivers help shape the biological diversity of the salmon studied. Using satellite images as well as field observations, they are also examining how the structural complexity of rivers affects food sources. For example, the southerly Kol River is fed by cold spring waters and contains mostly resident *mykiss*. Tundra rivers—such as the Utkholok River in the northern portion of the peninsula, which contains mostly anadromous *mykiss*—tend to be warmer, composed of brown-water systems that are high in tannic acid, which is produced by decaying organic matter. How salmon use and adapt to these river systems is the subject of emerging research. ■

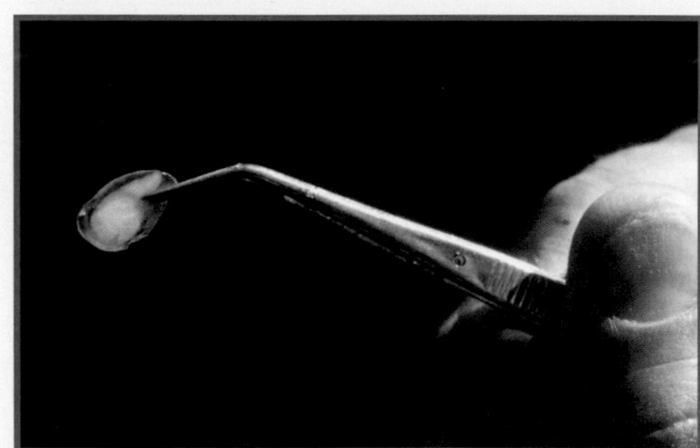

In 1989 American scientists reclassified *O. mykiss* from the *Salmo* genus to *Oncorhynchus*. Russian scientists, relying on hard-structure biological morphometric data, including skulls and otoliths (above), continue to classify rainbow-steelhead as *Parasalmo mykiss*.

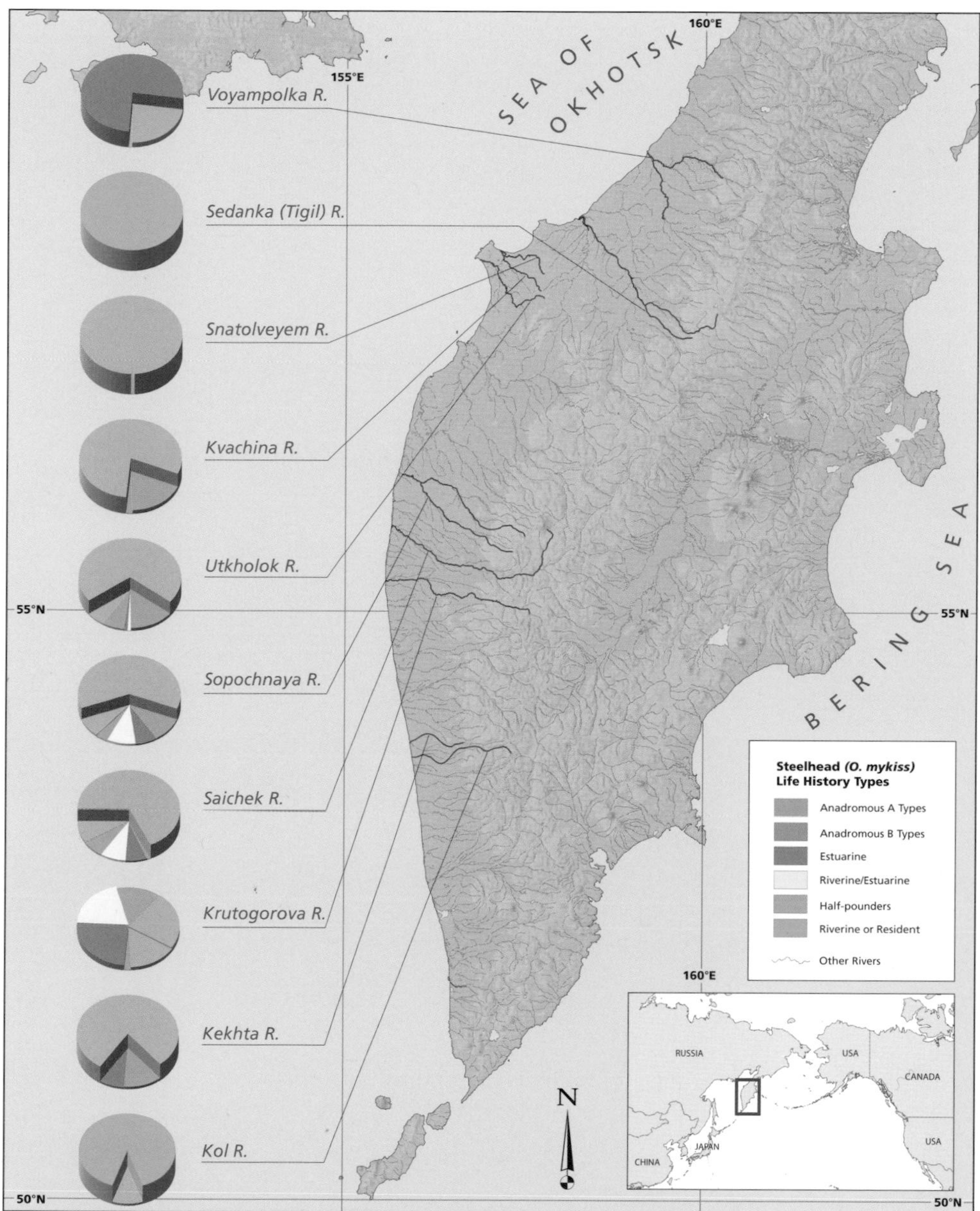


**KAMCHATKA RAINBOW-STEELHEAD LIFE HISTORIES** The rivers along the Kamchatka peninsula feature rich life history diversity within the genus *Oncorhynchus*, particularly within *O. mykiss*. Water temperature, channel morphology, as well as riverbed composition (e.g., volcanic or tundra) may contribute to the differences in life history types from river to river, ranging from the Kol in the south, where *mykiss* are largely resident, to the Snatolveyem in the north, where nearly all are anadromous steelhead. Russian scientists are the first to document this degree of life history variation within one species of *Oncorhynchus*.

*ONCORHYNCHUS MYKISS (PARASALMO MYKISS)*
Siblings from the same redd may later exhibit six distinct life history types, from resident (rainbow) to anadromous (steelhead).

Riverine/Estuarine
1.3 kg (0.4–2.5 kg range)

Estuarine, which may grow to around 90 cm.
2.1 kg (0.6–3.2 kg range)

Anadromous A migrate directly to the ocean, where they spend several years.
5.7 kg (2.5–10.5 kg range)

Anadromous B spend a summer at sea, return to freshwater for overwintering, and migrate back to the ocean for several years.
4.9 kg (1.0–9.3 kg range)

Anadromous B–half-pounders, immature entering migration.
0.2 kg (0.1–0.31 kg range)

Riverine or resident, which may be as small as 35 cm as a mature adult. This particular rainbow trout was photographed on the Shantar Islands.
1.4 kg (0.4–2.7 kg range)

# 2

# The People

*How five political provinces and generations of indigenous peoples built cultures and economies around Pacific salmon*



A First Salmon ceremony takes place in Oregon's Celilo Indian Village, arguably the oldest continuously inhabited community in North America.

**Where there are salmon, there is a** salmon culture. Salmon are food, and food draws people; people modify waterways and landscapes, which in turn affects the salmon. The cycle has repeated for thousands of human and salmon generations throughout the North Pacific at least since the close of the Pleistocene era, some 11,000 years ago.

## NORTH PACIFIC INDIGENOUS PEOPLES

Throughout the salmon ecoregions of the North Pacific, before the merchant class and the industrial economy began to drive resource consumption beyond sustainability, more than 130 discrete native cultures relied—to lesser and greater extents—on salmon for survival. Some inland and tundra tribes subsisted on caribou, which feed on lichens and berries that grow in soil enriched by minerals provided by spawned-out salmon. Other tribes along forested ridges high above rivers had limited access to fishing but may have supplied downstream neighbors with wood for building platforms, or plant fibers for weaving nets, in exchange for a sure supply of protein year-round.

For tribes who inherited prime fishing grounds along rivers and coastal systems, salmon was a cultural centerpiece: the men were dedicated to fishing and maintaining gear and tools; the women, to cleaning, smoking, and drying the fish; and the community, to the ceremony and ritual that upheld and sustained robust salmon populations on which the people depended.

A Yakama Indian dipnets near the Dalles Dam on the Columbia River. In the 1940s a Yakama tribesman was convicted for fishing without a license, an event that spurred the Native American civil rights movement.

### MYTH, RITUAL, AND CEREMONY

As animists, indigenous peoples believed in the presence and power of spirits in daily life. Creation myths reinforced and commemorated the role of spirits. For example, for the Ulchi of the Amur River Valley, respect for the Salmon Spirit ensured abundance of food, a belief validated in this creation myth. Long ago, a husband, facing scarcity of food, presented to his wife the lone salmon he had caught that day. When the wife began to cook it, the salmon spoke, asking the woman to spare its life in exchange for its scales. The wife agreed and made broth of the scales. When the husband returned home that evening, he discovered that the pot was teeming with fish. The husband and wife feasted on salmon and never again wanted for food.

The Quinault people, based in Taholah, Washington, were granted sovereignty by executive order in 1859.

Among the ways to demonstrate respect for spirits was through ceremony, and the celebration of the First Salmon returns are practiced by salmon people across the North Pacific. Ritual surrounds how the salmon are caught, presented, and welcomed by the community. The Ainu, for example, have dozens of traditional ways of preparing salmon to eat. Once the first salmon has been eaten by all in attendance, the bones, head, tail, or fins would then be returned to the river so that the salmon would encourage its relatives to enjoy the village's hospitality.

In addition to high ceremony, tribes practice ritual in daily fishing. For example, the Ulchi would not speak loudly around water. Women could not approach the fishing areas or touch equipment. The Ainu did not wash in the river but withdrew buckets and did laundry elsewhere. Their special tool for clubbing caught salmon, *isapakikni,* was made exclusively of willow wood. In the Karuk tribe of California, wood was also used for clubbing—except at the mouth of the Klamath River, where only rocks were used.

**ECONOMY AND ECOLOGY** Native peoples used the entire fish in daily life: the bladder was made into glue; vertebrae and cheekbones were dried to make toys; skin was used to make clothing and shoes. What's more, native peoples used the entire ecological system to fulfill fishing needs while maintaining natural processes. In basket-making, hillside communities might burn off brush and plant iris to use its fibers; coastal tribes would weave with stinging nettles; in the floodplains, tribes might use Dog Bane, a plant similar to milkweed. For the softest baskets worn close to the skin, the Japanese linden tree was a preferred material. Willow, grape vines, and cottonwood were fine choices for lattice work and woven materials.

To begin the salmon ceremony the eldest Ainu woman is given the salmon to present to the spiritual leader (Sapporo, Japan).

Along the Klamath River in California, the Karuk tribe used long dip-net poles to fish off falls, made from a specific kind of Douglas fir. After a forest disturbance (e.g.; logging, fire, or road building) stands grew back in thick clusters, and only scant few trees received sunlight through the canopy. The others grew in the shade and became thin and tall—around 5 centimeters wide and 12 meters long. Eventually their branches dropped off. The Karuk would cut down these thinner trees for use as fishing poles, fullfilling their own equipment needs while thinning the canopy so that the healthier trees could survive.

Indigenous peoples modified vegetation in their favor through prescribed burns, which were often celebrated in ritual. Frequent burning reduced fuel loads, preventing catastrophic wildfire. During droughts, seasonal burns could reduce transpiration levels and cause the release of water at springs, increasing surface water flows at times and places critical to salmon survival.

**SOCIAL CONSTRUCTS** Tribal fishery methods were as unique and distinctive as the other elements of native culture. Fishing gear was made and modified according to local social and ecological conditions. Simple spears and gaffs were used to harvest individual fish along the edges of water bodies or from canoes. To catch salmon along wider gravel-dominated river channels, tribes combined weirs with basket traps, dip nets, and seines. They designed materials to reduce bycatch of other species, which necessitated social organization and catch norms that protected the runs from overharvest. For example, from the First Salmon ceremony in May to the War Dance/World Renewal celebration in September, Karuk dip-netters worked in pairs. It is the job of the netter's helper, the clubber/fish killer, to properly identify fish species at all times, as they would release summer steelhead and, often, female chinook and coho.

William We-ah-lup, an ancient indigenous person of the Puget Sound, smokes salmon and salmon eggs on a beach in 1906.

The act of making and using gear required a highly ordered social infrastructure; for instance, a tribe's religious leader or a village headman might orchestrate communal weir construction. And among tribes, in order to ensure equitable allocation and spawning success, indigenous people coordinated fishing activities so as not to take more than their fair share of the upstream salmon migration.

# Indigenous Peoples of the North Pacific, c. 1880

**Native peoples and salmon coexisted** for millennia throughout the North Pacific. Precontact, the connection was intimate, as the culture of a people revolved around the seasonality and abundance of its salmon runs. Daily life, nourishment, clothing, ceremony, trade, migration, prosperity, and survival were shaped by the presence of salmon.

The interdependence between salmon and people clarifies when we visualize the distribution of native ethnic and linguistic groups living along the seashore and valleys and compare it to the distribution of Pacific salmon. As we can see in the map to the right, the level of diversity among native peoples was much greater on the eastern side of the North Pacific. The landscape was glacially carved and mountainous in the north (above the Columbia River basin) and segmented into riverine microzones in the south (southern Oregon and California). People who inhabited a streamside village in a deep canyon or secluded valley might have had little or no contact with neighbors upstream or over the ridge because of hard-to-navigate terrain. As a result, despite the existence of regional trade networks and some localized resource centers, small pockets of distinctive and long-standing peoples and cultures uniquely adapted to their setting, much like the highly individualized local salmon populations. In most areas tribes could be supported by local salmon runs without having to venture beyond their own watersheds; other resources (e.g., oak trees that provided edible acorns) were also strictly localized.

The topography of the western side of the North Pacific is less diverse, particularly across the expanses and mountain ranges of the northeastern Russian Far East. Here indigenous cultural groups occupied vast swaths of landscape that stretched across several major river basins. Salmon exhibit greater homogeneity in this region of tundra and taiga. Exceptions to this general rule of homogeneity are the Amur River basin and the "Ring of Fire." In the latter, the landscape of volcanoes, earthquakes, and tsunamis—in Kamchatka, the Kuril Islands, and Sakhalin and Hokkaido—has a mountainous and varied topography, a diversity of salmon, and somewhat more heterogeneous indigenous cultural groups.

During the past 150 years, the boundaries of North Pacific indigenous nations have blurred through assimilation, disease, migration, and forced removal of indigenous groups by new settlers. Similarly, biodiversity among salmon populations has diminished as a result of overharvest, habitat degradation, dams, and hatchery mitigation programs.

## VISUALIZING INDIGENOUS CULTURES: HOW WE BUILT THIS MAP

This map was developed in collaboration with specialists from the Smithsonian Institution's Department of Anthropology. Dr. Igor Krupnik provided us with the data for the native nations living in the western half of the Pacific; Dr. Ives Goddard provided us with similar information for the eastern half of the North Pacific. Several elements require explanation.

We chose the circa 1880 timeframe because it reflects the general situation that existed prior to the main influx of outside settlers—Russians, Americans, Canadians, Chinese, Japanese, and Koreans—into the areas populated by the indigenous fishing-dependent nations of the region.

It is important to note that not all ethnic and linguistic groups were traditional users of salmon. For example, the headwater people such as the Mongol in east Asia and the Blackfoot in British Columbia had no documented association with salmon and no words for salmon; in fact, the Mongols had no knowledge of fishing.

The basic ethnic and linguistic groups shown in this map are culturally and linguistically distinct but internally homogeneous. Indigenous groups in northeast Asia are usually recognized more as cultural entities, while the North American groups are historically defined by language; but in these areas culture and language typically go hand in hand. Each group featured has a distinctive combination of attributes, including social structure, ecological adaptation, religious and cosmological beliefs, ceremonial practices, and language. In some groups, however, more than one dialect or related language was or is spoken. ■

By studying the relationship between indigenous peoples and salmon before widespread contact with European settlers, we have a clearer understanding of the symbiosis and connection between humans and fish. Above, Inuit fishermen spear at a fish trap in 1916.

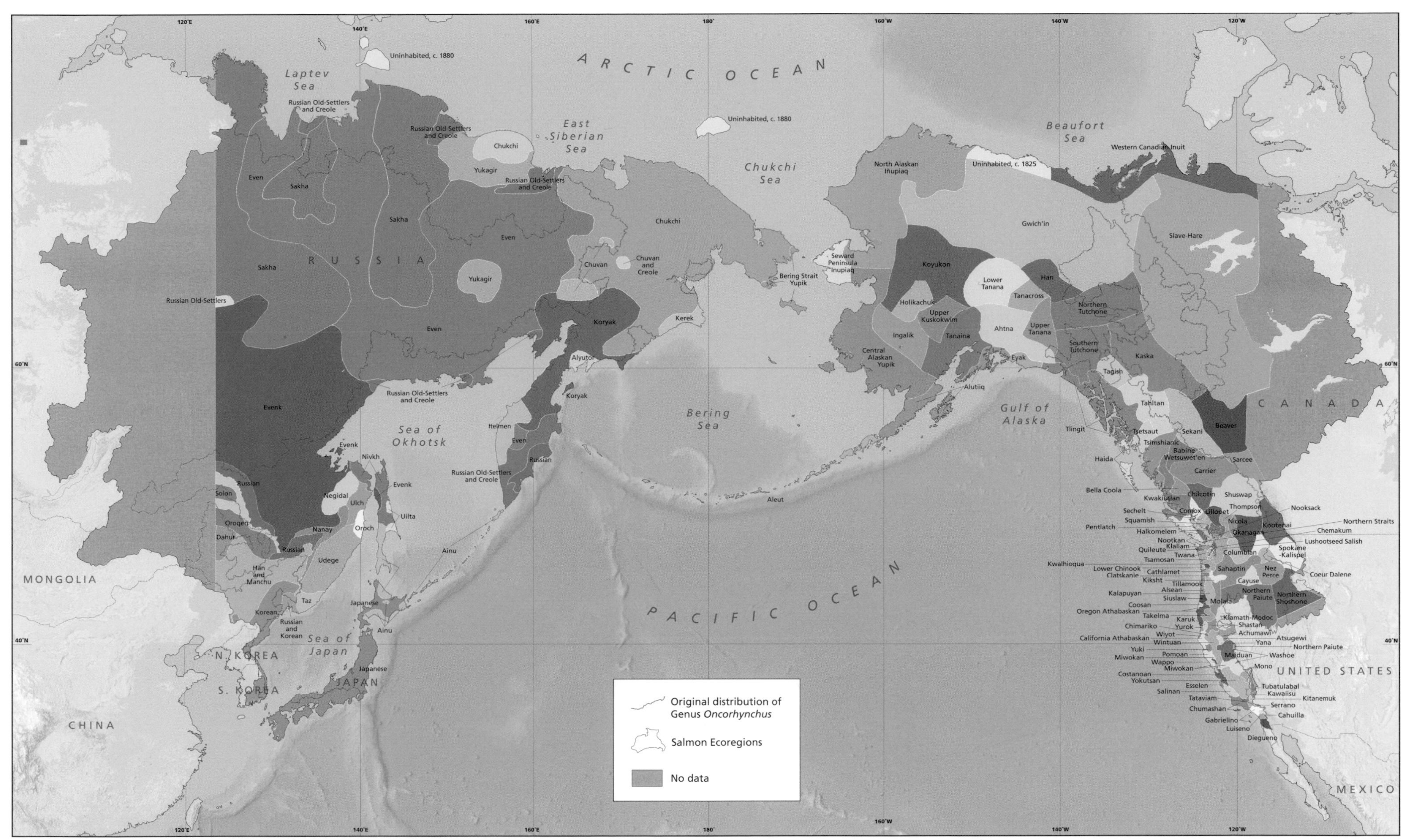


**INDIGENOUS PEOPLES OF THE NORTH PACIFIC, c. 1880** Note the abrupt edges of this map around the 125°E parallel and east of Canada's Mackenzie River: our researchers provided data only where native people fished for salmon. (The inland red line delineates original *Oncorhynchus* distribution, page 9.) By mapping indigenous culture heterogeneity and revisiting our salmon diversity map (page 11), we see how diversity of people and fish reinforce one another. Regions of high salmon speciation (the Amur basin, Kamchatka, and the North American coast) demonstrate vast cultural differences. Elsewhere, salmon and people exhibit greater homogeneity.

NOTE: The colors in this map were chosen to distinguish between language groups. They do not represent ranges of comparable data or severity of condition. Country borders are from present day to provide some comparative geographical reference to 1880.

# Native Americans and Salmon Coevolution

THE PRECEDING PAGE OFFERED A MACRO view of the relationship between tribal diversity and salmon diversity at a coarse North Pacific scale. Here, to demonstrate the extent to which salmon societies were linked to the salmon resource, we portray the relationship at the micro level, specifically as one group of Native Americans and one population of salmon connected to one particular terrestrial ecoregion straddling Oregon and California.

In the early 20TH century, ethnographers developed the theory of cultural core provinces, footprints of where indigenous people lived and shared material and cultural traits. The first map on the right delineates in purple the northwestern California core cultural province, ranging from Cape Blanco in the north to Punta Gorda in the south. At the margins of the northern region is the territory of the Cow Creek band of the Umpqua tribe; to the south, the Yuki tribe. However different, these tribes share a core cultural province, having each borrowed aspects of the region's material cultures.

In the second map, we have delineated in red the Southern Oregon-Northern California (SONC) coho evolutionarily significant unit (ESU, see below). This unit—a measure of biological relatedness essential to salmon sharing genetic traits and distinct from their neighbors—was identified by biologists with NOAA Fisheries (National Marine Fisheries Service) who were conducting population viability assessments. In the process, these scientists distinguished clusters of localized salmon runs on the basis of genetics (reproductive isolation) and life histories (local adaptations). These yardsticks represent important components of the species' evolutionary legacy.

In the third map, we have outlined in green the approximate boundaries of the Klamath-Siskiyou terrestrial ecoregion, the biotic province that describes this area (see page 54).

By overlaying these maps, we can see how the boundaries reinforce one another. For thousands of years, the Native Americans who lived in the Klamath-Siskiyou ecoregion and the coho salmon within this ESU developed concurrently. Settlements, economies, and cultures clustered around the fisheries along the coast, estuaries, and rivers.

Researchers have been unable to accurately and decisively define tribal cultural traits and boundaries, which have been blurred through forced removal and genocide. Similarly, natural salmon ESUs may have been blurred through urbanization, hatchery proliferation, resource extraction, and other anthropogenic threats presented to salmon throughout their range.

## EVOLUTIONARILY SIGNIFICANT UNITS

Biodiversity manifests as variety within and among species. In the United States, the 1973 Endangered Species Act became the first law to protect species biodiversity, mandating the U.S. Fish and Wildlife Service to conserve species and "distinct population segments" deemed to be at risk of extinction. But scientists soon begged the question, what constitutes a distinct species, and how can we identify discrete population segments that may be at risk?

The question was answered only as recently as 1991, when Dr. Robin Waples of the U.S. NOAA Fisheries (National Marine Fisheries Service) Northwest Fisheries Science Center in Washington offered a set of guidelines that, if met, would constitute conservation units meriting protection under the law. If a population were "substantially and reproductively isolated from other conspecific populations" and represented "an important component in the evolutionary legacy of the species," then it would qualify as an "evolutionarily significant unit" (ESU). The two-part test takes into account genetic, ecological, and life-history variability exhibited by individual populations, each with its own unique contribution to the diversity of the species.

Today, although there are seven *Oncorhynchus* species—including cutthroat trout (*O. clarki*) in Washington, Oregon, California, and Idaho (WOCI)—there are more than 50 salmonid ESUs, each with life history strategies and risk factors unique to their native streams. This relatively fine filter allows managers and conservation strategists to tailor efforts to the needs of locally adapted salmon populations. For example, coho populations in central California face different challenges from those in Southern Oregon-Northern California and from those in the Lower Columbia River and southern Washington. ■

Although they may look identical at this stage, alevin from one population will later exhibit life history strategies markedly different from other populations within the same species. These differences ultimately define the complex population structure of *Oncorhynchus*.

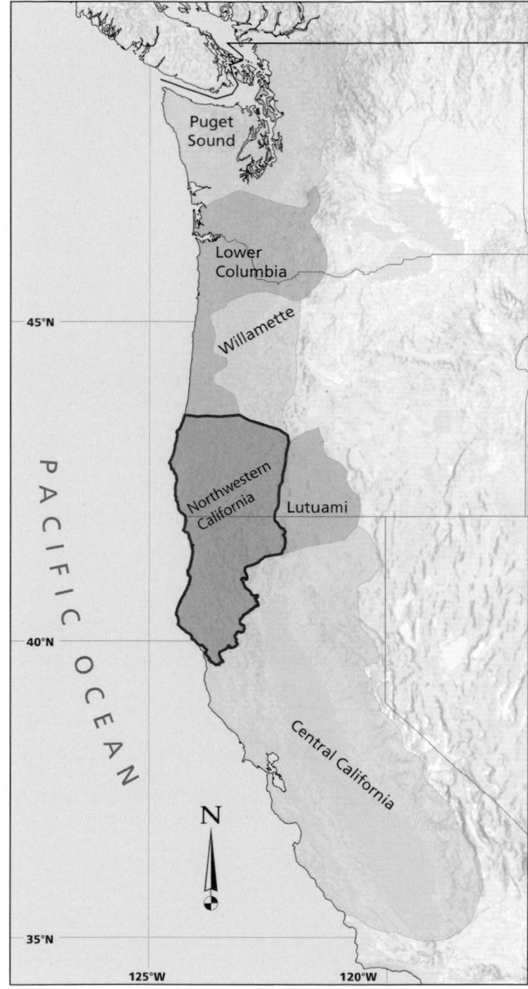


Native American Core Cultural Provinces

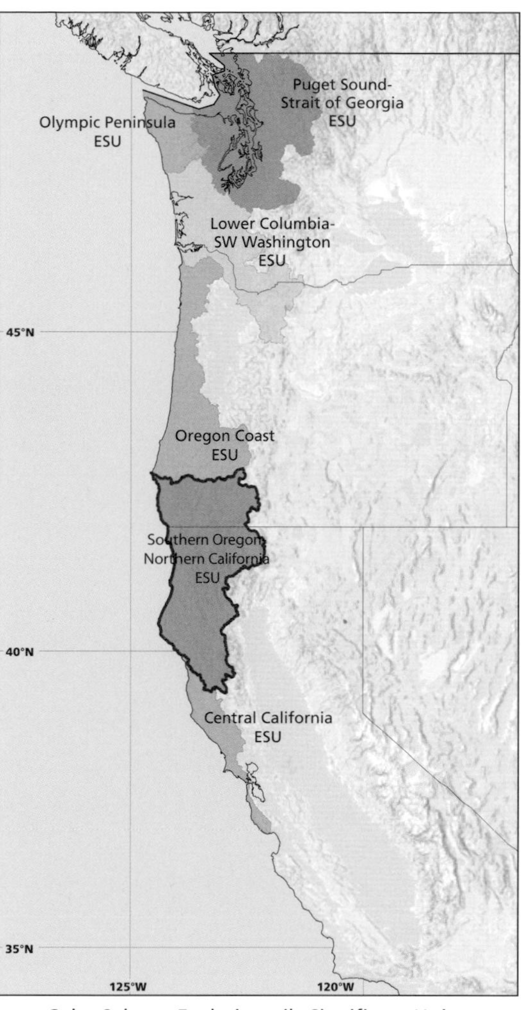


Coho Salmon Evolutionarily Significant Units

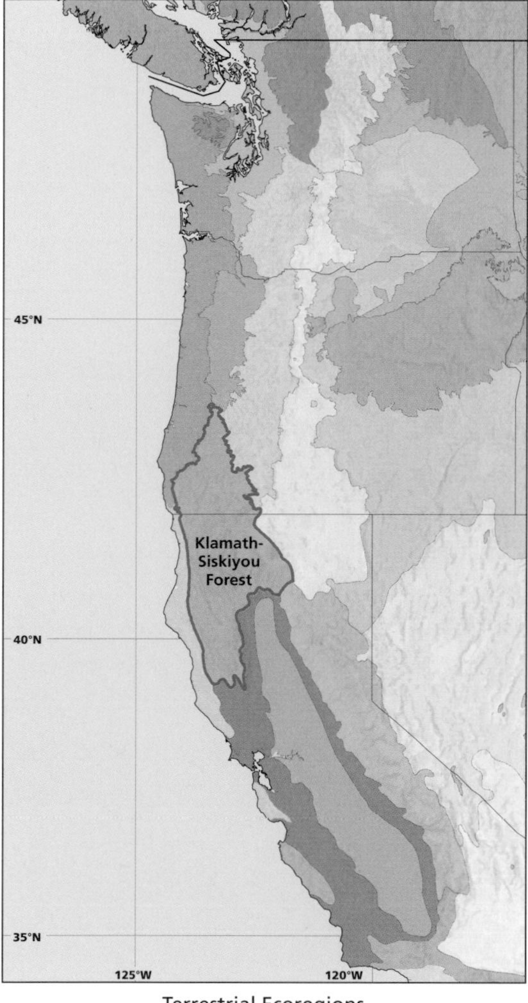


Terrestrial Ecoregions
(see page 54)

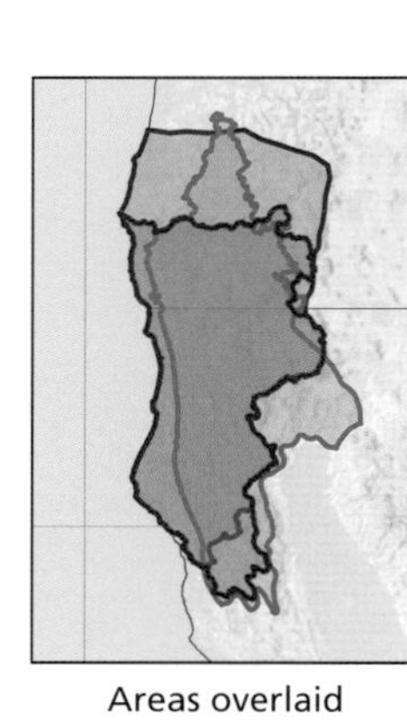

Areas overlaid

**NATIVE AMERICANS AND SALMON COEVOLUTION** In this trio of maps, we examine one region through three lenses: people (the Northwestern California core cultural province, including the Karuk and Yurok tribes); fish (the Southern Oregon-Northern California coho evolutionarily significant unit, including discrete salmon populations within it); and place (the Klamath-Siskiyou terrestrial ecoregion). By overlaying these data layers, we can envision how Native Americans, salmon, and landscape influence and reinforce one another, as humans and fish adapt to the unique setting of their catchment. The Klamath Range to the north, the Sierra Nevada to the east, and the Sacramento-San Joaquin estuary to the south served to isolate biological units of both people and fish, which adapted uniquely to the basin. The Eel, Klamath, and Rogue rivers formed the heart of this catchment and bonded living creatures within it.

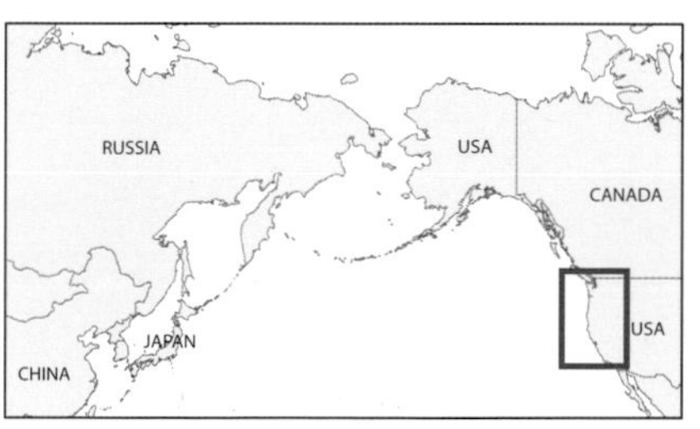

**NATIVE FISHERIES MANAGEMENT**

In the mid-19TH century, as European settlers migrated to the American West seeking opportunity and as the Meiji Restoration ushered a new merchant class into Japan, native peoples were marginalized, their lands overtaken, and their cultures diminished by lack of access to natural resources, including salmon. The extent to which this damage to indigenous people and their communities has been mitigated varies widely by tribe, region, and nation.

**JAPAN** The word for salmon in the language of Japan's indigenous people is *shipe,* which means "the staple food for the Ainu people."

The Ainu culture has depended on salmon for thousands of years. Until approximately 500 years ago, the Ainu fished freely and traded with the Japanese, but colonization and trade wars chipped away at the Ainu's access to salmon. By the early 1700s indigenous rights had eroded under economic and political pressures. The Ainu were proclaimed landless in 1872. The government relocated the Ainu away from fisheries and rich agricultural areas, and nonnatives hunted and fished without restriction on former Ainu lands. In 1883 the last major river of Ainu cultural importance was closed to harvest. By the late 19TH century, governing bodies outlawed Ainu fishing; and in 1899 the Ainu were decreed fully assimilated and proclaimed a nonpeople through the Hokkaido Former Aborigines Protection Act.

Nearly a century later, in 1997, this act was officially replaced by the Law for the Promotion of the Ainu Culture, which requires that local and national governments support the traditions and culture of the Ainu people. The legislation arose from a landmark court decision, based on a lawsuit filed by two landowners who opposed expropriation of Ainu farmland during the construction of the Nibutani Dam along the Saru River, which began in the 1960s. The court ruled that the local government's decision to build the dam—which had submerged the Ainu's sacred land and rendered it unfit for salmon habitat—was illegal because it violated Article 27 of the International Covenant on Civil and Political Rights. (Article 27 obligated Japan to recognize the Ainu as an indigenous people of Japan and to protect the Ainu's distinct culture.)

Although fishing is not permitted in Japan's rivers, the governor of Hokkaido may make exceptions for native tradition and ceremony. Here in Sapporo, Ainu celebrate the season's first salmon.

During ceremony, Japan's indigenous people place salmon in a woven basket, demonstrating traditional Ainu fishing techniques.

Today the Ainu make up a small percentage of modern-day Hokkaido's population, and salmon remains a dietary staple and a cultural symbol. As throughout the rest of Japan, salmon fishing is banned on Hokkaido's inland waterways, leaving the Ainu without their most important traditional resource. The Nibutani Dam decision required the governor of Hokkaido to grant the Ainu fishing rights for ceremonial and traditional purposes. These rights are typically given several times a year, though the licenses narrowly limit catch and prohibit commercial use of salmon. There have been proposals in recent years to grant the Ainu permanent indigenous rights to salmon fishing, but these efforts have been unsuccessful thus far.

**RUSSIA** Assimilation was the official policy regarding indigenous peoples in the former Soviet Union until its collapse. Today more than a dozen discrete groups of indigenous peoples inhabit the Russian Far East, often in remote regions. Some depend predominantly on salmon and retain many elements of traditional subsistence culture.

Several Russian laws protect the rights of indigenous minorities to maintain cultural traditions, including subsistence hunting and fishing. Articles 69 and 73 of the Constitution of the Russian Federation acknowledge the role of the federal government in protecting the cultures and traditions of native peoples. However, it is widely held that these constitutional principles, and the various laws that have been passed in the same spirit, are not enforced.

Salmon dry in a Markovo native tent in Anadyr, Russia, among the northernmost regions in the distributional range of *Oncorhynchus*.

New legislative provisions have established Territories of Traditional Nature Use, which are approved by the Russian Federation and guarantee indigenous groups the right to participate in resource management decisions and to continue traditional fishing practices without having to comply with state fishing regulations. To date, the success of these laws has been limited, and many hurdles must be overcome before they can present meaningful opportunities for Russia's indigenous populations.

**ALASKA** With American statehood in 1959, Alaska's government only recently entered into organized political discussions with native peoples regarding rights to natural resources. Two major pieces of federal legislation—both prompted by negotiations around oil reserves—have shaped the Alaskan native communities. In 1971, to secure the land that would create the Trans-Alaska pipeline from Prudhoe Bay to Valdez, the Alaska Native Claims Settlement Act (ANCSA) gave Native Alaskans title to 44 million acres and the authority to create 13 regional, four urban, and more than 200 village corporations. This total compensation package—which paid indigenous peoples US$962.5 million for ceded land rights—affected 80,000 Native Alaskans.

Further negotiation, largely guided by United States and Russia's ownership issues over Prudhoe Bay oil reserves, led to passage of what is known as the Alaska lands bill in 1980. Whereas the legislation was to include settlement of native claims to land, the concept of tribal sovereignty was left largely untouched. That said, the Alaska

Gear construction was tailored to local needs and materials. Above, fish traps built by Native Alaskans along the Chilkoot River (1894).

National Interest Lands Conservation Act (ANILCA) includes provisions that strip Native Alaskans of their tribal hunting and fishing rights in return for state and federal protection of the rural subsistence way of life in Alaska. Although all Alaskans can fish for subsistence, Native Alaskans are given limited preference through regulatory mechanisms, including customary and traditional determinations of culturally important fishing grounds. In these areas, Native Alaskans may have priority over other users in times of resource scarcity. However, the details of these matters remain unsettled because federal interests have taken over subsistence programs and salmon management on federal lands.

**BRITISH COLUMBIA** Canada's native fishery management laws are young and developing. In 1982 Canada's Constitution Act affirmed the existing aboriginal and treaty rights of Canada's First Nations. In 1990 this legislation was tested in the Supreme Court in what came

In 1993 Marie Smith Jones, the last full-blood Eyak, extends her thanks to fishermen who protested logging on sacred Native American land near Cordova, Alaska. For millennia, the Eyak have depended on salmon for food and spiritual renewal.

to be known as the *Sparrow* decision. This ruling held that aboriginal rights protected under Canada's Constitution included access to fisheries and the right to procure fish for food, social, and ceremonial purposes. As a result, limits were placed on the ability of the national Department of Fisheries and Oceans (DFO) to regulate aboriginal fisheries. The *Sparrow* decision did not settle the matter of the constitutionality of First Nations' rights to commercial fishing; in fact, the question of what the government recognizes as inherent tribal rights to fish remains unanswered.

Meanwhile, some specifics have been addressed. In 1992 the DFO introduced the Aboriginal Fisheries Strategy (AFS), outlining the agency's responsibilities in negotiating agreements with First Nations for harvest opportunities. Agreements may be tailored to the specific needs of First Nations; rights granted may take precedence over all other uses of fisheries, with the exception of conservation. Today the DFO maintains a commitment to provide commercial fishing opportunities for First Nations and to encourage economic development and community sustainability, although the DFO is presently renegotiating the AFS.

In 1993 the DFO introduced the concept of Communal Fishing License Regulations (CFLRs), giving the DFO supervision over all fisheries while providing fishing access to First Nations. As a result, First Nations who wish to fish for salmon may be granted an Aboriginal Communal Fishing License following negotiations with the DFO. The DFO may establish certain restrictions on aboriginal licenses only for conservation purposes. In the event that agreement cannot be reached, a single Communal Fishing License is granted. Since 1992, 125 agreements have been signed on a yearly basis. Under many of the agreements, First Nations have comanagement responsibilities for some or all aspects of fisheries. Allocation is annually renegotiated with the DFO, and First Nations are provided with resources and training to aid the DFO in stock assessment and data collection. Some First Nations have been particularly successful in forging cooperative management agreements with the DFO; agreements with First Nations on the Fraser, Skeena, and Nass rivers and on Vancouver Island have been in operation for nearly ten years.

The Tulalip Tribes celebrate their First Salmon ceremony in Marysville, Washington. The Tulalip Reservation, a home for several Puget Sound tribes, was established in 1855 with the signing of the Point Elliott Treaty.

**WASHINGTON, OREGON, CALIFORNIA, IDAHO (WOCI)** To Native Americans, maintaining healthy salmon populations has ensured cultural survival for thousands of years. But with European settlement came clashes over natural resources, including salmon. The 1787 Northwest Ordinance described the theory of governance over the land west of the Ohio River, offering protections for Native American

During a tribal opening ceremony, a member of the Swinomish tribe, in LaConner, Washington, clears a net full of nearly 4,000 pink salmon.

rights and traditions: "The utmost good faith shall always be observed toward the Indians; their lands and property shall never be taken from them without their consent; and, in their property, rights, and liberty, they shall never be invaded or disturbed, unless in just and lawful wars authorized by Congress; but laws founded in justice and humanity shall from time to time, be made for preventing wrongs being done to them, and for preserving peace and friendship with them." The practice of governance, however, would not necessarily uphold this optimistic decree, which ultimately was not enforceable.

In WOCI many tribes reserved fishing and hunting rights when they ceded their land through treaties with the United States in the 19TH century. For tribes in the Columbia River basin, western Washington, and elsewhere, the right to fish in traditional and customary places was reserved in treaties made in the 1850s that were actually documents signed by native peoples under false pretext. These rights have been legally challenged numerous times, and in each case tribal rights have ultimately prevailed.

Beginning in the 1960s and 1970s, several federal court decisions in Oregon and Washington reaffirmed treaty rights to fish, including in the Columbia River basin. The 1969 *U.S. v. Oregon* decision reaffirmed treaty rights to fish, although these rulings apply only to Washington and Oregon treaty tribes that negotiated agreements in the mid-19TH century.

The *Boldt* decision, handed down by Federal Judge George Boldt on February 12, 1974, is often held up as a landmark solution to shared stock conflicts between native and nonnative Americans. Decided with respect to Puget Sound fisheries, the decision decreed an even split of the salmon harvest between tribes and nontribal interests. Furthermore, it validated interpretations of treaties among Sovereign Nations "as the native peoples understood them" and retained the notion of "usual and accustomed places."

The majority of the tribes in Washington and Oregon, in fact, had no treaties, treaties that were not ratified, agreements created by executive orders, or treaties that did not address fishing rights. Called nontreaty tribes, these Native Americans experienced a parallel struggle in establishing inherent tribal rights to fish. Today's nontreaty tribal members must fish according to state regulations, not by the rules of their tribal governments. (There are exceptions, such as the Hoopa and Yurok tribes on the Lower Klamath in California, which forged their own agreement with the federal government in 1988 to apportion natural resources.) Recognition of cogovernment resulted in comanagement, and many native groups have established their own fisheries management authorities. In recent years partnerships among tribal, federal, and state agencies have evolved into formal cooperative management agreements.

The Columbia River basin offers a vivid example of the jurisdictional complexities inherent in the management of native fishing rights. At the headwaters of the Columbia River in the mountains of British Columbia and Alberta, Canadian First Nations are still deeply enmeshed in resolving treaty and aboriginal rights to fish. Downstream in the United States several nontreaty tribes on the river and its tributaries have significantly diminished rights to the salmon harvest in Washington and Oregon. However, in the lower reaches of the Columbia River basin, four treaty tribes—including Yakama, Warm Springs, Nez Perce, and Umatilla—exercise their full rights to comanagement and ownership of the fishing resource.

Darrel Jack uses a single-pointed fish pugh to transfer catch, in the Washington portion of the Columbia River.

## THE INDUSTRIAL ECONOMY AND FISHERIES MANAGEMENT

Of the seven countries that rim the North Pacific, we omit China and North and South Korea from our study because we have so little data for salmon from these jurisdictions. The remaining jurisdictions have enjoyed their own histories and human connections with salmon. Within each—Japan, Russia, Alaska, British Columbia, and WOCI—salmon have had different importance and meaning and varying economic and cultural value. From gear to contemporary management infrastructure, core political provinces have constructed very different worlds around the same Pacific salmon.

**JAPAN** With the overthrow of the Tokugawa Shogunate in 1868, Japan underwent an abrupt transition from a hierarchical empire to an industrial economy. The Meiji Restoration fast-tracked Japanese commerce and entrepreneurship. An emergent and powerful class of fish merchants exploited Hokkaido's salmon wealth and funded expeditions in Russian waters to the Sea of Okhotsk. The arrangement was agreeable for a time: the salmon-rich Russians had no source of salt for preserving and scant shipping resources, while the Japanese fishing fleets provided good infrastructure and ready markets. The 1880s witnessed Japan's peak wild salmon catch, but in search of more fish, the Japanese ventured farther into Russian waters. Conflict brewed and escalated to help incite the Russo-Japanese War in 1905. The Japanese occupation of Sakhalin from 1907 to 1945 debilitated many salmon streams, which were altered by the process of splash-damming, employing stream power to transport logs downstream.

Having exhausted its own wild resources, Japan explored the viability of hatcheries. During the 1980s hatchery-supported, shore-based fisheries overtook distant-water wild capture fisheries in economic importance. Today salmon remain essential for local economies on Hokkaido and northern Honshu, but farmed salmon imports are beginning to erode the stability of traditional fishing communities.

The five core political provinces within our study have their management differences, but none stands apart like Japan. About 95 percent of the salmon produced in Japan are hatchery derived, and catch is viewed in terms of biomass. Nearly all scientific research is performed in hatcheries. Rivers in Japan are managed by the construction ministry, with the exception of several dozen conservation rivers. Sport fishing in freshwater for salmon, with the exception of juvenile masu, is illegal.

What makes Japan fisheries management unusual is that it has been and continues to be community-based and sets practices from the bottom up. Japan has no nationwide fishery regulations; it is only beyond the 200-mile exclusive economic zone that the national fisheries ministry sets management regulations. Fisheries Cooperative Associations (FCAs) operate most hatcheries and, in most cases, take all the chum and pink salmon returning to spawn every season.

For centuries, in a practice that dates back to the feudal system when one lord oversaw a region's fisheries, this trade has been the ballast for small coastal communities. Fishing provided not only economic stability but also national security; because Japan had no naval fleet until the early 20TH century, fishing villages were valued

Fishermen pull in a net full of salmon near Ishikari City, Hokkaido, Japan. Commercial fishing in Japan, mostly for pink and chum, is managed at a community scale by the local Fisheries Cooperative Associations.

as the first line of defense and as outposts for communication that served as an early-warning system for attack. Fishing also served as a form of food security for an island nation that had limited land for tending large herds of livestock.

Today cooperatives serve the societal role of corporations to which fishermen pledge loyalty. The association of all the fishing cooperatives is Zengyoren, which works in tandem with the Ministry of Agriculture, Forestry, and Fisheries.

Sport fishing has gained tremendous popularity in Japan in recent years, and, as a result, entrepreneurs, scientists, and government officials are investing more time and resources into research for river health. Activism is a contemporary development in Japan, and concerned citizens have mobilized to oppose channelizations and dam building.

**RUSSIA** The salmon history of the Russian Far East has been joined to Japan's, for better and for worse. The past century has witnessed muted and overt conflict and nearly continual negotiation over fishing rights in the Seas of Japan and Okhotsk and the Bering Sea. The Japanese maintained dominance over infrastructure and technology; out of necessity, the Russians often benefited from Japanese capabilities, and the fishing boundaries between the countries blurred prior to World War II.

In the Soviet era, fisheries loomed in institutional importance, providing the infrastructure and all social services for coastal and streamside townships. Most Russian Far Easterners were part of the working class, and many were fishermen who rarely partook of salmon, considered a delicacy. Instead, salmon caught in the Russian Far East were shipped west to the cultural elite. The fishermen and their families instead ate char, herring, shad, pollock, and other less valuable species of fish—a daily reminder of the economic necessity of highly prized salmon in the lives of Russians.

Workers gut salmon at a plant on Sakhalin Island, Russia, where fisheries remain one of the largest sectors of the local economy.

After World War II, the Russians took southern Sakhalin back from the Japanese. Most of the hatcheries in Russia, primarily pink and chum, are on Sakhalin. The Russians upgraded the infrastructure in the 1980s with help from the Japanese, who maintained an active gill net industry off Sakhalin. After the collapse of the Soviet Union, fisheries erupted in chaos and catch limits were rarely enforced. The industry sped toward privatization as fishermen received stock in their former cooperatives. Without subsidies, remote companies failed. Stock was further centralized in the hands of the enterprises' former heads.

Salmon continue to be the backbone of the Russian Far East economy in many places, particularly in Kamchatka, Primorye, and Sakhalin. A barrel of fish, smoked or brined, can provide for the average family through the winter; more importantly, the proceeds from a few buckets of salmon roe can support a family for one year. As a result, harvest pressures on salmon spawning grounds are tremendous, particularly near townships.

Officially, the government owns the salmon resource in contemporary Russia: RybVod is the institution that oversees fisheries management and salmon enhancement programs; the Fisheries Research Institutes (the NIRO system) develop forecasts; regional fisheries councils apportion catch to individual companies after quotas are apportioned by the Moscow Federal Fisheries Committee to each of the territories.

Russia has neither numeric escapement targets nor biodiversity mandates, except for the United Nations Development Programme (UNDP) Kamchatka Salmonid Biodiversity Program, launched in 2003 and funded by the UNDP and the Global Environmental Facility. This effort is dedicated to helping Kamchatka's fishery agencies shift management strategies from maximizing biomass to preserving biodiversity.

**ALASKA** Since 1993 Alaskan fisheries have captured around 40 percent of the average annual North Pacific salmon catch. Fishing has always been a cornerstone of the economy, which is representative of the maverick pioneering spirit of this vast wilderness state. Personal-use fisheries continue to be an important food source for many Alaskans, especially outside of the three major urban centers and for native peoples. It would seem that the human hand on salmon populations has been comparatively gentle in Alaska, but this was not always so.

Until statehood in 1959, Alaska was a U.S. territory; as such, its processors, canneries, and, sporadically, hatcheries were managed by the Bureau of Fisheries in Washington, DC, to the dismay of many Alaskans. The salmon canning industry began in Alaska in 1878. Development was slow at first: by 1900 there were 42 shore processing facilities, but by 1929, Alaska had 159 active canneries. Fluctuating abundance and encroachment by foreign fleets spurred regulation of the Alaskan fisheries from Washington, DC, although little was done and even less was enforced. World War II brought about another period of intense catch as the U.S. secretary of the interior relaxed fishing regulations to maximize production for the war. The labor force remained predominantly nonterritorial, and powerful fishermen's labor unions barred or limited entry into the fishery by local fishermen.

Limited opportunities, declining salmon runs, and overharvest (due to the use of fish traps instead of small boats) spurred the drive for Alaska statehood. The new state created its own Department of Fish and Game and assumed control of all of Alaska's fisheries, except for Bristol Bay, which was the most lucrative. Embedded in the state's constitution was a provision for sustainability. Salmon fishery management was turned over to biologists, who by 1965 had set formal escapement goals for the major fisheries. Limited entry and vessel licensing systems were established and enforced, particularly during periods of low return, such as from 1968 to 1972.

Depleted stocks prompted the state to invest in new hatchery systems. A decade later, these were sold to local cooperatives that continue to manage them today. Because of the relative health of Alaska's river systems, the presence of hatcheries is controversial; some argue that hatcheries are warranted only in a place like Ketchikan, a major timber town that has greatly altered streamside forests.

The success of Alaskan fisheries is due in part to management sovereignty as well as to

## MANAGEMENT FOR BIODIVERSITY IN THE UNITED STATES

If we may compare concepts, salmon biodiversity and fisheries harvest management are "processes" that have worked in opposition. Biodiversity creates a complex mosaic of genetically distinct populations resilient to change: while some are declining, others flourish, and the species persists in the face of multiple stressors. In sum, biodiversity maximizes individuation to ensure species longevity.

Fisheries management, on the other hand, focuses on maximizing the harvestable resource. In practice, fisheries managers reduce the complexity of salmon life history into something more tractable: managed aggregate "stocks" composed of multiple separate populations. But harvest at this coarse scale can put weak populations at risk. Managing for biodiversity begs the question: how do we reconcile the need for harvest management and biodiversity preservation?

The 1973 U.S. Endangered Species Act prompted inquiry into resolving questions of stock scale; by the 1990s scientific advancements enabled identification of at-risk populations using more strict biological criteria. But even these units may be too coarse for biodiversity preservation.

New molecular genetic tools can resolve biological mechanisms at the most fundamental scale and, in essence, describe evolution in action. For example, it was long held that rainbow trout and steelhead were reproductively isolated, but studies using molecular genetic markers to establish parentage demonstrated that these subspecies do interbreed and produce offspring with both life history forms. By managing these "stocks" separately, agencies had missed the mark; but by recognizing these two life history forms as a single interbreeding population, agencies can now better manage for biodiversity of *Oncorhynchus mykiss*. ■

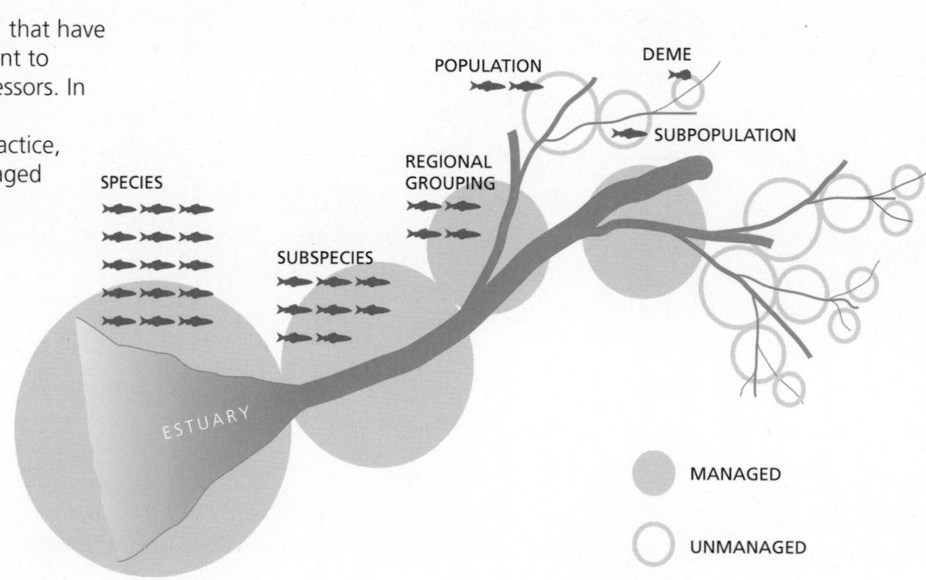

Fishermen lay fireweed to symbolize rebirth during a protest at a logging operation in Prince William Sound near Cordova, Alaska.

several major pieces of legislation. The Alaska Legislature passed the 1973 Limited Entry Act, which ended the practice of open access and stemmed the growing number of fishermen. The 1976 Magnuson Fishery Conservation and Management Act forced foreign fleets to move outside of the Exclusive Economic Zone.

The wiles of politics aside, Alaska's fisheries have been vulnerable to the effects of climate variability (see page 49). Scientists have identified three major climatic periods in the 20TH century that negatively affected Alaska stocks: the 1910s, 1940s, and 1970s. The most recent lowpoint coincided with salmon market collapses and devastated most of the commercial fishing fleet in Alaska. Legislation has attempted to compensate for climatic fluctuations, which further muddied the complexities of harvest quotas.

**BRITISH COLUMBIA** In the 1920s Canadian scientists were the first to document that salmon return to natal streams. Among the five core political provinces in the Pacific Rim, Canada has been consistently farsighted when it comes to honoring and protecting biodiversity. But separate governance of natural resources—fisheries at the federal level and habitat at the provincial level, with management often at odds—has contributed to strained relations between authorities that continue today.

Although Canada has fished with a softer hand on salmon runs than have other Pacific Rim nations, British Columbia did not escape a wave of westward immigration in the 19TH century, which had grave consequences for the health of salmon populations. Logging and mining impaired many river catchments, and urbanization and overfishing along the Fraser River led to serious declines of salmon runs. Habitat alteration was a significant factor; in the early 20TH century, salmon harvest in British Columbia took an economic backseat to resource extraction, which had deleterious effects on rivers and catchments. Canada also explored hatcheries in the 1930s but rejected them in favor of natural production.

Within the last decade, First Nations were constitutionally guaranteed access to salmon; therefore, fishery managers must do all they can to get salmon upstream, focusing on the upstream subpopulations in managing mixed-stock fisheries. Declines in production as a result of shifting climate regimes in the 1970s prompted the adoption of hatcheries under the Salmon Enhancement Program. With further declines, in 1985 the DFO was established at the national level to protect Canada's oceans and inland waters. Research and management were linked and became relatively streamlined. Commitment to salmon issues was reiterated in 1998 with the government's creation of the Pacific Fisheries Resource Conservation Council (PFRCC), an independent body that reports to the provincial and federal governments annually on the status of British Columbia salmon stocks and their habitats.

With the exception of Vancouver and the lower Fraser River Valley, Canada's habitat remains relatively pristine. Because Canada's cold, deep waters and fjords are ideal for aquaculture, in the 1980s Norwegian salmon farmers began building net-pen farms along the island coastline of British Columbia, largely for nonnative *Salmo salar* (Atlantic salmon). This controversial industry has created jobs and boosted supply of salmon, but its effects on the health of humans, the environment, and wild salmon are the subject of intense scrutiny. Research in Europe has demonstrated that farmed salmon can spread disease and parasites such as sea lice to wild salmon populations, but the relationship has not

British Columbia is a major force in global fish farm markets. Aquaculture is the province's biggest agricultural export.

been unequivocally proven in the North Pacific. Other concerns include questions of carrying capacity and changes in the food web as Atlantic salmon escapees have been observed spawning in northern British Columbia's rivers. Since 1993 the PFRCC has fostered the Salmon Aquaculture Forum to improve dialogue and seek solutions on net-pen farms (see page 36).

**WASHINGTON, OREGON, CALIFORNIA, AND IDAHO (WOCI)** We describe the management history of WOCI as a unit, but this region is far from homogeneous: states manage salmon separately and independently of the federal government.

The balance between First Nations and salmon was unsettled by early European immigrants to North America. But that balance was irrevocably upset in the mid-19TH century with the advent of the industrial economy, which was fueled by unchecked natural resource exploitation. Fur traders and trappers, farmers, and gold miners flooded the American West, often transiently, churning up landscapes and moving on for more prospecting. California attained statehood in 1850; Oregon in 1859; Washington in 1889; and Idaho in 1890.

The first cannery was built on California's Sacramento River in 1864, and the industrial salmon economy was launched. Caught off-guard by the economic free-for-all, legislators could not make harvest laws fast enough to protect stocks, and gross overharvest quickly followed. Incredulous observers wrote of the enormous salmon runs, estimated by the Northwest Power Planning Council to be as much as 16 million fish at their peak in the Columbia River basin. Harvest kept in step and depleted runs, and the federal government saw the need for artificial propagation, building the first federal hatchery in 1872 on the McCloud River, in the Sacramento River basin in California.

The first cannery appeared on the Sacramento River in 1864. Less than two decades later, in 1883 the river would produce 200,000 cases of salmon a year, packaged by 21 canneries.

A purse seiner casts its circular net—which may be anywhere from 7 to 70 meters long—and draws it closed to trap fish. Until the 1960s nets were gathered by deckhands; today it is done hydraulically.

Gasoline-powered boats were introduced around the turn of the century, and so began the practice of trolling, which revolutionized the fishing industry. New technology to preserve salmon transformed small-scale subsistence fisheries into large-scale commercial fisheries. A class war developed as individuals mobilized against companies and lobbied legislatures to ban the use of corporately owned fish wheels and traps. Individual gear groups battled each other intensely to ban gill nets and purse seines.

In the 1930s dams constructed by the New Deal's Works Progress Administration (WPA) aggravated losses; hatcheries, built as a mitigation strategy, supplemented salmon-drained rivers. Economic and ecological battles increasingly broke out over salmon resources at the local, regional, state, and federal levels. Coalitions of special interests—farmers, scientists, fishers, tribes, and more—required mediation and advocacy. After World War II, governance over fisheries issues remained fractured and scattered across many agencies at the federal level.

The 1970s presented two watershed events that have since shaped WOCI fisheries. In 1974 the *Boldt* decision allocated 50 percent of salmon available for catch to treaty tribes and presented avenues for collaboration between tribes and nonnative peoples.

The 1976 passage of the Magnuson Fishery Conservation and Management Act created the Pacific Fishery Management Council, one of eight regional organizing bodies through which conservation and harvest regulations would be set.

The machine of WOCI agency management has expanded into an unwieldy, decentralized, and fragmented behemoth. States work within their departments of fish and wildlife. They must also coordinate with the multilayered landscape of other fish management authorities and with adjoining states through interstate agreements; with native tribes; and between two departments at the federal level—National Oceanic and Atmospheric Administration (NOAA Fisheries, which manages anadromous fish) and the U.S. Fish and Wildlife Service (which manages resident fish). States oversee fishing activities in waters that are within three miles of their coastline. Further out in the ocean, yet still within the Exclusive Economic Zone of the United States, between three and 200 miles offshore, the federal government is responsible for management and oversight, acting through NOAA Fisheries and the U.S. Coast Guard.

## INDUSTRIAL FISHING GEAR

The way people fish reflects a society's cultural organization. In general, today's salmon fishing fleets also mirror the cultures from which they set sail.

The ethos of the North American fisherman has been entrepreneurial, individualized, and market-driven, and the diversity of gear types underscores these characteristics. Gear such as fish wheels or Alaska-style net traps operated by cannery operators or other private landowners were eliminated from North American fisheries due to political pressures favoring small operators from the 1920s through the 1950s. Fleets were tailored to fish ecology; trollers, for example, were suited for hunting nonschooling salmon (solo coho, chinook, and steelhead), which are more efficiently harvested by hook and line. In fishing for schooling fish such as pink, chum, and sockeye, North American fishermen used purse seines and gill nets, which could be handled by small crews.

On the western side of the Pacific, fisheries have traditionally been cooperative ventures. The gear they use—gill nets, beach seines, and net traps—require large crews. These collective operations efficiently harvest schooling chum and pink salmon, the predominant western Pacific species.

In the 1960s the use of nylon monofilament nets spurred the practice of high seas driftnet fishing. Nets up to 65 kilometers long and 12 meters deep were strung across swaths of ocean, ensnaring fishes, marine mammals, and seabirds, regardless of size or species. Let out at night when the filaments were essentially invisible, the drift nets were used largely by Japanese, Taiwanese, and South Korean fleets in the open ocean. Until 1976 drift nets were permitted outside the 12-mile zone offshore; thereafter, within the 200-mile EEZ (unless otherwise dictated by prior treaty agreement).

Due to the destructive nature of this practice, the United Nations General Assembly passed a resolution on June 30, 1992, calling for a moratorium on the use of drift nets. In spite of strong enforcement, the practice continues: in the North Pacific Ocean, 26 possible cases of unauthorized large-scale driftnet fishing—including seven interceptions of active driftnet fishing—were reported in 2003. ■

**Gillnetter**
Typical length 25 to 35 feet (8 to 11 meters); average 30 feet (9 meters)

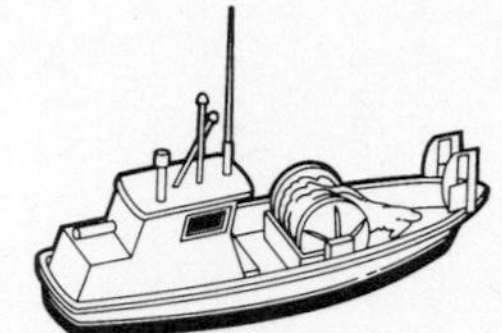

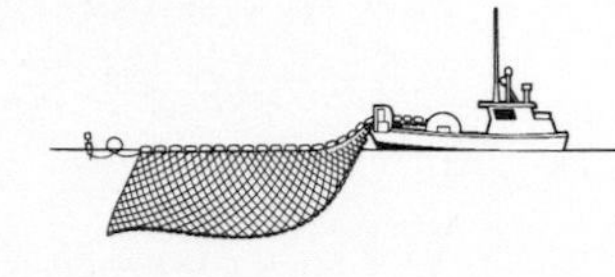

**Troller**
Typical length 35 to 65 feet (11 to 20 meters); average 50 feet (15 meters)

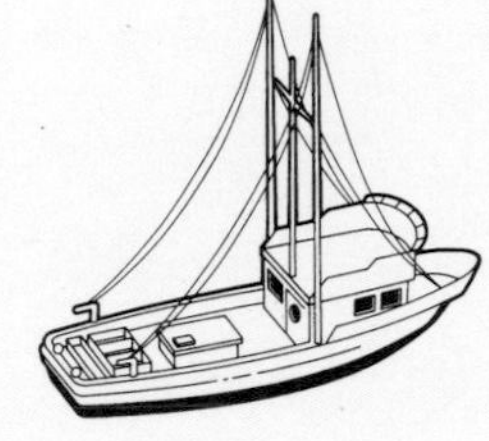

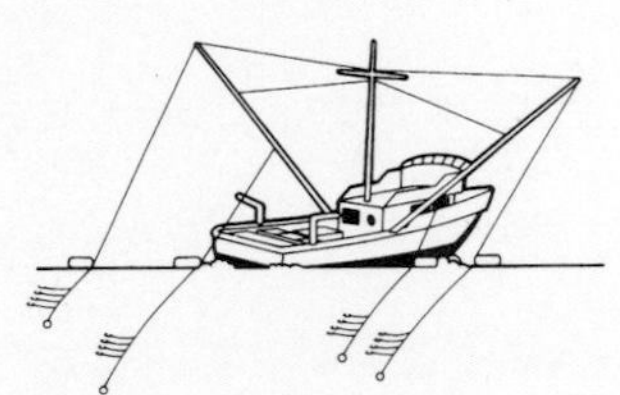

**Seiner**
Typical length 50 to 80 feet (15 to 24 meters) average 70 feet (21 meters)

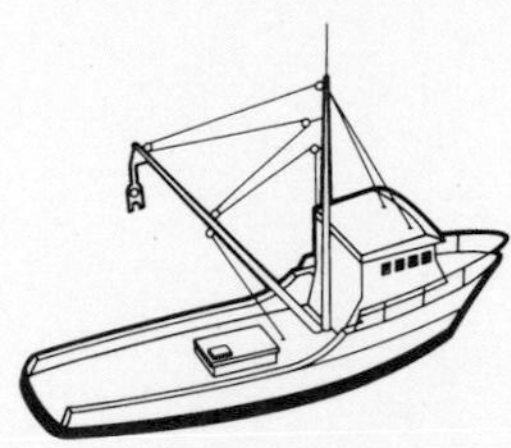

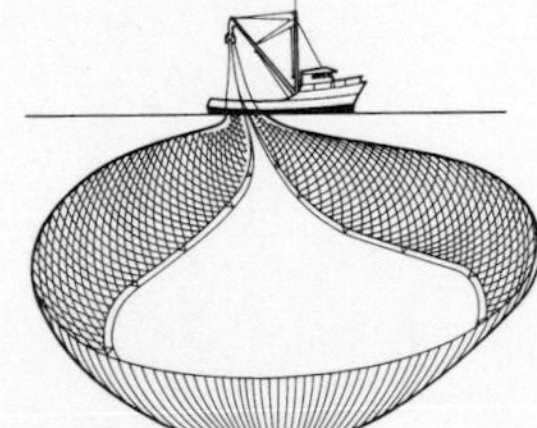

# Catch Composition

**The map to the right summarizes** average North Pacific salmon catch by jurisdiction during a recent nine-year period. Until the establishment of the International North Pacific Fisheries Commission, which was succeeded by the North Pacific Anadromous Fish Commission (NPAFC), data were aggregated (if at all) at a national level, using different metrics: Alaska, British Columbia, and WOCI quantified catch in millions of fish; Russia and Japan, in metric tons. When converted to millions of fish, chinook may be underemphasized, and pink may be overemphasized, due to their relative weights. Furthermore, accurate catch data from the western Pacific have been elusive. Prior to 1945, Russian and Japanese catch was frequently double-counted, especially pink salmon in Japanese-held lease areas on the rivers of Kamchatka and Sakhalin. Masu catch has proven to be exceptionally difficult to follow: in Japan's mixed-stock fishery, pink and masu catch are often reported together. In Russia, where masu populations are small, it is a personal-use fishery, and although quotas exist, reporting is unreliable.

Alaska and Russia net the highest catch numbers; Japan is third. With the exception of masu, there is greater species variety in the east than the west. Japan proves unique in its concentration on chum. The most ubiquitous species is pink salmon, which makes up most of the catch in Russia and Alaska. In the eastern half of the Rim, sockeye overtakes chum as the second most productive fishery. Both chinook and coho become increasingly significant commercially from west to east.

Steelhead has its own history. In years of low productivity it became a game fish in WOCI. After the *Boldt* decision in 1974, only subsistence fishing was permitted; steelhead was not sold commercially again until the 1990s—and only by Native Americans. Today British Columbia catch is negligible but WOCI increases slightly.

What portion of catch is wild and what portion is hatchery derived? The graph below demonstrates periodic declines in overall catch in the years around 1930, 1950, and 1970. During clusters of low annual returns, fisheries managers increasingly turned to hatcheries to supplement depleted wild stocks, until the last major catch downturn in the years around 1970, when most Pacific Rim nations adopted artificial propagation to support a portion of commercial fisheries. Since then, overall catch has increased. Alaskan hatcheries account for more than one-quarter of chum catch and around one-third of pink salmon catch; in Japan, more than 95 percent of salmon are hatchery derived.

Note that this graph plots salmon catch—wild and hatchery—in millions of fish. Measuring by number instead of weight overrepresents pink salmon catch and underrepresents chinook catch.

Since the 1970s widespread implementation of hatcheries and favorable ocean climate conditions have resulted in steadily increasing overall catch for most abundant salmon stocks.

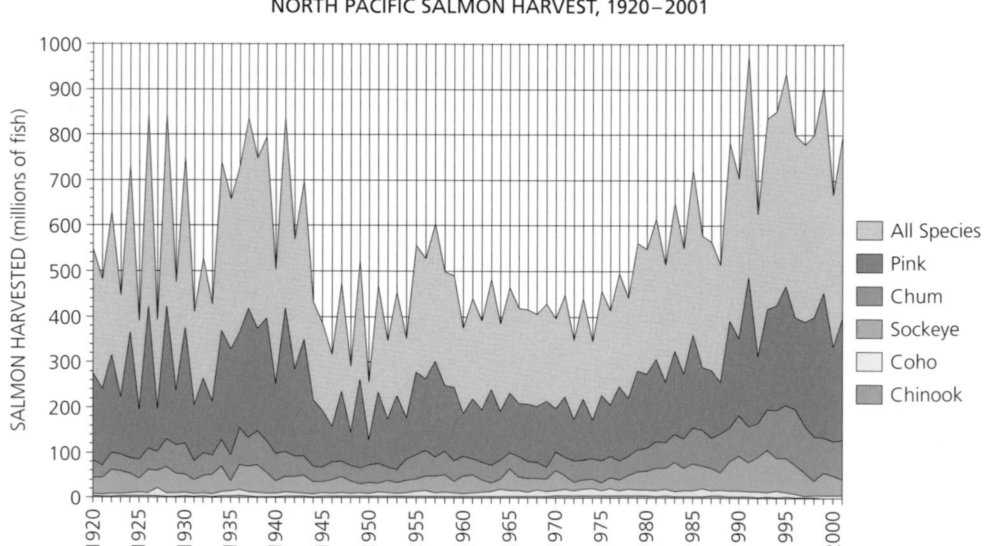

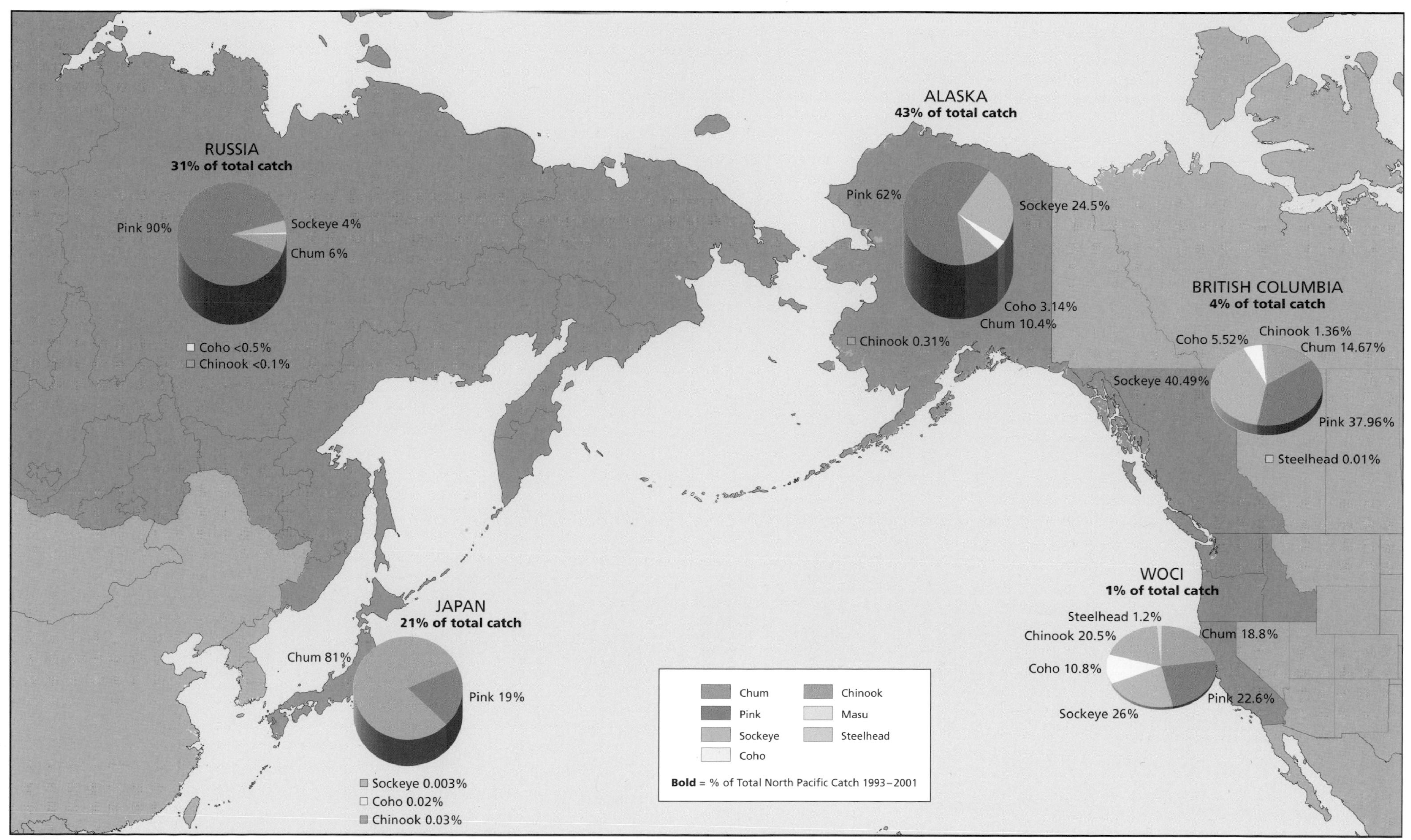


**NORTH PACIFIC CATCH COMPOSITION BY JURISDICTION (NINE-YEAR AVERAGE)** Among the five political jurisdictions, total salmon catch (both wild and hatchery) varies widely. Noteworthy is the trend in catch composition: from pink and chum in the west to a more equitable species distribution in the east.

Our numbers do not reflect masu catch in Japan and Russia; in some regions, masu represented a significant portion of the catch. Data were not available for personal-use, sport, or tribal fisheries. We report no data for South Korea, a new member of the North Pacific Anadromous Fish Commission.

NOTES: Data are reported for 1993–2001; data for 2000–2001 were preliminary when accessed.
California catch data: only chinook commercial catch is reported (1993–2000). These data were not available for 2001.
WOCI catch data: only commercial catch is reported. These data do not include Idaho catch; nor do they include "subsistence", sportfishing, or tribal catch.

# Hatcheries

In the years following the construction of the first salmon cannery (California, 1864), overfishing and habitat loss prompted fisheries managers to seek solutions to stock depletions—and soon the first salmon hatchery was constructed (California, 1872). Within a few years, the Japanese adopted artificial propagation and would later build hatcheries in Sakhalin during the Japanese occupation; by 1930, Russia built hatcheries in the Amur River basin. Yet commitment was mixed: Alaska closed its hatcheries in 1933; British Columbia, in 1936.

Ultimately, habitat loss, periodic run failures, and increased demand for salmon prompted aggressive new investments in hatchery programs. By 1970 hatcheries were widely implemented and today are used throughout the North Pacific—from plywood streamside incubation boxes to simulated natural lakes with engineered gravel-based spawning channels. Percentages of hatchery and wild salmon vary by region. In 1995 in Alaska hatchery fish accounted for 79 percent of total catch in Prince William Sound, but only 6 percent in Kodiak Island and the Alaska peninsula. Today in WOCI up to 80 percent of salmon are hatchery derived; in Japan, around 95 percent.

The benefits of hatcheries are compelling: they may offset losses in abundance in naturally spawning stocks and reduce harvest pressure on wild populations; they help stabilize commercial harvest; and they can serve as laboratories for the study and preservation of biodiversity. Hatcheries also provide a solid economic base for salmon-dependent communities, including native peoples.

Yet these benefits are counterbalanced with significant scientific uncertainty regarding freshwater and ocean carrying capacity, particularly within a trans-Pacific context (see below). Interbreeding and brood stock transfer among rivers can challenge wild population viability and genetic integrity. Hatchery production can mask ecological problems at the heart of declines in wild populations. Artificial propagation can deprive rivers of marine-derived nutrients (see page 54) essential to functioning freshwater ecosystems. Unfortunately, isolating ecological and biological variables to measure the impacts of hatchery fish on wild populations is extremely difficult, and so efforts to determine hatchery success or failure remain inconclusive.

Two legislative debates—whether to count hatchery fish under endangered species legislation (see below) and whether to allow surplus hatchery fish to spawn in the wild—have fulminated in recent years, underscoring the fact that hatchery management is among the most controversial issues in fisheries today.

## THE EFFECTS OF HATCHERY SALMON ON WILD POPULATIONS

In 2004, when the U.S. federal government announced that it would count hatchery salmon along with wild salmon to determine species status under the Endangered Species Act, it created an uproar among fisheries managers. Yet parties on both sides of this polarizing legislation agreed on one thing: interactions between hatchery and wild fish, both in the ocean and in fresh water, are poorly studied and not fully understood.

Each year since 1991, around five billion salmon fry are released into the North Pacific ecosystem: 40 percent come from Japan; 30 percent from Alaska; 13 percent from Russia; 8 percent from WOCI; and 7 percent from Canada.

Unlike wild spawning salmon, which emerge and migrate to sea over a period of weeks, hatchery fish are often held in pens until they reach optimal size and are then released all at once. Such releases can overwhelm smaller and less numerous wild salmon in a competitive food and forage environment. Hatchery fish can eat salmon eggs, alevin, smolts, and juveniles; they can interbreed with wild fish, interfering with genetic integrity. Some studies have demonstrated that hatchery fish do not perform as well as their wild counterparts in natural settings and may attract predators sensing weakness. Transfers of brood stock frequently fail and may inhibit the success of native salmon in these locales.

Finding common ground and conducting further research is essential. Limiting releases and separating wild populations from artificially reared stocks—both geographically and chronologically—offers promise in mitigating problems in carrying capacity and food web disruption. ■

Quantifying the effects of hatchery fish, such as these coho fry from Washington, on wild salmon remains an uncertain science. Scientists are studying how wild and hatchery fish respond, behaviorally and genetically, to competition for food in the ocean and in freshwater.

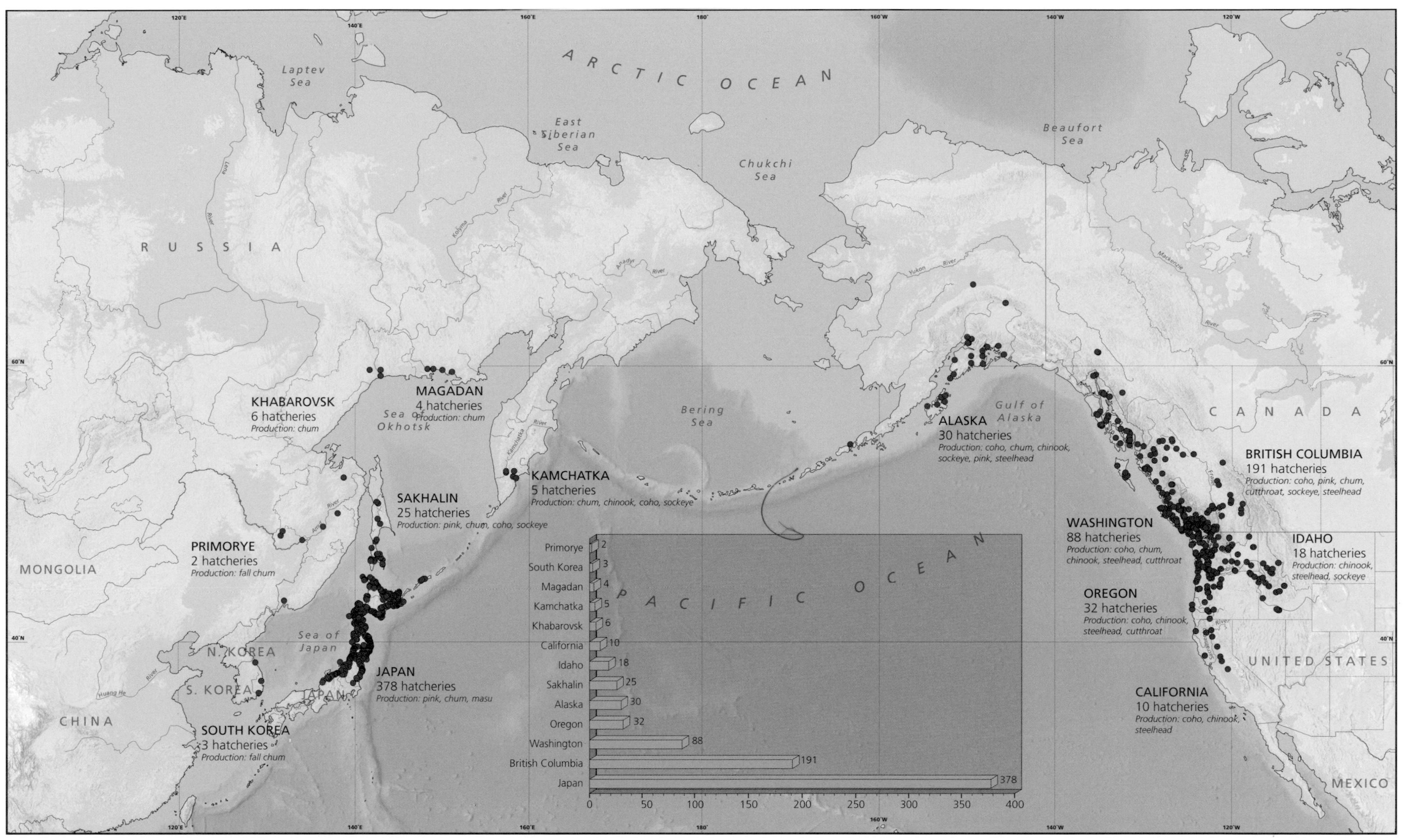


**HATCHERIES** Although the map above plots numbers of hatcheries around the North Pacific (our data are representative as of 2003), the effects of hatcheries on wild salmon populations and their habitats are widespread and not captured by points on a map. The numbers of smolts released can vary greatly among hatcheries. Hatchery brood stock collection and smolt-release sites are often located on different tributary systems, a practice that spatially extends the effects of hatcheries on wild salmon and habitat.

NOTE: Because of the layering of data points, dots representing hatcheries may not necessarily appear to add up to the actual number of hatcheries reported above.

# Fish Farming

**A GLOBAL INDUSTRY, FISH FARMING RAISES** salmon as livestock in near-shore net pens, where the fish spend their lives without migrating. Regardless of geography, one species of salmon—Atlantic salmon (*Salmo salar*, a different genus from *Oncorhynchus*)—is preferred in farming. Domesticated, it fares well in artificial conditions.

A typical salmon farm may include an average of 20 cages floating near the shore, each around 15 square meters and holding around 20,000 fish. Farmed fish consume food in the form of pellets, often composed of anchovy, herring, and sardines; one farmed fish will consume several times its weight in wild fish throughout its lifetime.

Researchers are exploring health concerns posed by fish farms to native fish and their habitats, which have been well documented in Europe. Density of fish within net pens can breed parasites and disease, especially as mortalities decay in the bottom of the pens. Escapees may compete with wild fish for food and habitat. Farm waste can raise nitrogen levels and reduce dissolved oxygen, which may harm local marine life. Finally, because net pens are fixed, the local habitat does not have the opportunity to recover from use as the ocean does during fallow periods following salmon migration.

The effects of farmed salmon on human health are also receiving attention. Farmed salmon are dyed so their flesh color is similar to wild salmon meat. They may be injected with antibiotics, as livestock are. They are fed pellets made from higher-order fish, which have more bioaccumulated toxins than do animals of lower trophic order consumed by wild salmon. A 2004 study published in *Science* reported that farmed salmon may contain up to ten times the cancer-causing toxins as wild salmon; but the U.S. Food and Drug Administration responded by noting that contaminant levels were well below federal standards, and that the health benefits of salmon consumption, farmed or wild, far outweighed the risks. The subject remains extremely controversial.

Among Pacific Rim jurisdictions, China, Japan, British Columbia, and Washington State host fish farms. Alaska, however, has a moratorium on salmon farms, and Russia has neither the capital nor the locations to invest in aquaculture.

Given the certainty that wild fish catch cannot meet global demand and that farmed fish can greatly increase world supply and mitigate harvest pressure, the aquaculture industry will continue to grow. Therefore extensive research is called for to determine the sustainability of this practice, to prevent escapes, sewage releases, and other ecologically harmful events, and to improve farms through new technologies.

## THE ECONOMICS OF PEN-REARED SALMON

Within the last three decades—most dramatically within the last ten years—the aquaculture industry has transformed global fish markets. In 1980 salmon farming produced around 12,800 tons of salmon, which was less than one percent of the global farmed salmon product in 2001. In 2002 the six largest companies in the industry controlled half the $3 billion global market. Today salmon farms operate around the world, with Norway, Chile, British Columbia, and Scotland as the major producers.

The year 1997 marked a dramatic turning point for the world salmon supply as pen-reared salmon surpassed wild salmon in annual production. Just four years later, in 2001, wild salmon catch amounted to just 62 percent of total farmed salmon production. Aquaculture has made salmon available year-round—and with increased global supply, the international market price for wild-caught salmon has halved in the past decade. The Alaskan fishing industry, the world's largest producer of wild salmon, has been hardest hit by the downturn in the salmon economy. In 1980, Alaska supplied nearly half the world's salmon; today Alaska supplies less than 20 percent of the world market, and the value of that salmon has plummeted. In 1992 Alaska produced 306,000 metric tons of salmon, and catch value peaked at nearly $600 million. In 2001, although wild Alaskan harvest surpassed 1992 figures, catch value plunged to $216 million.

Yet salmon farming has spurred economic growth in other sectors. In British Columbia, the farmed salmon industry provides more than 3,500 jobs directly and indirectly, half to First Nations and women. Aquaculture is the province's biggest agricultural export sector, as the wholesale farmed seafood products industry generated around CA$390 million. ■

Fish farms provide jobs and income to local residents, particularly in rural areas, as well as global stability in salmon markets. But increased supply and year-round availability have gravely depressed wild salmon prices, threatening community livelihoods around the North Pacific.

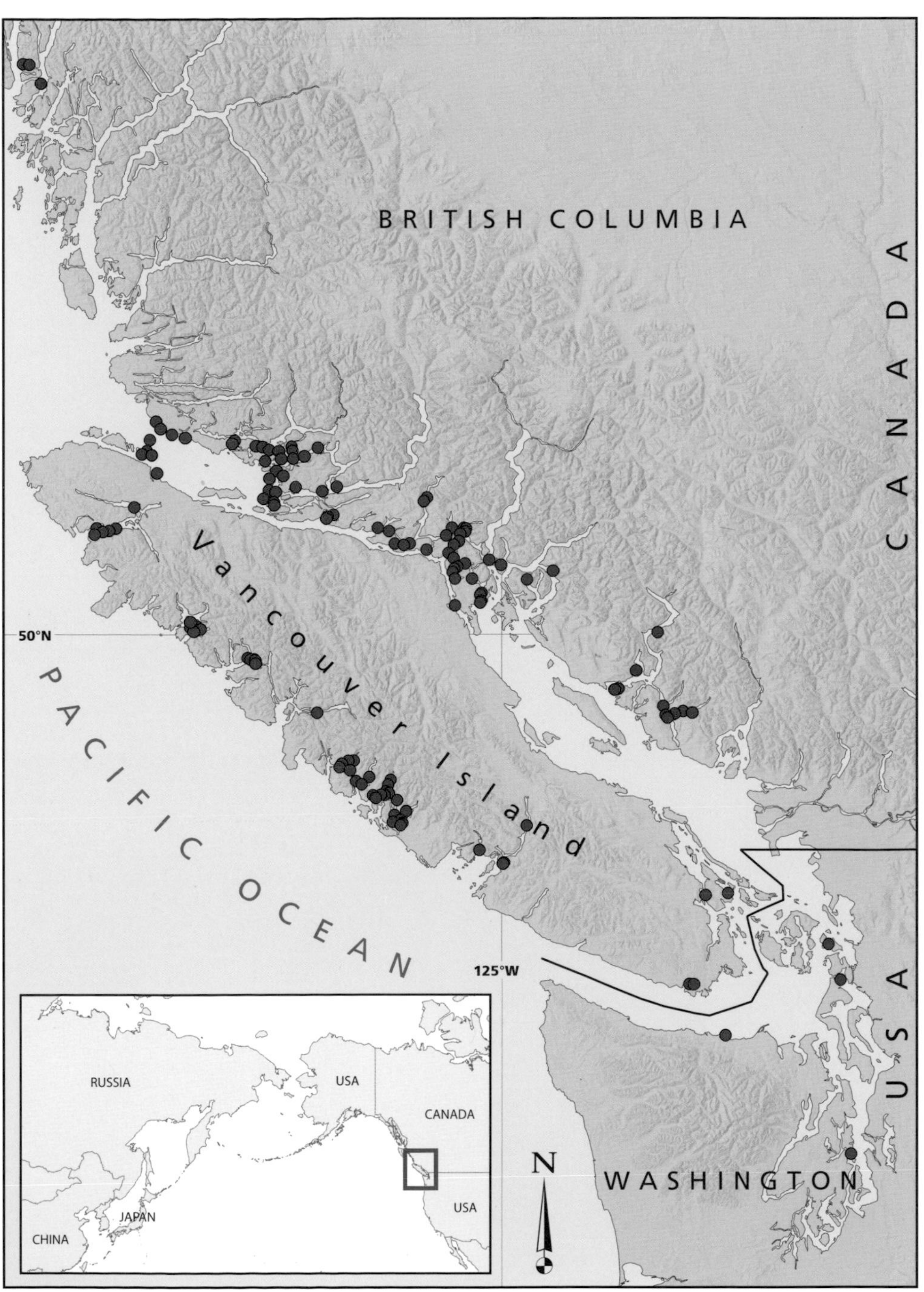


CANADIAN AQUACULTURE
The coastlines between Vancouver Island and mainland British Columbia—along the Strait of Georgia, Queen Charlotte Strait, and Johnston Strait—host the highest concentration of salmon farms in the North Pacific.

## FISH FARMS OF THE FUTURE

In 2002, when faced with dramatic declines in pink salmon returns to the Queen Charlotte basin—the body of water between British Columbia and the northeastern tip of Vancouver Island—many investigators attributed the mortalities to sea lice infestation, which may have been caused by a high density of fish farms in the region. (North Atlantic studies have documented the relationship between fish farms and parasites.)

Others disagreed, maintaining there were no studies linking fish farms and sea lice outbreaks in this region. Noting that pink returns to the region two years later (progeny of the 2002 spawners) were unusually high, they maintained that if a relationship existed between farmed fish and sea lice, it was complex at the very least.

Nonetheless, as a result of declines in the 2002 pink runs in an area commonly known as the Broughton Archipelago, fish farming came under intense scrutiny, and the British Columbia provincial government committed to improving industry practices. Today the government enforces stringent standards and oversight in 52 provincial and federal policies and statutes to mitigate and minimize ecological and biological concerns.

Throughout all aquaculture markets, science offers promise for industry improvements. Biologists are exploring the effects of limiting farmed stocks to native species and reducing netpen densities. Improvements in fish feed have reduced the amount needed to raise farmed salmon, which is 42 percent lower than what it was in 1972. Increased use of vegetable-based proteins for animal feed is also reducing dependence on wild fish populations from which pellets are made. (At the same time this reduction may correspond to a drop in omega-3 fatty acid, one of the main selling points of a fish-rich diet.) The doses of antibiotics, used to stem infections in farmed fish, amount to less than 0.5 percent of what was used in the early 1990s. Vaccines may also reduce disease within farmed populations. Sterilization could eliminate some of the threats posed by escapees.

Although many argue otherwise, technological opportunities can offer solutions. Parasites and disease transfer may be avoided by isolating farmed salmon through the use of near-shore closed containment pens or "walled" underwater tanks. Land-based tanks, made of cement or fiberglass, are another option. By recirculating water, these tanks eliminate the deposit of fish-farm waste and disease-causing microbes into the ocean.

Finally, following models in the wood product and tuna industries, market-based incentives such as certification may greatly encourage best practices in fish farming. ■

# Salmon Trade

**Overnight shipping, flash-freezing,** and growing consumer demand make the international salmon trade a multibillion-dollar business. In 2001, for example, Americans bought more than US$55 billion in seafood, and salmon continues to be a favorite seafood option.

More than 60 percent of the world market supply of salmon is farmed. In 1997 farmed salmon and trout surpassed wild production internationally, with most of the farmed fish coming from Norway, Chile, and the United Kingdom. Although it has few remaining populations of wild Atlantic salmon, Norway is the world's largest producer of farmed salmon. Fjords providing clean, cold water, and swift currents that wash away net-pen pollutants are considered ideal for production. Norway's dominance of the world's net-pen industry extends beyond its own borders, as many multinational aquaculture conglomerates in Chile, the United Kingdom, and British Columbia, for example, are owned by Norwegian companies. Denmark is also an extension of the Norwegian farmed salmon industry: it imports salmon from Norway and exports it to the United States, a legacy of a threatened U.S. trade embargo against Norway for perceived violations of international whaling laws in the 1990s.

Although it is not yet among the world's top 10 importers, Russia is emerging as an important market. Even while the Russian Far East is a major salmon producer, the country's western cities are increasingly turning to Europe for salmon, because transportation, processing, and cold storage capacity in the Russian Far East are inadequate. Between 2001 and 2002, for example, Norwegian exports to Russia quadrupled.

The world's largest salmon importer is Japan, which drives the international salmon market. Where sockeye was once preferred in Japan, Chilean coho and Atlantic salmon (both farmed) are current favorites.

The increased availability of farmed salmon has devastated the wild salmon market. In Japan, for instance, the average price per 100 grams of salmon was nearly 160 yen (around US$1.96) in 1990; by 2001, the price had fallen to around 130 yen (roughly US$1.47). Alaska fishermen have also experienced dire effects. In 1998, for example, the ex-vessel price for sockeye was US$2.37 per pound (454g); by 2002 it had fallen to US$0.55—a decrease of more than 75 percent. It is not only the fishermen and their communities that are economically harmed but also agencies, which suffer drastic budgetary cuts when their industry is devalued; losses in funding hinder agencies' abilities to manage wild stocks.

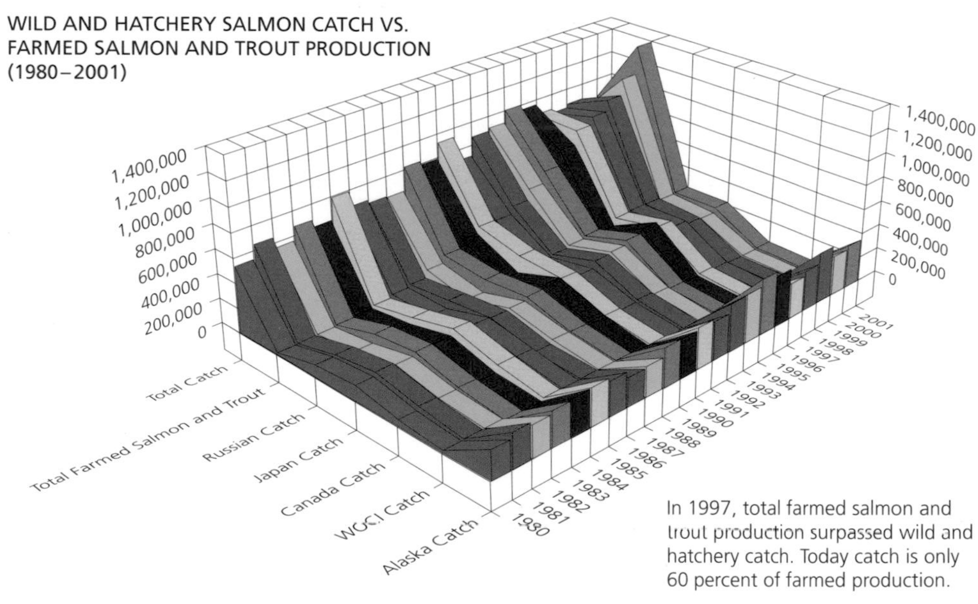


WILD AND HATCHERY SALMON CATCH VS. FARMED SALMON AND TROUT PRODUCTION (1980–2001)

In 1997, total farmed salmon and trout production surpassed wild and hatchery catch. Today catch is only 60 percent of farmed production.

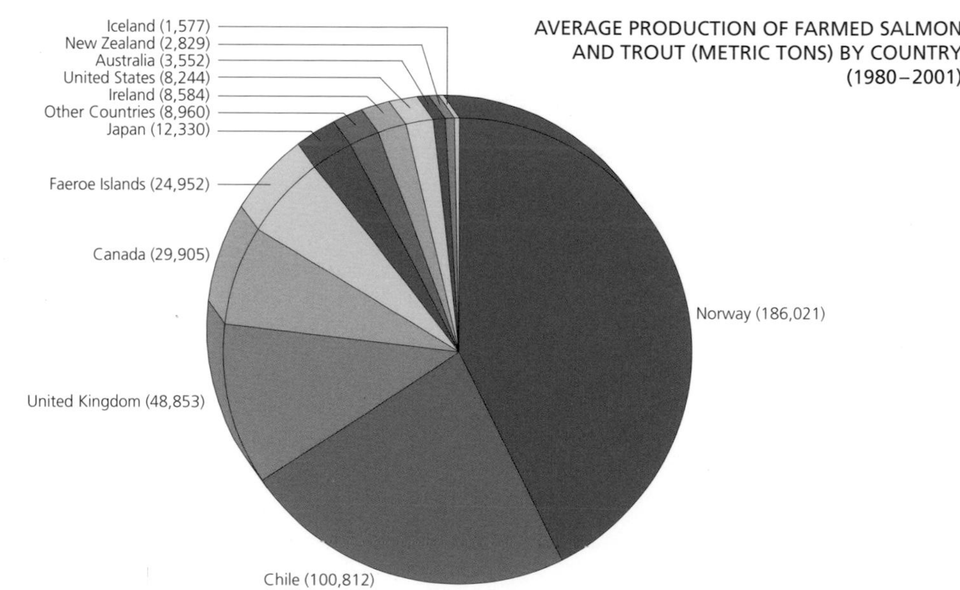


AVERAGE PRODUCTION OF FARMED SALMON AND TROUT (METRIC TONS) BY COUNTRY (1980–2001)

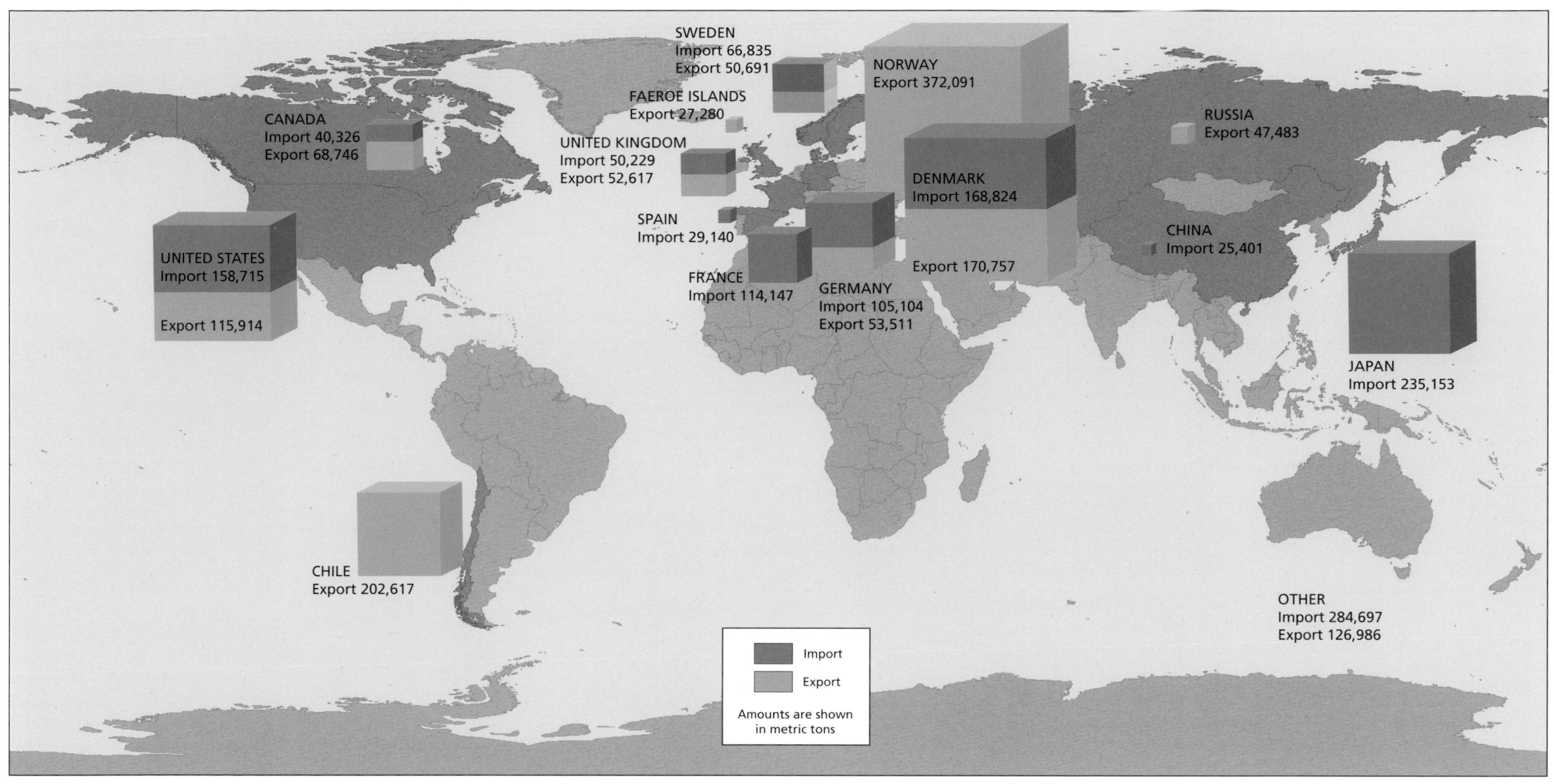


**SALMON TRADE (WILD, HATCHERY, AND FARMED)** Above is the only world-view map in this atlas. It demonstrates that the salmon trade story is best told at a global scale. Note also that we have shifted our perspective from Pacific-centric to Euro-centric, which splits the Pacific Ocean in two. Global spatial data is most frequently represented this way.

Several countries import as much salmon as they export, which reflects market preference and the economic value of different species. In reading this map, it is critical to note that around 40 percent of the world's salmon supply is either wild or hatchery derived *Oncorhynchus*, and the proportions vary by location. The remaining 60 percent of the global supply of salmon is farmed and of that, most is Atlantic salmon (*Salmo salar*). Once prevalent along the western Atlantic coastline from Quebec to New York, and throughout many western European countries, native Atlantic salmon populations remain healthy only in a small number of rivers in Norway, Greenland, Iceland, Scotland, and western Russia, which remains the last real stronghold of *Salmo salar*. Elsewhere and for commercial purposes, *Salmo salar* is farmed. In Chile, one of the world's top salmon producers, Atlantic salmon is an introduced species. The proliferation of net-pen farms has made salmon available year-round, spurring demand and lowering prices.

# Marine Jurisdictions

**The freedom-of-the-seas doctrine,** a 17th century common law, granted nations rights to ocean resources along their shorelines (traditionally within three nautical miles). Beyond that, all nations shared in the resources of the high seas. But as European settlers transformed the landscape of North America and Canada and as the Meiji Restoration ushered in a new class of Japanese fish merchants, so too was the seascape transformed, from shared resource to commodity.

The Russian-American Convention of 1824 opened the North Pacific to Russian and American citizens for fishing, trading, and navigation. The next year the Anglo-Russian treaty of 1825 decided terrestrial borders in the Alaska panhandle, with the exception of the Dixon Entrance off the southeast coast, which is disputed to this day. Rancor developed, eventually prompting the Russian-American Treaty of 1867. This treaty—the first such maritime claim over open oceans—finalized the U.S. purchase of large portions of what is now Alaska from the Russians, effectively dividing the Bering Sea. Even today the two countries interpret Bering Sea maritime boundaries differently.

Yet tensions continued. After the Russo-Japanese War, Japanese access to Russian waters in the northwest Pacific was zoned by treaty in 1907. Until World War II the Japanese dominated the fisheries of Kamchatka, Sakhalin, and in the Kuril Islands, where sovereignty continues to be a matter of intense negotiation. Today the enduring presence of Japanese driftnet fisheries off the coasts of the Kuril Islands and Kamchatka remains highly controversial.

In 1945 President Truman extended U.S. jurisdiction to the continental shelf, and other countries followed suit. Nonetheless, nations strove for accord. In 1952 the International North Pacific Fisheries Commission was created to unite Pacific Rim fishery interests and research. In 1993 it was annulled and recreated as the North Pacific Anadromous Fish Commission (NPAFC), with broadened membership that included signatory nations Japan, Russia, Canada, the United States, and, as of 2003, the Republic of Korea. Today the role of the NPAFC is limited to enforcement and research coordination; it does not oversee management or habitat use.

New technologies spurred competition for ocean resources and the creation of exclusive economic zones. In the 1970s Chile was the first to delineate its own Fisheries Conservation Zone; in 1976 the United States followed suit with the Magnuson Fishery Conservation and Management Act; Canada, Russia, and Japan asserted boundary lines thereafter.

## THE LAW OF THE SEA

The United Nations Convention on the Law of the Sea is arguably the most comprehensive international environmental accord in existence, governing all aspects of ocean space. The Law of the Sea, ratified by 145 of the United Nations' 195 independent members, oversees procedure on exclusive economic zones, transit passage, revenues from resource extraction, research, pollution, dispute law, and many economic and political activities related to uses of ocean resources.

The seeds of the work for the Law of the Sea were planted in 1967 by Malta's ambassador to the United Nations, who called for recognition of the conflict brewing in the world's oceans and for an international governing body to take charge of all marine matters. By 1973 the third United Nations Conference on the Law of the Sea was convened in New York and continued for nine years. The product, finalized for signature in 1982, was a comprehensive treaty for the oceans. In 1994 the Law of the Sea was fully in force with 60 countries on board.

The United States and Canada were not among the signatories in 1982, nor did they ratify the treaty because they opposed a provision that regulates offshore seabed mining and establishes a multinational court that would oversee taxes, permitting, pricing, and research for marine activities. (An agreement was eventually reached to resolve outstanding issues.)

The United States has now signed—but not ratified—the Law of the Sea treaty. Other than the provisions dealing with deep-sea mining, the United States regards the Law of the Sea as customary international law and complies with its provisions. In 2003 Canada became a signatory of the Law of the Sea. ■

Treaty negotiations may take years. Above, stakeholders craft the terms of the 1985 Pacific Salmon Treaty, regarding interception fisheries in Canada and the United States. It was renegotiated and amended in 1998, with the Pacific Salmon Commission as the implementing body.

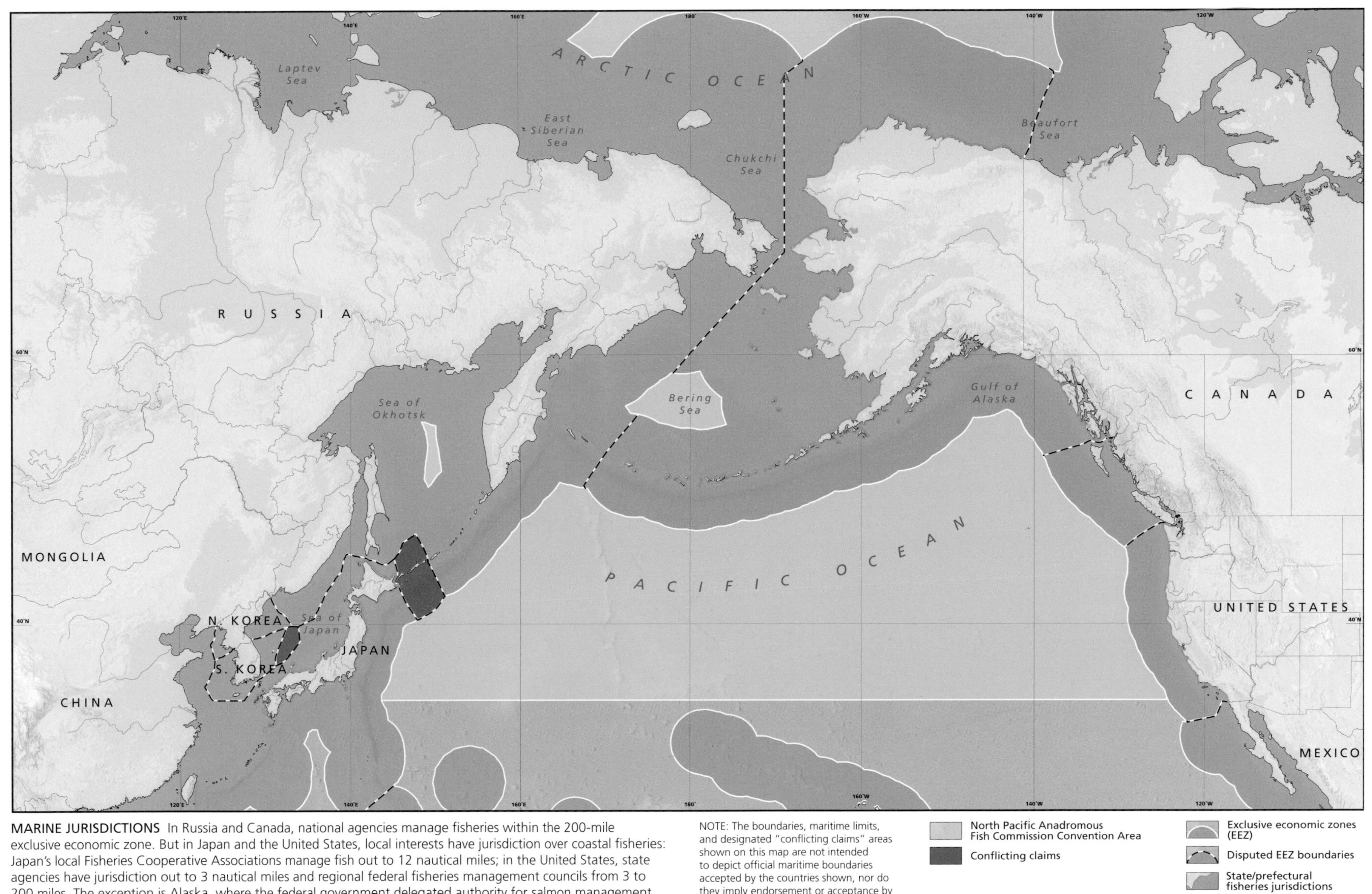


**MARINE JURISDICTIONS** In Russia and Canada, national agencies manage fisheries within the 200-mile exclusive economic zone. But in Japan and the United States, local interests have jurisdiction over coastal fisheries: Japan's local Fisheries Cooperative Associations manage fish out to 12 nautical miles; in the United States, state agencies have jurisdiction out to 3 nautical miles and regional federal fisheries management councils from 3 to 200 miles. The exception is Alaska, where the federal government delegated authority for salmon management from 3 to 200 nautical miles to the state. Tribal comanaged stocks are not depicted in this map.

NOTE: The boundaries, maritime limits, and designated "conflicting claims" areas shown on this map are not intended to depict official maritime boundaries accepted by the countries shown, nor do they imply endorsement or acceptance by the United Nations or governing bodies of Pacific Rim nations.

# Shared Stocks

**Marine jurisdictions outline the** boundaries of natural resources "ownership" at sea. But salmon are moving targets and cross jurisdictional borders throughout their life cycles. Because they are as essential to commerce as they are to an ecosystem's biological integrity, salmon have been the focus of bitter treaty and boundary skirmishes for centuries. For example, Japanese fishing in Russian waters added sparks to the incendiary relationship between these growing countries, and territorial conflicts eventually erupted into the Russo-Japanese War in the early 20TH century.

Shared stocks are those that are explicitly managed by more than one country due to migratory pathways, where national borders cross river basins, or where there are conflicting claims over territory. The Russians and the Chinese have wrangled over chum runs on the Amur; the transboundary rivers of southeast Alaska and Canada—the Taku, Stikine, and Alsek, among others—have served as battlegrounds over salmon runs. Just as transboundary rivers provoke conflicts, so too does catch in shelf waters.

Canada and the United States share many salmon runs where national borders cut across straits and fjords; the Strait of Juan de Fuca is just one example. Relations between the two countries worsened during the 20TH century when Washington fishermen harvested a disproportionate amount of Canadian-bound pink and sockeye. In the 1970s Canadian fishing interests deliberately overfished the Fraser River fishery in order to force an agreement with the United States.

At the same time, Native Americans were suing the states of Washington and Oregon over native fishing rights. By 1985 the heated conflicts prompted the creation of the Pacific Salmon Treaty, dedicated to conserving salmon and dividing harvest equitably between the United States and Canada. The Pacific Salmon Commission was then created to implement the treaty providing regulatory advice and recommendations to both countries.

Equitable distribution of catch is a hotly discussed topic within the shared stocks debate; species protection is another. Midway up the Amur River, for example, waters are flanked by Russian forest on one bank and developed Chinese communities on the other. Both countries are suffering the losses of this important fishery due to competitive overfishing and upstream agriculture and pulp mill proliferation on the Chinese banks; however, cross-border relationships are improving and perhaps will one day result in an Amur salmon treaty.

## MAJOR PACIFIC SALMON TREATIES

- **1824: Russian-American Convention**—opened North Pacific to citizens of both nations for fishing, trading, and navigation
- **1875: St. Petersburg Treaty**—Sakhalin Island claimed by Russia, granting Japanese permission to fish along shores; Kuril Islands claimed by Japan
- **1892: U.S.-British Joint Commission**—first formal attempt to address international allocation and fishery management, focusing on transboundary issues between Canada and the United States
- **1907: Russo-Japanese Salmon Treaty**—created a lease system giving Russians preference over the Japanese in large rivers, creating lease areas
- **1939: First U.S.-Canadian bilateral salmon sharing convention in the Eastern Pacific**—designed to solve transboundary wars over Canada's Fraser River basin sockeye
- **1952: Tripartite Convention, "International Convention for the High Seas Fisheries of the North Pacific Ocean"**—limited Japan's high seas harvest of salmon originating in U.S. and Canadian rivers; gave Canadian and U.S. fishermen rights to harvest within each other's waters; created International North Pacific Fisheries Commission
- **1982: U.N. Convention for the Law of the Sea III**—governed use of the world's oceans; asserted coastal state-of-origin jurisdiction over salmon
- **1985: Pacific Salmon Treaty**—created Pacific Salmon Commission to prevent overfishing and apportion equitable harvest, for all areas with interception fisheries
- **1991: U.N. adoption of Resolution 46/215**—implemented moratorium on pelagic driftnets
- **1993: Convention for the Conservation of Anadromous Stocks in the North Pacific Ocean**—prohibited all high-seas harvest of Pacific salmon beyond national boundaries; created North Pacific Anadromous Fish Commission
- **1999: New Pacific Salmon Treaty**—allotted shared stocks based on forecasted abundance and escapement, with dedicated national funding for salmon run and salmon habitat protection and restoration

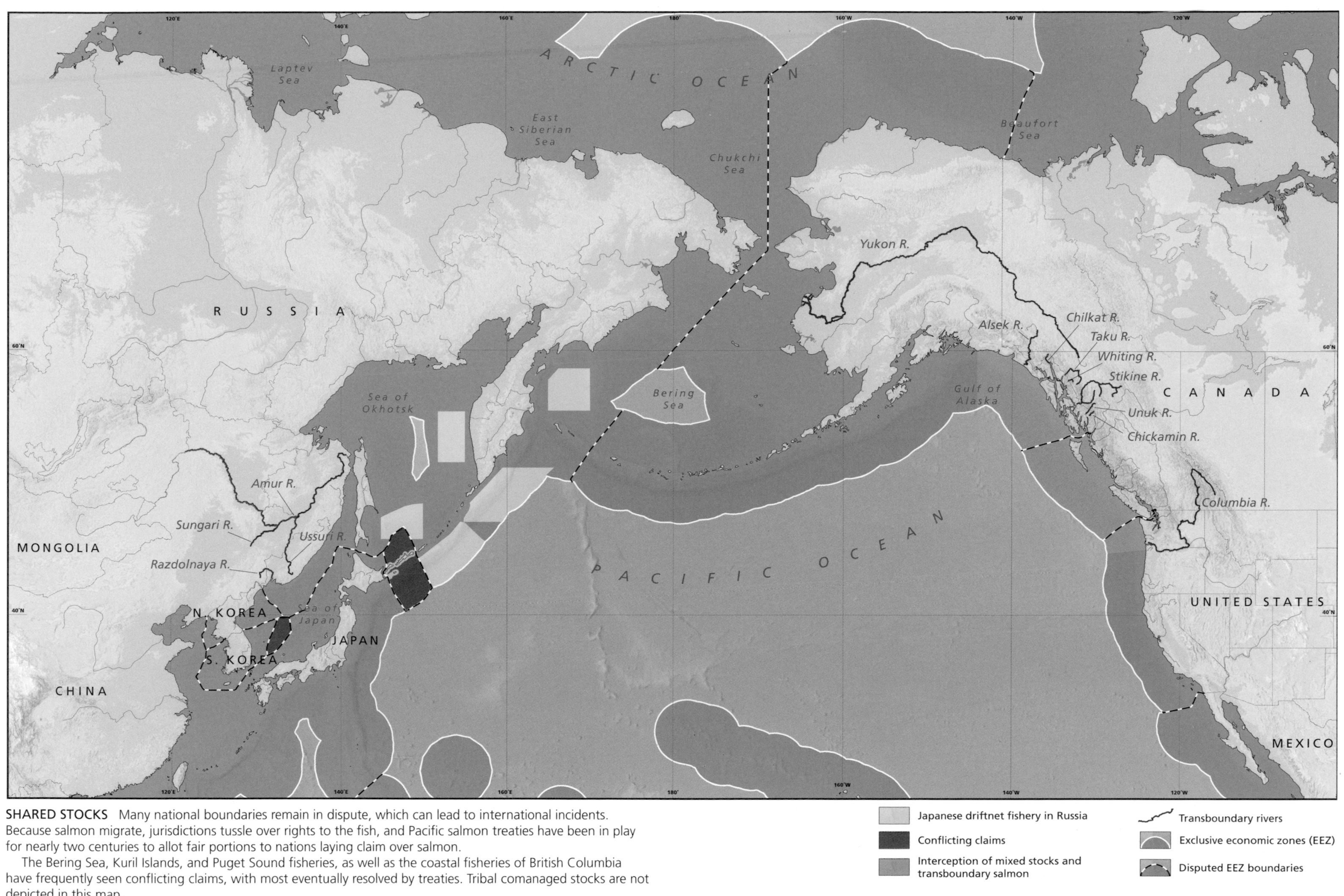


**SHARED STOCKS** Many national boundaries remain in dispute, which can lead to international incidents. Because salmon migrate, jurisdictions tussle over rights to the fish, and Pacific salmon treaties have been in play for nearly two centuries to allot fair portions to nations laying claim over salmon.

The Bering Sea, Kuril Islands, and Puget Sound fisheries, as well as the coastal fisheries of British Columbia have frequently seen conflicting claims, with most eventually resolved by treaties. Tribal comanaged stocks are not depicted in this map.

# Protected Areas

**The formal movement to establish** protected areas began in 1872, when president Ulysses S. Grant created America's first national park in Yellowstone, Montana. Twenty years later, in 1892, famed naturalist Livingston Stone took the notion a step further, proposing before a meeting of the American Fisheries Society the creation of salmon parks. Stone's vision was a century ahead of its time. With a few exceptions (see below), there are no fully protected wild salmon rivers in the North Pacific.

The map on the facing page depicts existing protected areas in the North Pacific using classification systems that describe degree of protection, devised by IUCN-The World Conservation Union. Category I denotes nature preserves and wilderness areas, managed mainly for science or wilderness protection. Categories II and IV represent national parks and habitat and species protection areas, managed mainly for ecosystem management and recreation purposes, and for "conservation through management intervention." Russia's special category of Nature Parks provides little enforced species protection.

Protected areas exist throughout the landscape of the North Pacific but may not be of particular value to salmon. In fact, most were established as bird sanctuaries along migratory flyways or in low-value, high-elevation forests. A number of protected areas fall along ridgelines (in Alaska, the Kuskokwim and Coast mountains and the Alaska and Brooks ranges). Such protections may ensure pristine headwaters but leave vulnerable the lowland areas used by salmon, which are most often targeted for development. More protected areas occur inland at northerly latitudes, in places where salmon populations remain relatively robust and where human populations pose the fewest environmental threats. With scant few exceptions, protected areas fall within the Category II and IV status, offering limited species protection. Regions with great salmon diversity—Kamchatka, Koryakia, British Columbia, and WOCI—leave most coastal areas unprotected.

Perhaps most important to migrating salmon are whole-basin protections. This map also illustrates the extent to which protected areas are fragmented. Within the context of our ecoregion template, only a few major salmon producing drainage basins are completely protected, including the Vengeri and Pursh Pursh rivers in Sakhalin's Vostochnii Zakaznik refuge and the Kol River on Kamchatka. Only four ecoregions, albeit in remote locations, are largely protected: the Kobuk River (38), the Noatak River (39), the Southeast Bering Sea Inner Shelf (44), and the Alaska Coastal Downwelling (47).

## SALMON CONSERVATION RIVERS

Russia and Japan have gone so far as to designate protected areas specifically for salmon. In Kamchatka a headwaters-to-ocean salmon refuge—believed to be the first of its kind—was established in 2004. It encompasses the entire length of the pristine Kol River. Flowing from the mountains to the coastal plain of southwestern Kamchatka, the Kol River is among the most productive rivers in the world, supporting annual salmon runs of more than five million fish. The United Nations Development Programme and the Wild Salmon Center identified the Kol River Salmon Refuge as a conservation priority because it contains seven species of native Pacific salmon, including rainbow-steelhead trout. In addition to protecting abundant salmon habitat and a rich ecosystem supported in large part by salmon, the refuge will serve as a natural laboratory, allowing scientists to study a pristine salmon ecosystem unaltered by human development.

Japan has also taken steps toward salmon conservation. On the northern island of Hokkaido, 32 rivers have been assigned protective status in order to bolster wild populations of masu. Although the Ministry of Construction manages watersheds in Japan, Hokkaido prefecture oversees the living "resources" within rivers. Masu protection rivers provide sanctuary for fish and habitat: recreational angling is prohibited; developers may not alter the natural condition of these rivers; and natural habitat features are being restored. Although it is illegal to fish for adult salmon in rivers and streams throughout Japan, fishing for juvenile and resident salmon is permitted in most of Hokkaido's rivers, except within these select masu conservation rivers. ■

Along the species-rich western coastline of Kamchatka, the Kol River Salmon Refuge was established by the local Kamchatka Administration in 2004. It is the world's first whole-basin river system dedicated to salmon protection.

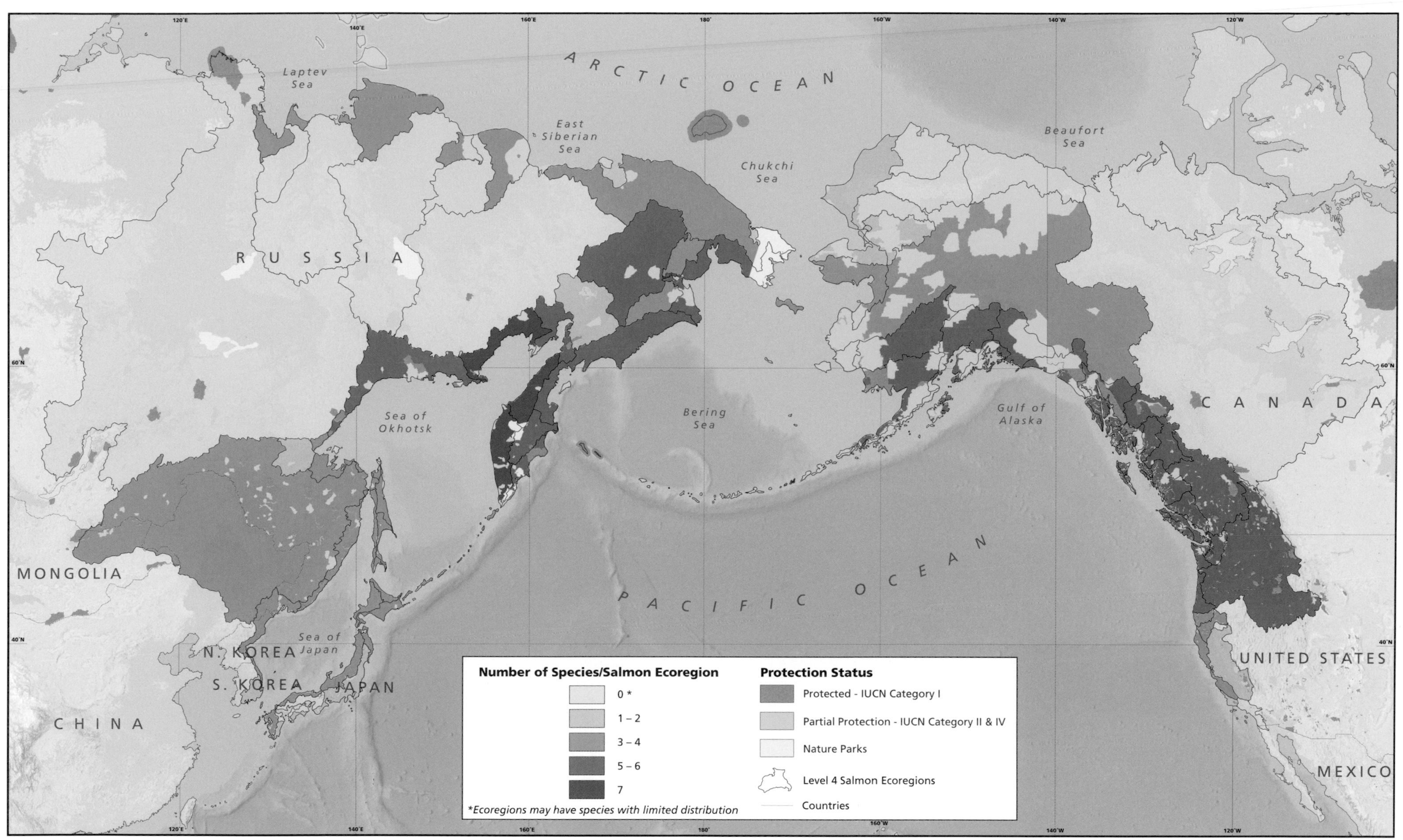


PROTECTED AREAS (COMPARED TO SALMON DIVERSITY, SEE PAGE 11) In general, protected areas have been established to conserve migratory bird flyways; they do not offer sanctuary for salmon at the basin scale. With few exceptions, these areas represent an incomplete patchwork of protections and tend to be inland and at high elevations—locations that are generally not used by salmon. Furthermore, protected areas may not be sited in areas where protections are needed most (such as coastal plains); nor are they necessarily located in the regions with the greatest salmon diversity, which remain vulnerable to threats.

NOTE: Nature Parks as defined in this dataset are located only in Russia.

# 3

# The Place

*How the North Pacific is sculpted by climate, geography, and biology*

■ Extent of Glaciation ■

■ Terrestrial Ecoregions ■

■ Sea Ice ■

■ Primary Production ■

■ Sea Surface Temperature ■

Six species of Pacific salmon spawn in the cold and fast waters of the Opala River in southwestern Kamchatka, 161 kilometers from headwaters to the Sea of Okhotsk.

**The Pacific Ocean is the world's** largest and deepest ocean. Composing about one-third of our planet's surface area (more than all the land area combined), the Pacific is vast but also nurtures creatures at the smallest of scales. The alchemy of air, water, ice, energy, and organic matter contained in the North Pacific and mixed over millennia produces a changeable environment for the world's greatest concentration of wild salmon—about one-half billion individuals each year. The focal point of productivity within the North Pacific ocean is the Bering Sea, one of the most nutrient-rich places in the world. On its eastern side, the continental shelf is the world's widest, with the gentlest slope; on the western side, it plunges to 4,000 meters—as deep as the world's greatest mountains are high.

The North Pacific contains nearly a dozen ocean provinces, each characterized by a unique mix of currents, sea surface temperatures, primary and secondary productivity, and a host of other physical traits. Each province is an essential component of the greater ecosystem that accommodates salmon during their various life history stages. For the most part, salmon arriving in the Bering Sea and the Gulf of Alaska use the region as a foraging ground, transitioning from a diet of small-bodied zooplankton (e.g., copepods) to their spring- and summertime buffet, including squid, jellyfish, and prey fish.

The glacially carved Chukotka coastline looms over the Bering Sea and ventures north to the East Siberian Sea in the Arctic. It offers some of the most northerly habitat for salmon.

Salmon link the North Pacific to the variegated landscapes along the periphery: dormant and active volcanoes in the Ring of Fire comprising Kamchatka, the Japanese archipelago, and the Kuril Islands; the tundra and taiga of the Russian Far East, with its dry continental climate, characterized by mountains of modest elevation and extensive lowland and highland plains; the rugged mountains and braided inland passages sculpted by heavy glaciation in Alaska, British Columbia, and Washington, with montane glaciers carving valleys farther south.

Russia's Kamchatka peninsula is home to some of the world's most active volcanoes, which have contributed rich nutrients to the landscape. The Space Shuttle *Endeavour* took this radar image of the erupting Kliuchevskoi volcano on October 5, 1994.

As landscapes vary, so does the ocean floor. Its ridges and troughs route ocean water, affecting currents, biological productivity, and climate.

**PRECIPITATION** The hydrology of the North Pacific is an immense machine. The timing, amount, and intensity of precipitation drive the flow regime of the riverine systems in which salmon spawn and rear. Sediments wash from the landmasses to the ocean during rainfall and snowmelt, representing a complex recipe that layers the ocean water column and feeds primary producers along the coast during the springtime. In the west, the monsoon climate brings heavy rainfall and peak precipitation in the form of summer rains from July through September; low water levels occur in January and February in Japan and Russia; and because most northern waters freeze over, a secondary snowmelt peak occurs in late spring.

In contrast, precipitation patterns are more variable in the eastern Pacific, driven predominantly by the east-west atmospheric flow. Peak precipitation occurs in fall and winter, with peak runoff typically occurring during the spring snowmelt. Low flows occur during the dry months of August and September. In the Coast Range and the Cascade lowlands, extending under various names from southeast Alaska through California, precipitation falls mostly in the form of rain. In the high Cascades and east slope, there is less precipitation but a greater proportion is in the form of snow. Peak river runoff tends to coincide with fall and winter rains on the mountain ranges' windward side, with spring snowmelt on the leeward side.

Precipitation and stream flow affect overwintering survival of salmon, which varies from species to species and from east to west.

Icy Bay, at the foot of Alaska's Mount St. Elias, was formed by the retreat of four glaciers within this past century. Here Tann Fjord rises above the Gulf of Alaska, the center of the *Oncorhynchus* range.

With the exception of the Kamchatka peninsula, where an unusually high proportion of stream flow is derived from groundwater, the western Pacific offers limited overwintering habitat for salmon; with few exceptions, rivers entirely freeze over. Therefore, chum and pink, with their shorter life histories and lesser dependence on freshwater, predominate. Masu, which overwinter for one season, find refuge in the rivers on the shores of the Sea of Japan and in the rivers fed by nutrient-rich groundwater in western Kamchatka. The eastern Pacific, however, offers rich opportunities for overwintering habitat and therefore better suits the longer freshwater residency life history types represented by sockeye, coho, chinook, and the rainbow-steelhead complex.

**CLIMATE VARIABILITY** In 1996 scientists studying the relationship between the North Pacific climate and salmon abundance theorized that approximately every 25 years, the climate from Alaska to California shifts from a warm regime to a cold regime. They called this Pacific Decadal Oscillation (PDO). Although PDO is evident in at least four discrete periods since 1890, recent paleo-oceanographic studies indicate that PDO has shaped the climate for far longer. Climatic variations with basin-wide effects on the North Pacific Ocean and its salmon, anchovy, and sardine populations appeared long before fisheries or other anthropogenic factors had any significant influence. While scientists are in accord regarding the existence of PDO, there is disagreement about the periodicity of the effects, geographic differences in climate effects across the salmon's

Salmon rivers may traverse seemingly inhospitable terrain, from rugged rock to arid plains. Here the Owyhee River, a salmon producer, runs through the high desert region of Eastern Oregon.

distributional range, and most significantly the driving factors and intermediate causal factors affecting salmon.

During the cold phase of the PDO cycle, conditions are favorable for salmon ocean-rearing and migration off the coasts of Washington, Oregon, and California. Air and sea surface temperature are lower than average from October through March, and snowpack, stream flow, and flood risk are high. At the same time, phytoplankton production in the Bering Sea and the Gulf of Alaska is diminished, which reduces carrying capacity for ocean foraging and migration and affects higher trophic productivity.

In the warm phase of the PDO cycle, the phenomenon is reversed: salmon production is boosted in Alaska and diminished in Washington, Oregon, California, and Idaho (WOCI). Variation in salmon productivity off British Columbia is more diverse, as the ocean province straddles two opposing regimes.

Alaskan tundra supports a wide variety of wildlife, including golden eagles and brown bears, which depend on salmon for food.

Climate cycles also can predict inland precipitation patterns and seasonality. Mild winters appear to foster greater algal biomass, increase insect populations, and minimize the destruction of salmon eggs by flood scour. Therefore periodicity of freshwater cycles may also contribute to salmon population variability.

Climate-driven cycles do not indicate species recovery or rebound. The PDO is only one manifestation of climate dynamics, and it can mask trends that occur independently of climate variability (e.g., overfishing). Although they both represent various forms of climatic variability, PDO differs from El Niño/Southern Oscillation (ENSO). ENSO periodicity, from 6 to 18 months, affects climatic conditions primarily in the tropics and secondarily in North America.

**RIVERS** In the broadest sense, a wild river may be described as the natural and productive flow of water—fed from above (snow and rain), from the ground (glacial and snowpack melt), and from below (subsurface and groundwater)—toward the ocean. Every river constitutes its own ecosystem. Most major rivers in the North Pacific are born atop mountains carved by glaciers or sculpted by volcanoes. Water droplets collect in furrows, and trickles funnel into pools that spill over, gaining momentum downhill to form headwaters. Flows converge, pushing and pulling currents in the direction of the ocean, which might be thousands of kilometers away. Smaller tributaries fall in step, perhaps fed by underground aquifers. As rivers amass volume and energy, they carry minerals, biological sediments, and other organic matter, churning gravel on the river bottom and scraping the banks. The composition of the rock and soil, the power and frequency of wind, storms, and precipitation, and the particular topography of place all contribute to the uniqueness of each salmon river.

As streams move from mountaintop to forest, taiga, tundra, valley bottoms, foothills, and rangelands, composition increases in complexity.

## CURRENTS OF THE NORTH PACIFIC

The ocean climate is determined by the interplay among major ocean currents, wind patterns, and smaller-scale mesofeatures such as fronts, eddies, gyres, and coastal up- and downwellings. All these factors change interannually, seasonally, daily, and hourly; furthermore, they have implications for water and air temperatures, food for salmon prey and predators, and migratory pathways.

Offshore winds bring cold, nutrient-rich waters to the surface, creating upwellings; cold currents are transfers of this upwelled water propelled by surface winds and fluid dynamics in the ocean. These currents include the Oyashio and Anadyr Currents in the western Pacific and the California Current in the eastern Pacific. Conversely, the eastward-moving Kuroshio-North Pacific Current pushes warmer waters clockwise, past Japan, Canada, and southeast Alaska through the Bering Current and the Arctic Ocean, via the Alaska and Subarctic Currents. These bring warm waters in years of high wind stress; with more wind, currents are faster, and the rate of gyre rotation increases.

Depending on their position and force, dominant current systems that sweep across the Pacific affect the intensity of the California Current and the Aleutian Gyre off the Bering Strait. In the Aleutian low pressure system, for example, extreme lows result in stronger winds and greater gyre and circulation, which affects ocean productivity. ■

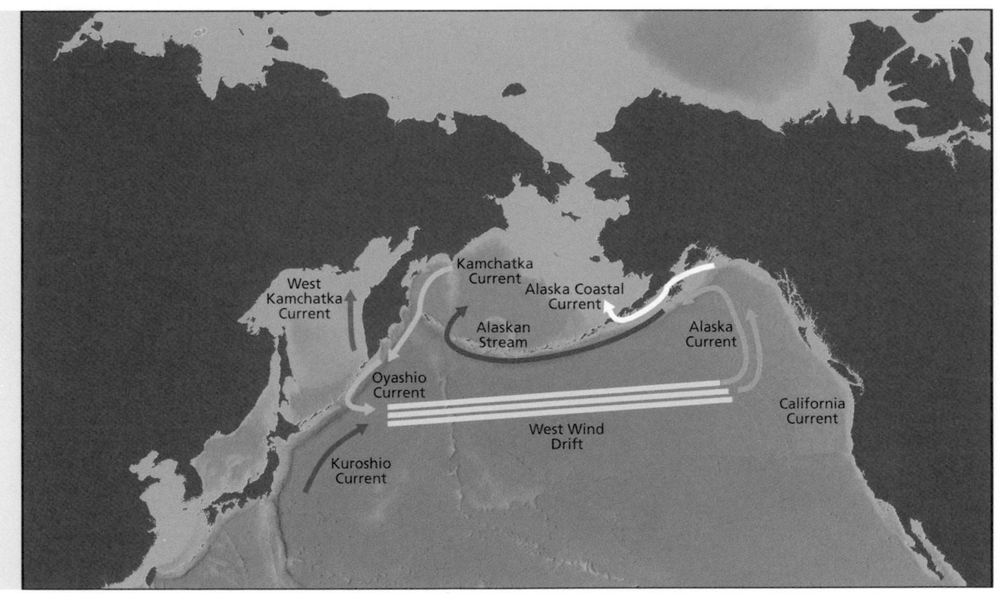

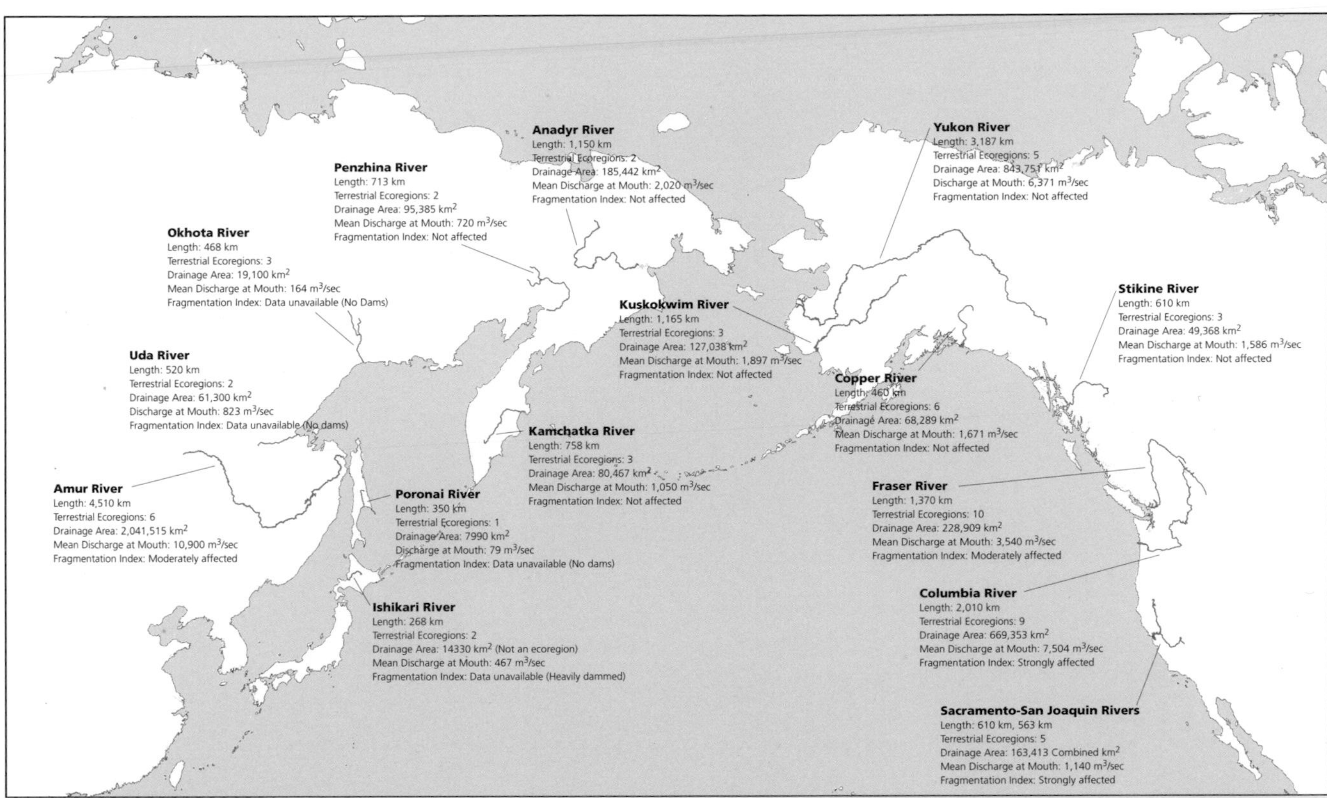


MAJOR SALMON RIVERS OF THE NORTH PACIFIC Here we offer information on some of the most significant Pacific salmon rivers. Drainage area is based on Level 4 ecoregions (see page 7). World Wildlife Fund terrestrial ecoregions describe discrete ecological areas (see page 54). Fragmentation index ranks degree of flow impingement by dams. "Strongly affected" means that less than 25 percent of the main channel river system remains free-flowing. "Unaffected" means that the mainstem has no dams and that tributary dams result in no more than a two-percent change in discharge.

There is a great difference between simple water flow and a life-giving, ecological mosaic within a setting composed of unique biological, chemical, and physical properties.

Robust vegetation, be it forest, tundra, or grassland, stabilizes stream banks through root systems. It shapes channel morphology through woody sediments. And vegetation serves an essential role in the riverine ecosystem through two major processes: the transfer of sunlight into energy (photosynthesis) by phytoplankton, algae, and aquatic plants; and the transfer of decomposed organic matter from vegetation to the stream waters. Organic matter may appear in a stream from fallen leaves, evergreen needles, branches, and trees, which are broken up by microbes, invertebrates, and the mechanics of a moving river. Animals—insects, fish, birds, and mammals—also contribute organic matter to rivers through death or excretion. Groundwater provides cool inflow, minerals, and dissolved organic matter brewed deep in the earth for hundreds or thousands of years.

From above, precipitation falls on the canopy, washes organic matter into the soil, seeps into the forest floor, soaks into rotting stumps and logs, and hastens the breakdown of organic materials. Naturally occurring fires can also shape the nutrient base of sagebrush and forest ecosystems, breaking up root systems, opening the forest canopy, and modifying stream flow, nutrient subsets, and sedimentation.

As water travels from headwaters to ocean, evapotranspiration cycles water back into the air, as absorption from the water surface and through plant exhalation into clouds, so that the water may be released again in precipitation to complete the hydrological cycle.

The major salmon rivers of the North Pacific differ profoundly in flow, productivity, and geomorphology, as do the estuaries and oceans into which they flow. Together they form one continuous and connected ecosystem. Water temperature, currents, and productivity are influenced by air and wind, the shape and quality of the landmasses, and the topography of the ocean floor. Currents of air, vapor, and water circulate, creating weather and climate patterns. Climate and geology determine rates of erosion and the morphological features of river valleys.

Human-introduced changes to habitat and waterflow—urbanization, water diversions, and logging—can disrupt these riverine ecosystems, upset the relationship between inland forests and river systems, and alter the process of evapotranspiration that connects land and air, affecting the salmon that use these waters.

# Extent of Glaciation

**Six million years ago, glaciation and** tectonic activity began to sculpt the northeastern coastline of the Pacific Rim, from Chukotka to California. Preceding the onset of the Pliocene epoch around five million years ago, this was a time of sharp cooling that brought about the first *protosalmon* (see *Oncorhynchus* family tree, page 3). As the Pleistocene era unfolded about 1.8 million years ago, increased cooling ushered in more frequent bouts of glacial advance. This series of ice ages may have produced as many as 10 distinct glacial episodes, as ice retreated and advanced.

At the last glacial maximum, about 18,000 years ago, the Cordilleran Ice Sheet covered most of western Alaska and nearly all of British Columbia. Throughout these glacial episodes, there were at least three major places on the eastern side of the Pacific where tundra or forest met ice and consistently promised safe harbor for *Oncorhynchus*: Kodiak Island, in what is now the Gulf of Alaska; northeast Queen Charlotte Island and likely parts of southeast Alaska; and the Brooks Peninsula on Vancouver Island's west coast. These refugia were home to several of the genus's core populations. At glacial maximums, these salmon populations remained discrete and developed specialized life history survival tactics for their particular basins. When warming set in, estuaries expanded, rivers reached inland, and headwaters shifted. Salmon explored new migration pathways, and species radiated outward from refugia, allowing salmon to spawn new generations with variable life history strategies.

Although Russia and Japan experienced far less glaciation, changing sea levels reshaped basins and land bridges throughout the western part of the Pacific Rim for millennia. At glacial maximums, water was locked in ice in what is now Alaska and Canada. Sea levels were lower, and landmasses connected the Kuril Islands, Japan, and South Korea. The sea level at its most recent nadir was likely around 125 meters lower than present-day levels. During these times of glacial advance, four major river basins persisted and perpetually shifted in the central and western Pacific: the Yukon (which stretched across what is now the Bering Sea), the Anadyr, the Amur, and the Songhua.

At low sea levels, the Seas of Japan and Okhotsk were shallow and likely enclosed, creating the unique conditions favorable to masu and char (also a salmonid). Some scientists hypothesize that the Sea of Japan was once a brackish lake, which offers insight into why masu salmon radiating out from this region may be predisposed to short ocean migrations and have such limited distribution.

## GLACIAL LEGACIES: MIXED SALMON POPULATIONS

Nearly all of British Columbia and southeast Alaska was covered by ice during the last glacial maximum, except for three major refugia. At times, isolated and distinct populations of salmon inhabited the waters in and around Queen Charlotte, Vancouver, and Kodiak islands, separated by towering mountain ranges. But as the ice receded about 11,000 years ago, the ground became marshy, and water inundated the coastline. Ice dams backed up rivers and rerouted entire river systems.

Headwaters exchanges helped create the tremendous life history diversity exhibited along the coastline from southeast Alaska to WOCI. At one point, the Upper Fraser River drained into the Thompson River and Lake Okanagan, which emptied into the Columbia River. So too did the Stikine and Skeena rivers, which backed up and flowed into what is now the Upper Fraser. And as ice melted in the Queen Charlotte refugia, salmon swam in with the rise of the Georgia Strait and into the Puget Sound. Today's salmon populations do not necessarily derive from the rivers where they reside; rather, they radiated out from regional geographic refugia by these processes. The result: two populations within the same species residing in the same river may exhibit divergent marine migration and life history behaviors because of discrete genetic coding.

The phenomenon applied across the Bering Land Bridge. At one time, the Beringia refugia was the *proto*-Yukon, which likely linked the Anadyr and the Yukon rivers in one large catchment. The fact that tundra and ice sheets connected Chukotka and Alaska may have a great deal to do with why today's populations of Yukon River chum and Anadyr River chum appear genetically similar. ■

The 120-kilometer-long Hubbard Glacier spills from Mount Logan through Canada's Yukon Territory, meeting the ocean at Disenchantment Bay in Yakutat, Alaska. Hubbard is advancing relatively quickly, nudging into the mouth of Russell Fjord, pictured above.

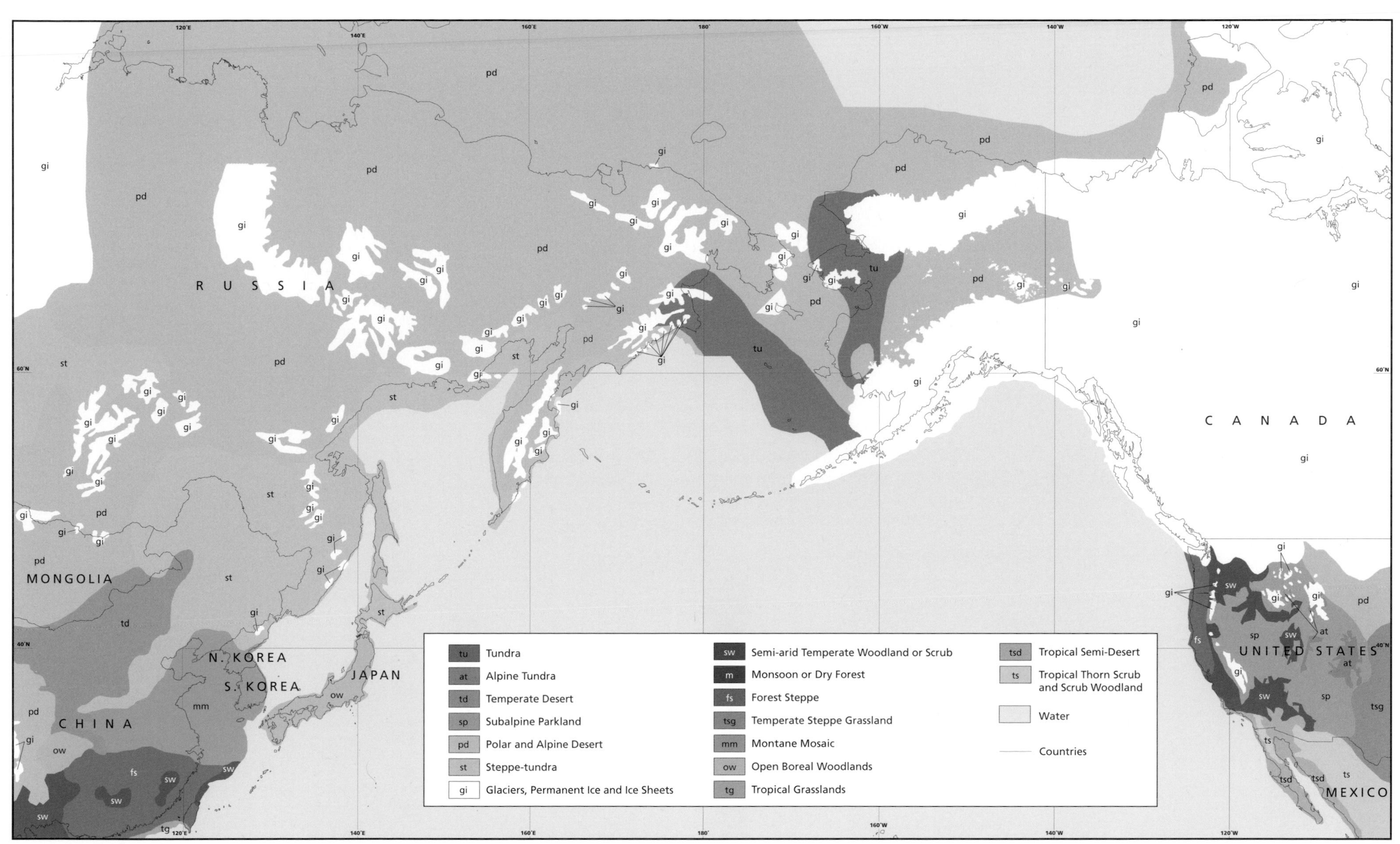


**EXTENT OF GLACIATION AT THE LAST GLACIAL MAXIMUM** Approximately 18,000 years ago, so much water was locked up in massive glaciers and ice sheets in Alaska and Canada that sea levels were around 125 meters lower than they are now. Glacial advance carved out basins, valleys, and fjords in southeast Alaska and British Columbia, sculpting new pathways for future salmon generations, which would colonize inland waters when glaciers eventually ebbed and rivers swelled. Although China, Japan, and most of Russia remained ice free, the landscape (largely tundra and polar or alpine desert) was nonetheless affected by the rising and falling sea levels.

# Terrestrial Ecoregions

**In 1995 World Wildlife Fund launched a** multiyear project to map the world's terrestrial biodiversity, separating units into "distinct assemblages of… natural communities sharing a large majority of species, dynamics, and environmental conditions." It termed the units of biogeographic regionalization "terrestrial ecoregions," mapped on the facing page.

The North Pacific Rim embraces eight major plant-based ecological communities, or biomes, from tundra to forest to shrubland. Within each biome, there may be several or dozens of distinct terrestrial ecoregions, through which salmon transition during their life cycles. Terrestrial ecoregions are influenced not only by geology and climate but also by biological phenomena such as salmon migrations (see below).

Through this prism, we note that the boreal forests and taiga of Kamchatka are similar to those in western Alaska. The temperate broadleaf and mixed forests of the Ussuri River straddling the border between China and Russia have much in common with Oregon's Willamette Valley forests. Likewise, the Hokkaido montane conifer forest shares many features with the leeward forests of the Cascade Range.

When we compare our salmon diversity map (see page 11) to this map, we see that salmon can thrive in vastly different landscapes. Kamchatka, Magadan, and western Alaska feature tundra and boreal forests and rich biodiversity; the Columbia River basin embraces similarly robust biodiversity within different climates ranging from deserts to temperate forests; the Amur River ecoregion features savannas, shrublands, and grasslands, a landscape similar to inland California.

**Tundra**

- Alaska-St. Elias Range tundra
- Aleutian Islands tundra
- Arctic coastal tundra
- Arctic foothills tundra
- Bering tundra
- Beringia lowland tundra
- Beringia upland tundra
- Brooks-British Range tundra
- Cherskii-Kdyma mountain tundra
- Chukchi Peninsula tundra
- Interior Yukon-Alaska alpine tundra
- Kamchatka mountain tundra and forest tundra
- Low Arctic tundra
- Middle Arctic tundra
- Northeast Siberian coastal tundra
- Ogilvie-Mackenzie alpine tundra
- Pacific Coastal Mountain icefields and tundra
- Taimyr Central Siberian tundra
- Trans-Baikal Bald Mountain tundra

**Temperate Grasslands, Savannas and Shrublands**

- California Central Valley grassland
- Canadian Aspen forests and parklands
- Daurian forest steppe
- Mongolian Manchurian grassland
- Montana valley and foothill grassland
- Palouse grassland
- South Siberian forest steppe

**Temperate Coniferous Forests**

- Alberta mountain forests
- Alberta-British Columbia foothills forests
- Blue Mountains forests
- British Columbia mainland coastal forests
- Cascade Mountains leeward forests
- Central British Columbia mountain forests
- Central Pacific coastal forests
- Central and Southern Cascades forests
- Da Hinggan Ling-Dzhagdy Mtns. conifer forests
- Eastern Cascades forests
- Fraser Plateau and Basin complex
- Great Basin montane forests
- Hokkaido montane conifer forests
- Honshu alpine conifer forests
- Klamath-Siskiyou forests
- North Central Rockies forests
- Northern California coastal forests
- Northern Pacific coastal forests
- Northern transitional alpine forests
- Okanogan dry forests
- Puget Lowland forests
- Queen Charlotte Islands
- Sierra Juarez-San Pedro Mártir pine oak forests
- Sierra Nevada forests
- South Central Rockies forests

**Temperate Broadleaf and Mixed Forests**

- Central Korean deciduous forests
- Changbai Mountains mixed forests
- Hokkaido deciduous forests
- Manchurian mixed forests
- Nihonkai evergreen forests
- Nihonkai montane deciduous forests
- Northeast China plain deciduous forests
- South Sakhalin-Kuril mixed forests
- Southern Korea evergreen forests
- Taiheiyo evergreen forests
- Taiheiyo montane deciduous forests
- Ussuri broadleaf and mixed forests
- Willamette Valley forests

**Mediterranean Forests, Woodlands, and Scrub**

- California coastal sage and chaparral
- California interior chaparral and woodlands
- California montane chaparral and woodlands

**Flooded Grasslands and Savannas**

- Amur meadow steppe
- Nen River grassland
- Suiphun-Khanka meadows and forest meadows

**Deserts and Xeric Shrublands**

- Great Basin shrub steppe
- Mojave Desert
- Snake-Columbia shrub steppe
- Sonoran Desert
- Wyoming Basin shrub steppe

**Boreal Forests/ Taiga**

- Alaska Peninsula montane taiga
- Cook Inlet taiga
- Copper Plateau taiga
- East Siberian taiga
- Interior Alaska-Yukon lowland taiga
- Kamchatka-Kuril meadows and sparse
- Kamchatka-Kuril taiga
- Mid-continental Canadian forests
- Midwestern Canadian Shield forests
- Muskwa-Slave Lake forests
- Northeast Siberian taiga
- Northern Canadian Shield taiga
- Northern Cordillera forests
- Northwest Territories taiga
- Okhotsk Manchurian taiga
- Sakhalin Island taiga
- Trans-Baikal conifer forests
- Yukon Interior dry forests

- Rock and ice
- Rivers, lakes and oceans
- Outside of Level 4 Salmon Ecoregions

## THE ECOLOGICAL IMPORTANCE OF MARINE-DERIVED NUTRIENTS

Salmon quite literally complete a connection between land and sea when they die. Essential to functional river systems, salmon carcasses transport nutrients and trace elements from the ocean—including carbon, phosphorous, and nitrogen—to upstream habitat. These marine-derived nutrients replenish streambeds, spur algal growth, and fortify aquatic and riparian foodwebs, from microinvertebrates to plants and wildlife. Headwaters hundreds of kilometers inland, fed by precipitation and snowmelt, would otherwise not contain some of these trace elements.

Salmon catch over the past 150 years has deprived headwaters of the marine-derived nutrients transported upstream by returning spawners. The loss of nutrients can be significant. In 1997 G. A. Larkin and P. A. Slaney posited that a salmon carcass contains approximately 3 percent nitrogen and 0.35 percent phosphorous. By comparing historic abundance and current escapement numbers for selected WOCI basins, Ted Gresh, Jim Lichatowich, and Peter Schoonmaker calculated the extent to which today's rivers have been depleted of nitrogen and phosphorous. Historic high runs in the Columbia River peaked at more than 100 million kilograms of Pacific salmon, netting 2 million kilograms of nitrogen and 270,000 kilograms of phosphorous. But at current return rates, biomass in the Columbia has plummeted to around 580,000 kilograms of salmon, which returns 18,000 kilograms of nitrogen and 2,000 kilograms of phosphorous to this river system—less than 1 percent of the historic contribution of marine-derived nutrients.

Research on marine nutrient dynamics is a new field. Native peoples, however, have long understood the importance of decaying salmon carcasses to the health of river systems and the life that depends on them. ■

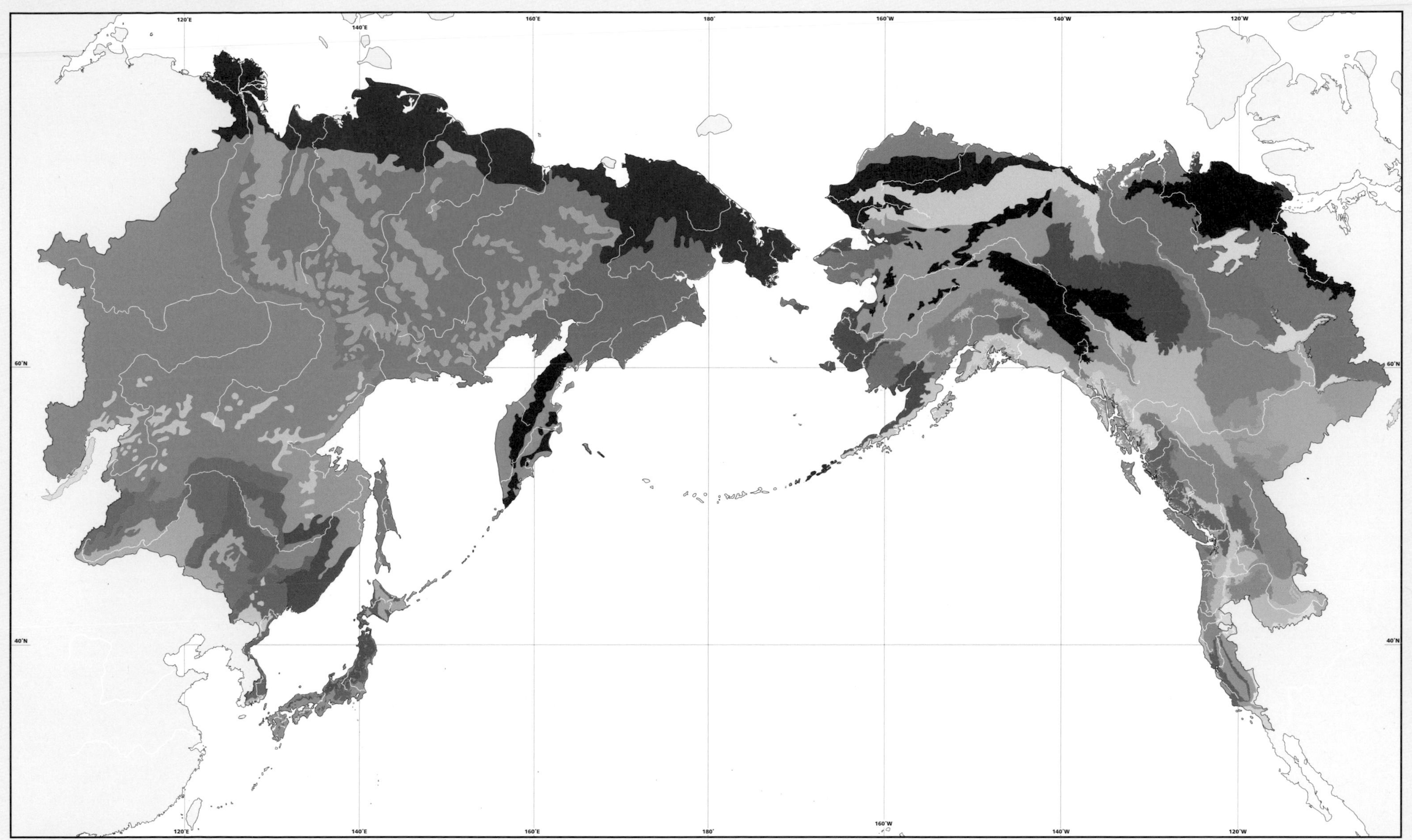


**TERRESTRIAL ECOREGIONS OF THE NORTH PACIFIC** There is tremendous diversity within coastal North Pacific drainage basins. Eight biomes, ranging from tundra to forest, comprise nearly 100 discrete terrestrial ecoregions. In some basins (e.g., the Amur, the Mackenzie, the Columbia), salmon may transition through more than a dozen terrestrial ecoregions; elsewhere, in areas equally rich in biodiversity (such as western Kamchatka or along the Vancouver coast), salmon may remain in just one terrestrial ecoregion.

# Sea Ice

**The timing, extent, and composition of sea** ice are major factors in the climate and nutrient transfer system of the Bering Sea. Sea ice acts as an intermediary between air and sea and as a conveyer for freshwater, nutrients, and salinity—all of which work together to create the unique conditions that determine the productivity of marine life from year to year and within a season.

Sea ice in the Bering Sea forms at –1.7°C, usually around November. Its formation locks up nutrients and, at the same time, results in a phenomenon called brine rejection, wherein the sea ice that floats on top of the ocean is less salty than the water underneath, which becomes more saline. Wind and current transport sea ice, which may thaw along coastlines to create smaller dynamic systems featuring the interplay of freshwater and sea ice.

The beginning of every salmon season is signaled by the melting of sea ice. That process is cued by the length of daylight, temperature, upwelling, currents that melt the ice from the bottom up, and the influence of inland snowmelt. The melting sea ice creates a soup of thawing nutrients and freshwater that then triggers the growth of phytoplankton, which is dependent on sunlight. In warmer years, sea ice is less extensive and retreats quickly in southerly regions. In colder years, the reach of ice is wider and duration is longer. Yet the temperature correlation is not necessarily so simple; an unusually warm spell in winter months may signal a rapid onset of ice melt, but daylight hours may be insufficient for photosynthesis.

More tangibly, sea ice can hinder fish passage. Typically water temperatures are colder along the shores of the Russian Far East than along the shores of North America at comparable latitudes. In colder years in the Sea of Okhotsk, for example, the presence of sea ice and prohibitively low sea surface temperatures in April and May can severely limit smolt outmigration. The relatively limited species composition of the Tuguro-Chumikan area in the southwest corner of the Sea of Okhotsk is a particularly strong example; this region retains the ice pack well into the summer months as a result of the counter-clockwise circulation pattern of the Sea of Okhotsk's cold current and frigid Siberian air masses, which keep temperatures low.

Interannual climate variations render the extent and nature of sea ice unpredictable. What we can predict is that even incremental fluctuations in sea and atmospheric temperatures due to global climate change will have dramatic effects on sensitive marine ecosystems that reverberate throughout the food web.

### THE EFFECTS OF SNOWPACK MELT

As sea-ice melt gives seasonal cues for the onset of marine phytoplankton production in early spring, snowpack melt signals the beginning of the freshet, the annual peak flow in freshwater systems. The processes differ throughout the Pacific Rim.

In North America, with the exception of Alaska, most of the precipitation falls during the fall and winter months. In Alaska, mountain snow melts later in the spring and summer; at moderate elevations, snowfall melts cyclically throughout the winter and into the springtime; at low elevations, precipitation falls usually as rain and is quickly absorbed into streams and seeps into groundwater stores during winter months. Therefore, inland tributaries at higher latitudes with mountain headwaters may expect high flows in late summer and early fall as glacial melt and snowmelt accumulate. Streams fed by lower elevation mountains and coastal ranges experience warmer temperatures and peak flows in summer.

Hydrology is markedly different in the western Pacific. Colder, drier winters in Chukotka and Magadan result in relatively low river flows in the spring. Furthermore, the Sea of Okhotsk often freezes over in the winter, with temperatures and cold currents from the north blocking fish passage to inland rivers well into the spring. In Japan, China, and southern Russia, the heavy rains of the monsoon season occur in the summer and early fall, increasing flooding, flow, and siltation.

With river flow and water temperature so dependent on the vagaries of climate, global climate change, however subtle, may have major ramifications for riverine ecology and, therefore, for salmon populations that have evolved to meet a basin's historical hydrological patterns. ■

In Washington's Olympic National Park, towering peaks, some reaching more than 2,400 meters, can receive more than five meters of precipitation each year. Above, the headwaters of the Soleduc River begin at the Seven Lakes basin.

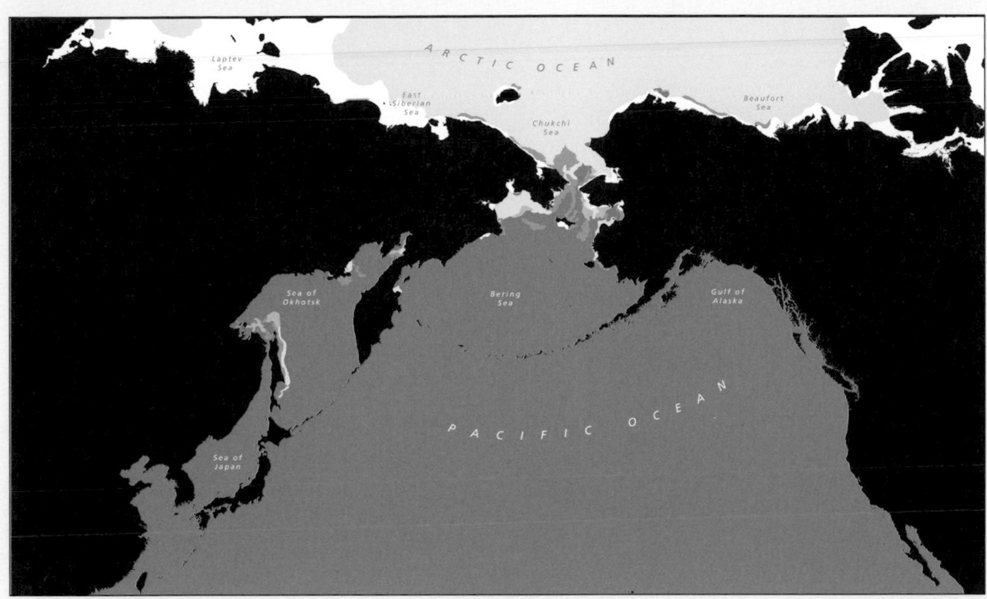


**May 1996**

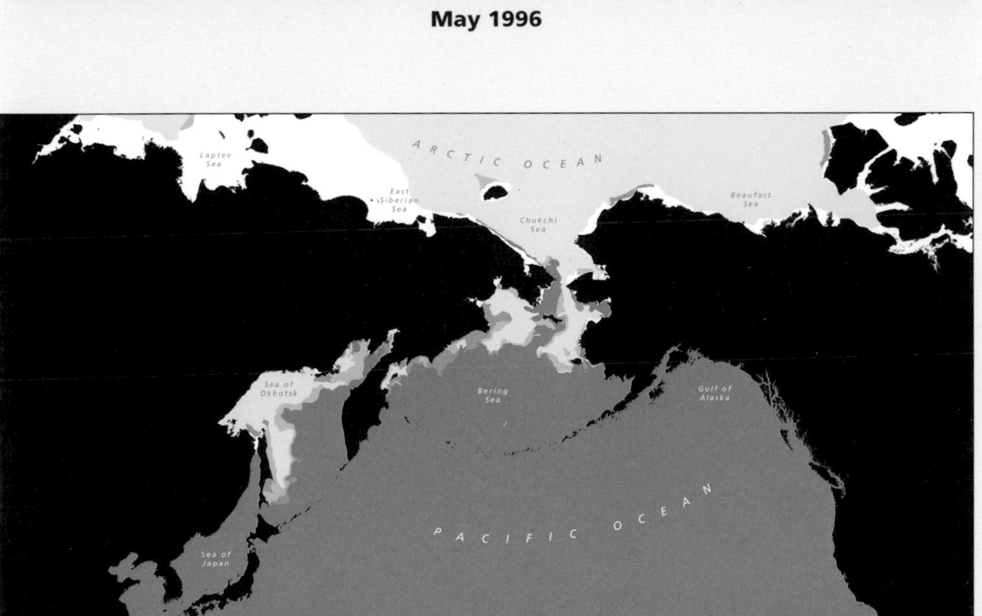


**May 2001**

FLUCTUATIONS IN SEA ICE

Because salmon use surface waters (except for chinook, which delve much deeper), the extent of sea ice has direct bearing on their range in a given season. Bodies of water particularly affected include the Arctic waters, the Chukchi and Bering Seas, and the Sea of Okhotsk, where migration pathways can be quite limited.

Sea ice melt can signal the beginning of smolt outmigration, cued by the temperature and length of daylight, upwelling and currents that melt the ice from the bottom up, and snowpack melt on land. Temperature will have bearing on the extent and quality of phytoplankton as the season progresses. As the maps to the left show, the extent of sea ice can vary dramatically from year to year.

**Precent Range of Sea Ice Concentration**

- 100% (solid ice)
- 70 – 100%
- 40 – 80%
- 10 – 50%
- < 10% (open water)
- Land

Rivers at higher latitudes in the North Pacific may remain frozen into late spring and early summer. Here, just south of the Bering Strait, ice clogs waterways near Alaska's St. Lawrence Island in May.

# Primary Production

**PHYTOPLANKTON ARE THE MICROSCOPIC BASE** of the marine food chain. Generally phytoplankton bloom on the continental shelf between February and May. As the days grow warmer, the snow melts, river discharge increases, and the seasonal upwelling moves nutrients from lower depths to the ocean's surface. The convergence of sunlight and nutrient-rich water spurs primary productivity.

By measuring levels of chlorophyll (the pigment that phytoplankton use to transform light into chemical energy), we can display the surface layer of phytoplankton in the sea. At the right time, place, and volume, phytoplankton can enrich the whole ecosystem by feeding organisms at successively higher trophic levels, leading to increased growth and survival of salmon and other top consumers. Algal bloom can serve as a yardstick for predicting salmon productivity.

Around the same time of year, freshwater juveniles transform into migratory smolts. Cued by temperatures and length of daylight, they swim downriver and enter estuaries and the ocean, where conditions are favorable for growth.

Future generations of salmon rely on the success of this carefully timed process. Variations in sea surface temperature, ocean currents, and upwelling affect primary and secondary productivity (zooplankton), and thus determine carrying capacity, the ocean's ability to produce and sustain biological productivity.

The maps to the right illustrate the apex of phytoplankton bloom in May, which largely synchronizes with smolt outmigration to nearshore ocean areas. The abatement of the bloom in July coincides with the smolts' transition to subadults as they move from shelf water to blue water offshore, shifting their diet to the larger prey that occupy higher trophic levels.

Interannual volume of algal bloom can vary by as much as 50 percent. In the maps on the facing page, for example, the density of the algal bloom in 2002 is much greater than in 1998, particularly in the eastern Pacific.

Global climate change will greatly alter primary productivity: even slight changes in weather conditions can have dramatic effects on the timing and extent of algal blooms—which, in turn, will have a ripple effect throughout the food web and affect salmon populations differently over time and space. Predicting how salmon respond to changing nutrient dynamics—which populations will benefit and which will suffer—is a subject garnering much attention today. To understand these complexities, researchers are studying marine rearing and migration pathways of specific salmon populations.

## THE CARRYING CAPACITY AND NUTRIENT RICHNESS OF THE BERING SEA

The 2.3 million square kilometers of the Bering Sea are separated from the Pacific basin and the Gulf of Alaska by the volcanic chain of more than 50 Aleutian islands; the Bering Sea's north edge is bordered by the Chukchi and Seward peninsulas. About half of the Bering Sea is a broad and shallow continental shelf, with an average depth of less than 75 meters. At its margin, the shelf drops off precipitously into seven submarine canyons (the world's largest) bordered to the south by two major underwater mountain chains.

For the most part, the Bering Sea is the heart of the Pacific salmon circulatory system. Smolts from east and west ride the currents up the coast to feed on the zooplankton, which feed on the marine algae that bloom with the springtime ice melt and warming temperatures. It's no wonder that two of the world's richest salmon grounds—Bristol Bay and Kamchatka—flank the Bering Sea.

Our new understanding of the Bering Sea prompts concern regarding ocean carrying capacity. There is evidence that we can overseed regional areas with aggressive hatchery release programs; for example, the release of pink salmon hatchery smolts in Prince William Sound all at once can have direct effects on the availability of food for all salmon and other marine life. Furthermore, because so much of the success of Pacific salmon depends on the integrity of the Bering Sea, there is increasing concern for how climate change will affect nutrient growth, sea surface temperatures, currents, and any number of variables that weave the intricate patterns of this unique ecosystem. ■

Copepods—zooplankton that feed on phytoplankton—are a major food source for salmon, particularly in the Bering Sea. There are more than 10,000 species of copepods inhabiting fresh and salt water throughout the world, from mountaintop to ocean canyon.

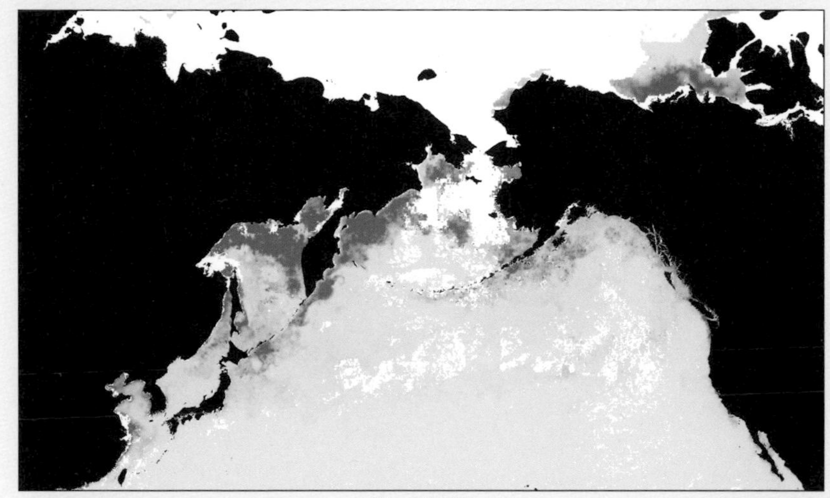

May 1998

CHLOROPHYLL PRODUCTION
Chlorophyll indicates the presence of phytoplankton, which is at least one trophic level removed from salmon (depending on life cycle stage) and functions as the base of marine food webs. The early spring window of algal bloom is narrow and can vary interannually. These fluctuations and the extent of the bloom may help explain changes in salmon growth and survival from one year to another. These maps reveal a significantly richer phytoplankton bloom in 2002 than in 1998.

(Note: These maps, representing mean monthly chlorophyll production, are composites of satellite images. White areas denote cloud cover or the presence of sea ice.)

May 2002

July 1998

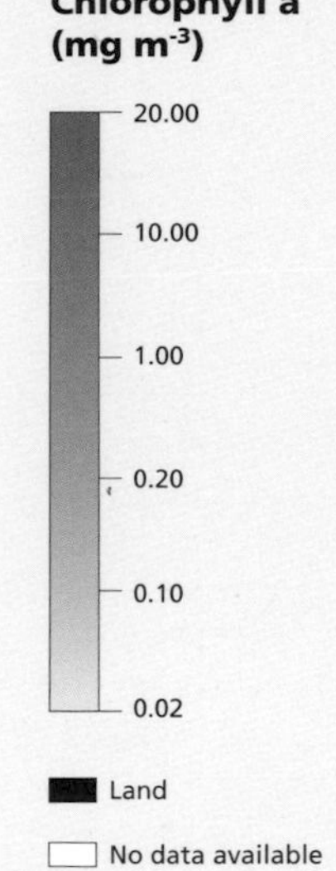

July 2002

Note: "Chlorophyll a" is a molecule present in all life forms that photosynthesize.

# Sea Surface Temperature

**Scientists routinely study salmon in** rivers, but knowledge of salmon in their ocean phase is less sure-footed. We do know that most salmon spend at least two-thirds of their lives at sea and gain more than 90 percent of their final weight there. We also know that most *Oncorhynchus spp.* remain close to the sea surface during their marine phase. (Chinook are the exception, venturing up to 100 meters below the surface.) And we know that salmon make particular use of the subarctic domain, bordered on the south by the transition zone, below which temperature and salinity in ocean waters dramatically increase.

At sea, salmon resolutely prefer cooler waters, and abundance drops by a factor of two at temperatures above 10.4°C. Furthermore, species appear to have specific preferences in the spring: for example, sockeye prefer waters cooler than 9°C; coho will not venture into waters warmer than 9.4°C; and pink and chum may tolerate waters as warm as 10.4°C. In order to remain in cool water as the season progresses, salmon distribution shifts toward the northern latitudes, which results in higher densities of fish.

Where the southern extent of cooler waters in May can reach below Hokkaido and off the coast of southern Primorye in the western Pacific and off the coast of Washington, Oregon, and California in the eastern Pacific, this thermal boundary—and salmon distribution—will shift to the north as the season progresses.

Sea surface temperature peaks in July and August in the North Pacific. In addition to seasonal temperature shifts, it responds to a host of other variables as well, including gravitational forces, ocean currents, wind, evaporation, and atmospheric circulation, to name a few. These factors combine to create constant and often unpredictable fluctuations in sea surface temperature that have a direct effect on the distribution and productivity of salmon populations. Phenomena such as PDO and El Niño/La Niña (describing variations of climate and temperature patterns and precipitation anomalies; see pages 49–50) constitute spatiotemporal patterns of sea surface temperature variability, which can have significant effects on the productivity of salmon populations.

In recent years, scientists have made great strides in identifying ocean modes or states that have direct relevance for the status and productivity of salmon in the North Pacific. Nonetheless the complexities of ocean dynamics continue to challenge oceanographers and climatologists in their efforts to accurately predict ocean productivity.

## STREAM TEMPERATURE AND THE EFFECTS OF CLIMATE CHANGE

Anadromous salmon spend most of their lives in the ocean, but the comparatively short time they reside in freshwater—as juveniles and again as spawners—occurs when they are most vulnerable. Salmon do best in freshwater streams with high flow and cool temperatures, from 15°C to 18°C and no more than 20°C for any extended time period. Temperature affects food and water quality, which indirectly affects salmon metabolism and vulnerability to pathogens and disease, particularly for returning spawners that expend tremendous amounts of energy in migrations to natal streams.

Scientists are working to assess how climate change will alter stream habitat. Their findings differ substantially throughout the North Pacific, depending on the models they use. Generally they have found that arid regions will experience more frequent bouts of drought, decreased stream flow, and increased water temperatures. Humid and temperate regions will be more likely to endure flooding and increased water flow. Seasonal changes may become more extreme, with increased runoff in winter and spring and decreased runoff in the summer.

A recent study of global warming effects on the Fraser River basin applied global circulation models to predict stream temperature and flow patterns between 2070 and 2099. Although total river discharge and average flow did not demonstrate much variance, peak flow timing was markedly different. With diminished snowpack at higher elevations and warmer summer temperatures, models predicted that average peak flow would shift 24 days earlier and decrease by 18 percent, increasing by a factor of 10 salmon's exposure to water temperatures above 20°C. ■

In the Fraser Valley in British Columbia, a creek flows into the Chilliwack River. Even slight climate changes can influence rate and timing of snowpack melt at headwaters, which can have dramatic effects on salmon runs downstream.

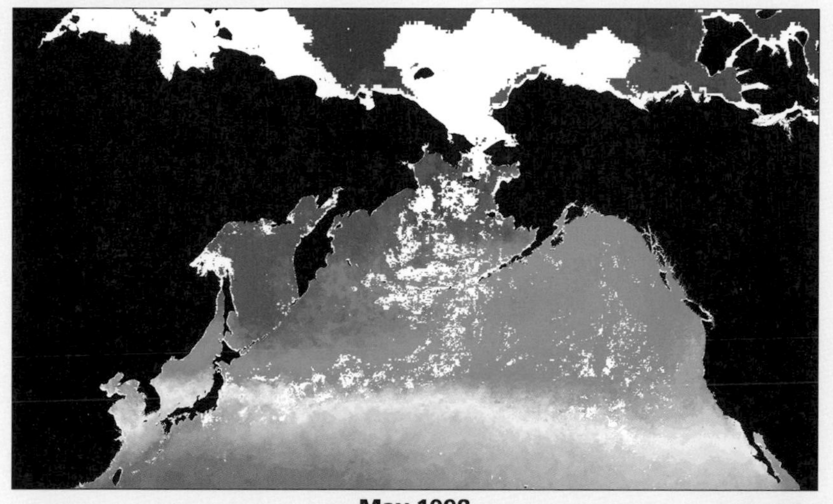
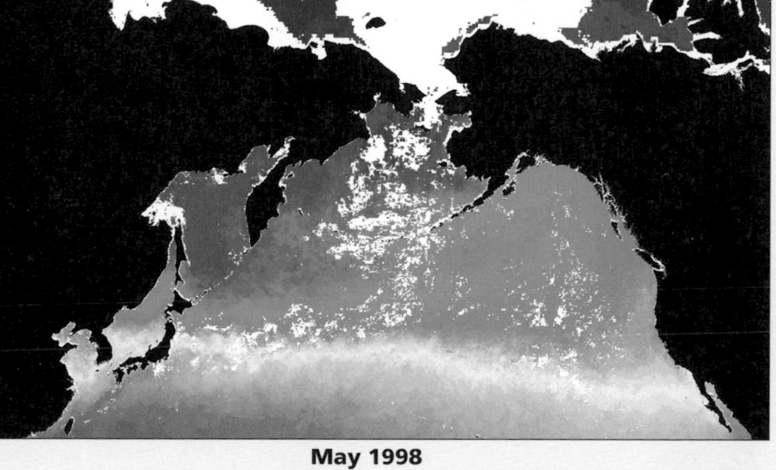

May 1998

SEA SURFACE TEMPERATURE
Depending on the species, the uppermost temperature salmon can endure in the marine life history phase is around 10°C–11°C. More importantly, temperature toleration changes continuously. For sockeye, summer temperatures above 13.5°C are lethal; in winter, sockeye prefer 6°C–7°C. Length of day and temperature affect marine distribution; climate fluctuations and warming in particular restrict suitable marine habitat even further.

(Note: These maps, representing mean monthly sea surface temperature, are composites of satellite images. White areas denote cloud cover or sea ice.)

May 2002

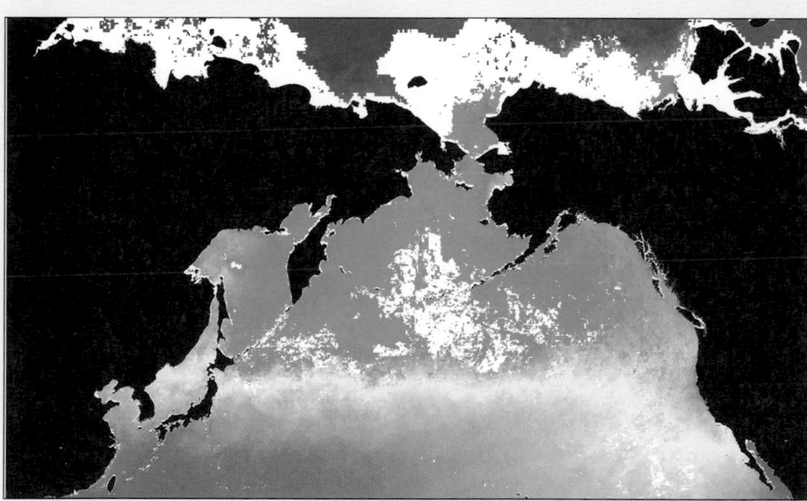

July 1998

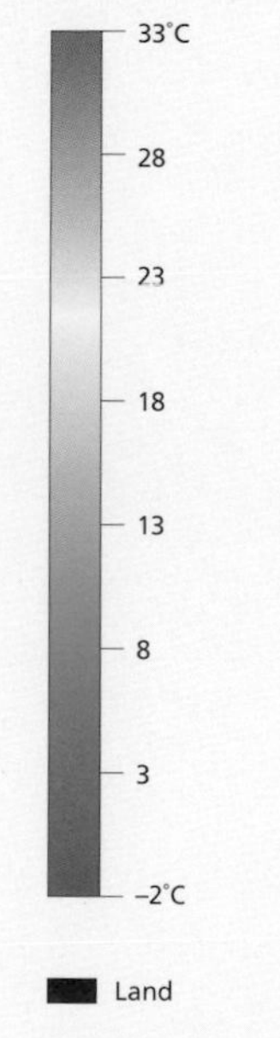


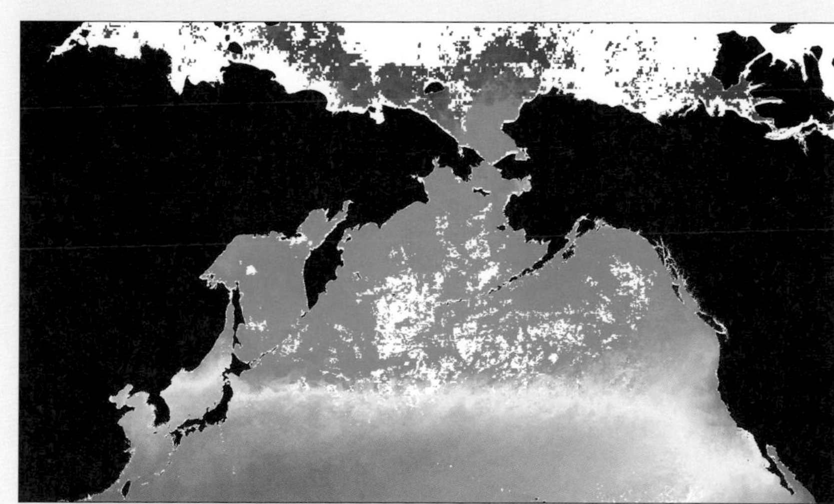

July 2002

# 4

# Distribution and Risk of Extinction

*Salmon in their setting: Stock status and trends*

■ Chum *O. keta* ■

■ Pink *O. gorbuscha* ■

■ Sockeye *O. nerka* ■

■ Chinook *O. tshawytscha* ■

■ Coho *O. kisutch* ■

■ Masu *O. masou* ■

■ Steelhead *O. mykiss* ■

A pink salmon carcass contributes rich nutrients derived from its ocean migrations to freshwater habitat, including essential nitrogen, phosphorous, and carbon.

**The information in this chapter—** where Pacific salmon are and how they are faring—is the fulcrum of this book. The maps on the pages that follow, like the others in this atlas, were built using geographic information systems (GIS) tools. Our georeferenced data spatially represent layers of information that have been researched by our GIS analysts and taken from any number of sources at multiple scales. Once the data were aggregated and analyzed, results from the analysis were used to construct maps, which were designed using several graphics software programs. Such status assessments have never before been presented at this spatial extent on this subject.

The maps of distribution and risk of extinction represent the outcome of a decade of work that began as the Pacific Rim Project. Launched in 1994 at Oregon State University with seed funding from congressional appropriations secured by Senator Mark Hatfield, this project was designed to assess distribution and threats to salmon across the North Pacific in order to give Oregon fisheries managers a greater context in which to assess depressed coho escapements during the early 1990s. This project was staffed by the book's author and fellow researchers Dan Bottom and Jeff Rodgers; Rodgers was among the first to create an accurate Mercator mapping of the North Pacific (see page 8). The Pacific Rim Project created a common geographic framework (ecoregions, see page 6) to represent data at a uniform scale.

Sockeye smolts navigate the Naknek River at the mouth of Alaska's Bristol Bay, which continues to be the world's greatest sockeye producer. Annual average sockeye catch in Bristol Bay ranges from 10 to 30 million.

As the project evolved and moved in 1999 to the Wild Salmon Center in Portland, Oregon, the author and Bottom and Rodgers conducted a survey to gather best expert judgment data on salmon distribution and risk of extinction from field biologists in the western Pacific, from Russia, Japan, China, and South Korea. We were successful for Russia, the Amur River, and the Heilon Jiang province of China. But our efforts in Japan were limited by the fact that its salmon are overwhelmingly hatchery derived, and therefore we confronted a lack of site-specific and species-specific knowledge of wild populations.

Along the United States–Canada border, chinook work their way upstream in the Chilliwack River in British Columbia.

**DISTRIBUTION** Until the publication of this atlas, the best available range maps for the anadromous forms of the genus *Oncorhynchus* were presented in the landmark publication *Pacific Salmon Life Histories,* edited by C. Groot and L. Margolis (1991), and, representing just the northeastern Pacific, in Robert Behnke's treatise *Trout and Salmon of North America* (2003).

Our distribution maps expand upon these earlier efforts by presenting for the first time georeferenced data for the western Pacific at a consistent measurable scale. We surveyed field biologists in Russia, Japan, China, and South Korea who provided best expert judgment data that we aggregated at a uniform scale.

To evaluate eastern Pacific salmon distribution, we relied primarily on datasets from provincial, state, and federal fisheries agencies, as well as previously published literature and best expert judgment.

Because these data were captured with different methods and at varying scales, we needed to use a consistent unit of measure across the North Pacific—yet spatial analyses of hydrological data at a global or continental scale can be difficult to organize, evaluate, and process. The U.S. Geological Survey had previously solved this problem by creating a HYDRO1k dataset, which we used to classify our information (see "How we mapped distribution," next page). This common spatial framework enabled us to depict current, limited, and historic spawning distribution across the North Pacific.

**RISK OF EXTINCTION** Our risk maps were also constructed using best expert judgment from biologists in Russia and North America.

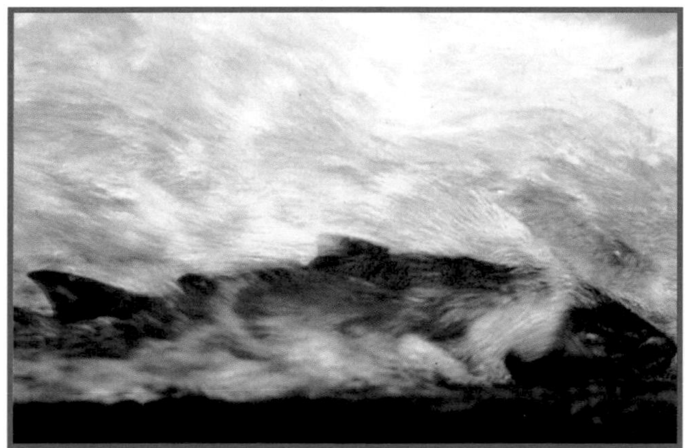
A returning chinook, here cruising in the foam of Washington's Klickitat River, can travel around 45 kilometers per day.

In addition, for North America, we depended on previously published literature: for Washington, Oregon, California, Idaho (WOCI) (Nehlsen et al. 1991; Huntington et al. 1996), for British Columbia (Slaney et al. 1996), and for southeast Alaska (Baker et al. 1996). Southeast Alaska stock status is reported by management areas, which could not be directly resolved to the boundaries of our ecoregions without access to the original georeferenced data; therefore, we report stock status for the region as a whole.

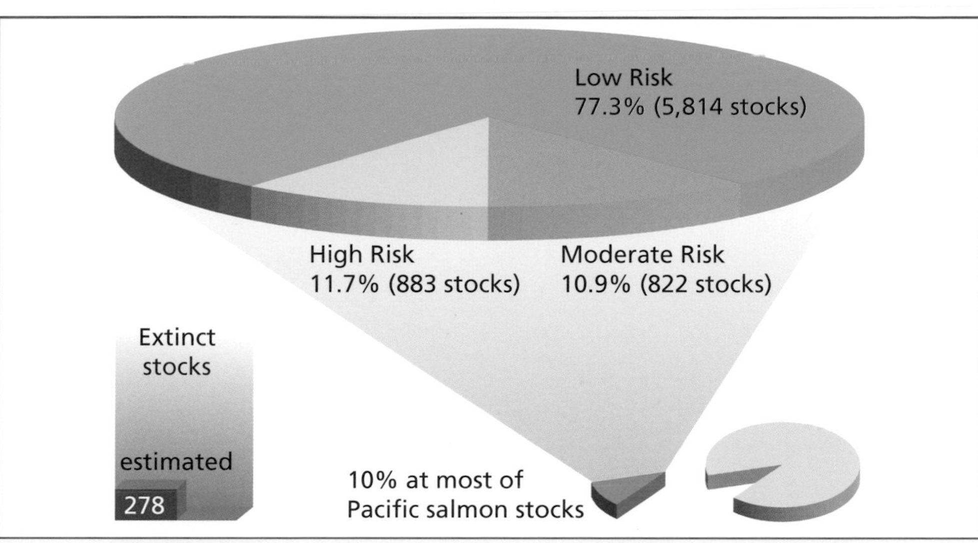

PACIFIC SALMON RISK OF EXTINCTION Representing at most an estimated 10 percent of *Oncorhynchus*, our data reveal that 23 percent of Pacific salmon are at risk. We also identified an estimated 278 extinct stocks.

We were unable to include data from Japan and central and western Alaska in our risk assessments. For Japan we did not pursue data from our colleagues because hatcheries dominate Japanese fisheries, and wild salmon data are insufficient for meaningful status assessments. In Alaska data holdings at the time were diffuse, and time and resource constraints made it impossible for us to pursue data collection and analysis.

Where spawning populations were viable, we assigned risk of extinction (as a percentage of the populations assessed) for each species by ecoregion; where applicable we also estimated the number of extirpations. However, we did not assess regions where populations were entirely extirpated or where little data were available. Lack of data prevented us from status assessments for South Korea; this was also true for cutthroat data throughout the North Pacific. Because so little is known about the status of salmon populations in 14 remote Arctic ecoregions, they were also omitted from this assessment.

Best expert judgment has its limitations, most notably perception bias, which can skew results. For example, very little data are available for masu salmon because they are low in abundance throughout their range. We turned to field biologist reports for data, which may reveal a systematic perception bias in our survey. For instance, where Kamchatka biologists rank their chum salmon populations at moderate or high risk of extinction, our Sakhalin respondents had contradictory best expert judgment data, ranking all populations at low risk of extinction. Although

## HOW WE MAPPED DISTRIBUTION

On the following pages, the spawning distributions for the seven species of Pacific salmon covered in this atlas represent our most recent effort to display current distribution at a consistent hydrological basin unit scale across the Pacific Rim. To create these maps, we assembled a dataset (GIS coverage) that shows current distribution for each species at the stream level.

- For Russia and the Far East, we used best expert judgment data from local biologists in each oblast or prefecture and previously published data
- For Alaska we primarily used the anadromous waters catalog, 2003
- For British Columbia we used the Department of Fisheries and Oceans (DFO) Fisheries Information Summary System, 2003
- For Washington-Oregon-California-Idaho (WOCI), we used StreamNet's distribution, 2003

After these data were compiled into one seamless coverage, the dataset was then intersected with a basin dataset, HYDRO1k, developed at the U.S. Geological Survey's (USGS) EROS Data Center. HYDRO1k is a geographic database providing comprehensive and consistent global coverage of topographically derived datasets, developed from the USGS's recently released 30 arc-second digital elevation model (DEM) of the world (GTOPO30).

The HYDRO1k dataset provides a standard suite of georeferenced datasets (at a resolution of 1 kilometer) that are valuable to all users who need to organize, evaluate, or process hydrological information on a continental scale. We use this dataset because it is the only one to offer uniform coverage for both sides of the Pacific Ocean.

We were also able to display areas of historic salmonid spawning distribution based on previously published literature, including status reviews prepared by NOAA Fisheries (National Marine Fisheries Service). We used the HYDRO1k dataset as the base layer. Then each basin or subbasin describing historic salmonid spawning distribution was manually edited using georeferenced 30m DEMs, 1:100,000 streams, 4th field watershed boundaries, and dams. ■

this is a flaw in the best expert judgment approach, we believe nonetheless that the data presented here provide a valuable benchmark, an initial effort to spur future quantitative study.

**FINDINGS** We had sufficient information to assess the status of 7,519 stocks from 41 of the 52 Pacific ecoregions. These included seven anadromous species of the North Pacific: chinook *(Oncorhynchus tshawytscha),* chum *(O. keta),* coho *(O. kisutch),* masu *(O. masou),* pink *(O. gorbuscha),* sockeye *(O. nerka),* and steelhead *(O. mykiss).* It is critical to note that even the quantitative data we use gravely underrepresent total numbers of salmon. Escapement trend estimates likely represent no more than 10 percent of known stocks in British Columbia and perhaps as few as 2 percent of the salmon stocks in southeast Alaska. Furthermore, data are skewed by fishery management needs, with valuable species more closely surveyed: sockeye data, for example, are more complete than coho or pink.

Based on published literature and our own findings, our maps demonstrate that salmon distribution is shrinking at the southern edges of the range across the North Pacific.

The northern edge of the range is represented primarily by pink, chum, and chinook. The northernmost populations in the East Siberian and Beaufort Seas are intermittent, appearing in significant numbers when climate conditions are favorable (e.g., warmer seas and less ice).

Our data also demonstrate that chum are the most widely distributed, and that masu have the smallest distributional range.

The giant rainbow trout of Lardeau River, a tributary of Kootenay Lake in British Columbia, can weigh up to nine kilograms.

Our findings on risk of extinction demonstrate that human-made threats present new and ever-increasing challenges to salmon populations. We determined that

- WOCI ecoregions have the highest concentrations of high-risk stocks;
- ecoregions in the Sea of Okhotsk and the western Bering Sea have a high proportion of moderate-risk stocks;
- the western Pacific has more ecoregions classified entirely in the low-risk category than does the eastern Pacific.

We found that on a species-by-species basis

- 7 percent of pink populations are classified at moderate or high risk;
- 24 percent of masu populations are classified at moderate or high risk;
- 29 percent of chum populations are classified at moderate or high risk;

- 30 percent of sockeye *and* coho populations are classified at moderate or high risk
- 36 percent of chinook populations are classified at moderate or high risk;
- 39 percent of steelhead populations are classified at moderate or high risk.

**MARINE DISTRIBUTION** Inland distribution is just one component of the salmon life cycle. Salmon may spend years at sea developing into mature adults, and we know very little about this period. What we do know is that many factors—temperature, primary and secondary productivity, climate variability, predation, and more—contribute to the habitat mosaic.

Our knowledge of Pacific salmon ocean distribution is based primarily on high seas tagging programs conducted between 1956 and 1989 under the auspices of the International North Pacific Fisheries Commission, which coordinated U.S., Japanese, and Canadian marine distribution research efforts. Sampling was conducted on a 2 x 5 degree grid system, rendering a block pattern; our data may consist of as little as a single observation during the period of record. Where appropriate, we added blocks representing migratory pathways between known high seas distribution areas and documented natal rivers, off the California coast and in the Sea of Japan, for example. Tag returns have never been robust and must be read as an incomplete representation. Because Japan had the most active high seas tagging program, the sampling effort focused largely on the western Pacific and in the central Bering Sea, the historic focus of Japanese high-seas fisheries.

The Skagit River system is the only one in Washington to host five species of spawning salmon, including the pink salmon pictured above.

New technology and dramatic improvements in research collaboration among North Pacific nations are rapidly improving our understanding the complexities of ocean distribution and migratory patterns, in the context of climate variability and stock-specific migration pathways. These investigations are mostly ship based and have been more extensive in the western Pacific, due to agency priorities.

Nuances of marine distribution are complex: which populations use which parts of the ocean at what time during their migratory period? How do we assess ocean carrying capacity, particularly as hatchery stocks are more abundant than wild fish? How do we conservatively manage fisheries during periods of low productivity? The answers to these questions are important, because salmon survival patterns over space and time reflect the status of lower trophic levels throughout the North Pacific Ocean ecosystem.

## HOW WE ASSESSED RISK

Risk assessment is based upon the best expert judgment of field biologists, with the exception of British Columbia and southeast Alaska, for which we used quantitative escapement data. Southeast Alaska stock status is reported by management areas. Because these regions did not follow the boundaries of our ecoregions, we report stock status for the area as a whole, combining portions of four ecoregions (Transboundary Fjords (50), Stikine River (51), Nass River (52), and Nass-Skeena Estuary (53)). Data for the British Columbia portions of the Nass-Skeena and Nass River ecoregions are reported separately. Lack of data prevented us from categorizing wild salmon status in North and South Korea, Japan, Arctic Ocean drainages, and central and western Alaska.

Extinction risk was assigned as follows:

**Unthreatened populations**

- are abundant relative to current habitat capacity;
- have not exhibited recent declines;
- have never been previously identified as at-risk.

**Populations at moderate risk**

- have experienced declines;
- exhibit stable numbers (more than 200) of spawners;
- return an average of one adult per spawner.

**Populations at high risk**

- have experienced declines;
- exhibit decreased number of spawners (fewer than 200 adults within the past five years), except for historically small populations;
- return fewer than one adult per spawner.

**Extinct populations**

- no longer reproduce in their historical range.

Note that the height of the data bars in the risk of extinction maps represents all assessed populations; colors represent the percentage of stocks assigned to different risk categories. ■

# Chum Distribution

**Chum are the most widely distributed of** Pacific salmon. They also undergo the longest migrations within the genus *Oncorhynchus.* Spawning in side channels of braided river reaches far upstream, they travel herculean distances—in Canada, as far as 3,500 kilometers up the Yukon River; in Russia, deep into the Amur River basin.

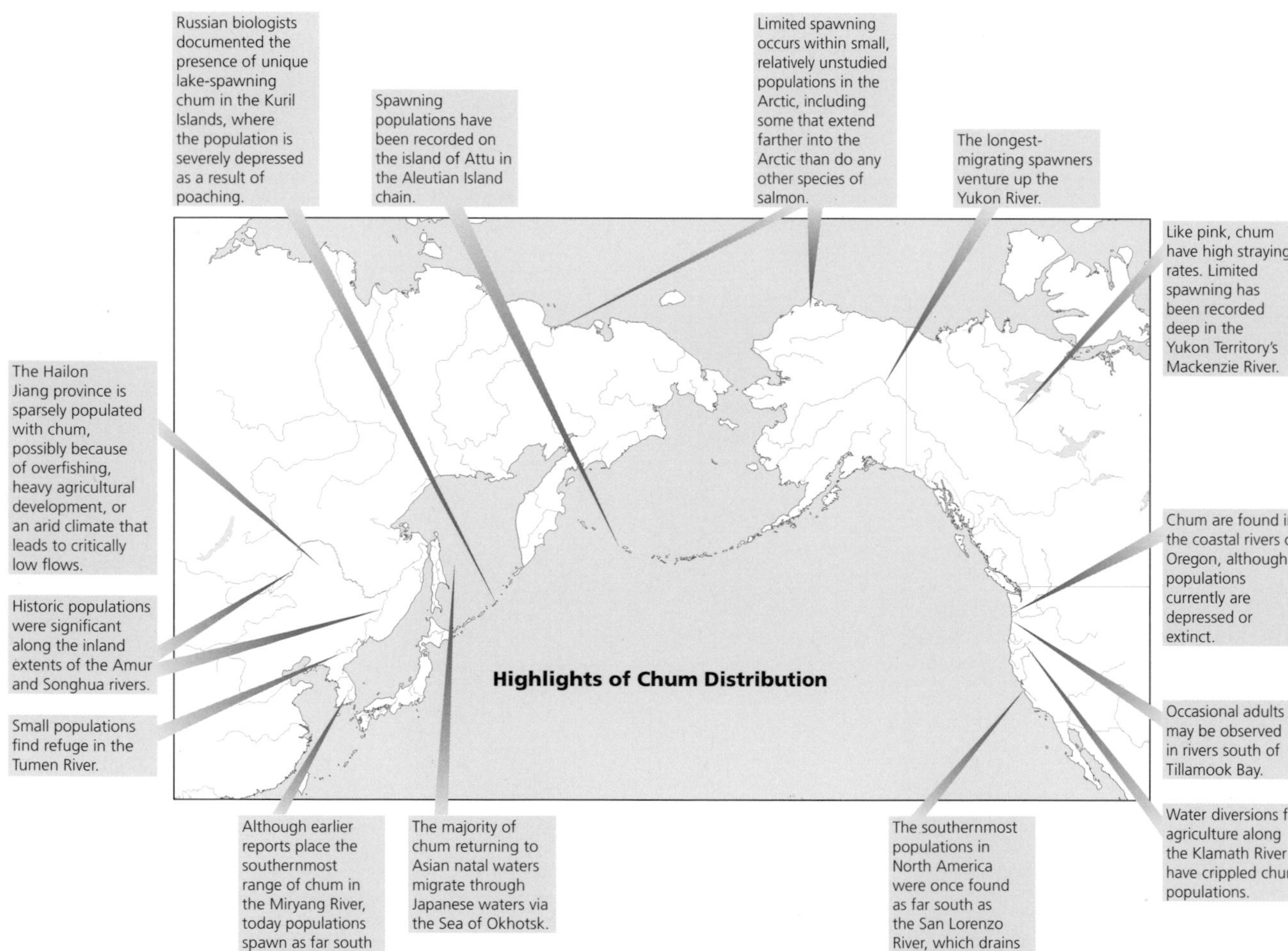


## CHUM SALMON

**SCIENTIFIC NAME**

*Oncorhynchus keta*

**ABUNDANCE**

Chum is the most abundant species in Japan. Elsewhere in the Pacific, chum are second in abundance only to pink salmon, except in Oregon and California, where they are less numerous than steelhead, chinook, and coho.

**SIZE**

Next to chinook, chum salmon are the largest of Pacific salmon, weighing 3–6 kg and measuring around 100 cm at maturity.

**LIFE HISTORY**

Chum live 3–5 years. They are most closely related to pink salmon and spawn in the lower reaches, side channels, and tributaries of gravel-bed rivers; unlike pink, however, fall chum are long-distance migrators. Most populations rely little on freshwater for development, beginning their ocean migration soon after emerging from the redd. Run timing varies; chum return to northern rivers from June to September and to more southerly rivers from August to November. A few Japanese chum salmon runs return as late as early February. Chum maintain a diet of large, gelatinous zooplankton while in the ocean.

**CULTURAL ROLE**

Chum are a staple for people in the western Pacific and for native peoples in Alaska and British Columbia. They are favored for drying by the indigenous Amur peoples because chum preserve better than the less oily pink salmon. In regions with abundant chinook or sockeye, chum are dried for use as sled-dog food—a critical fuel in northern latitudes in preindustrial times.

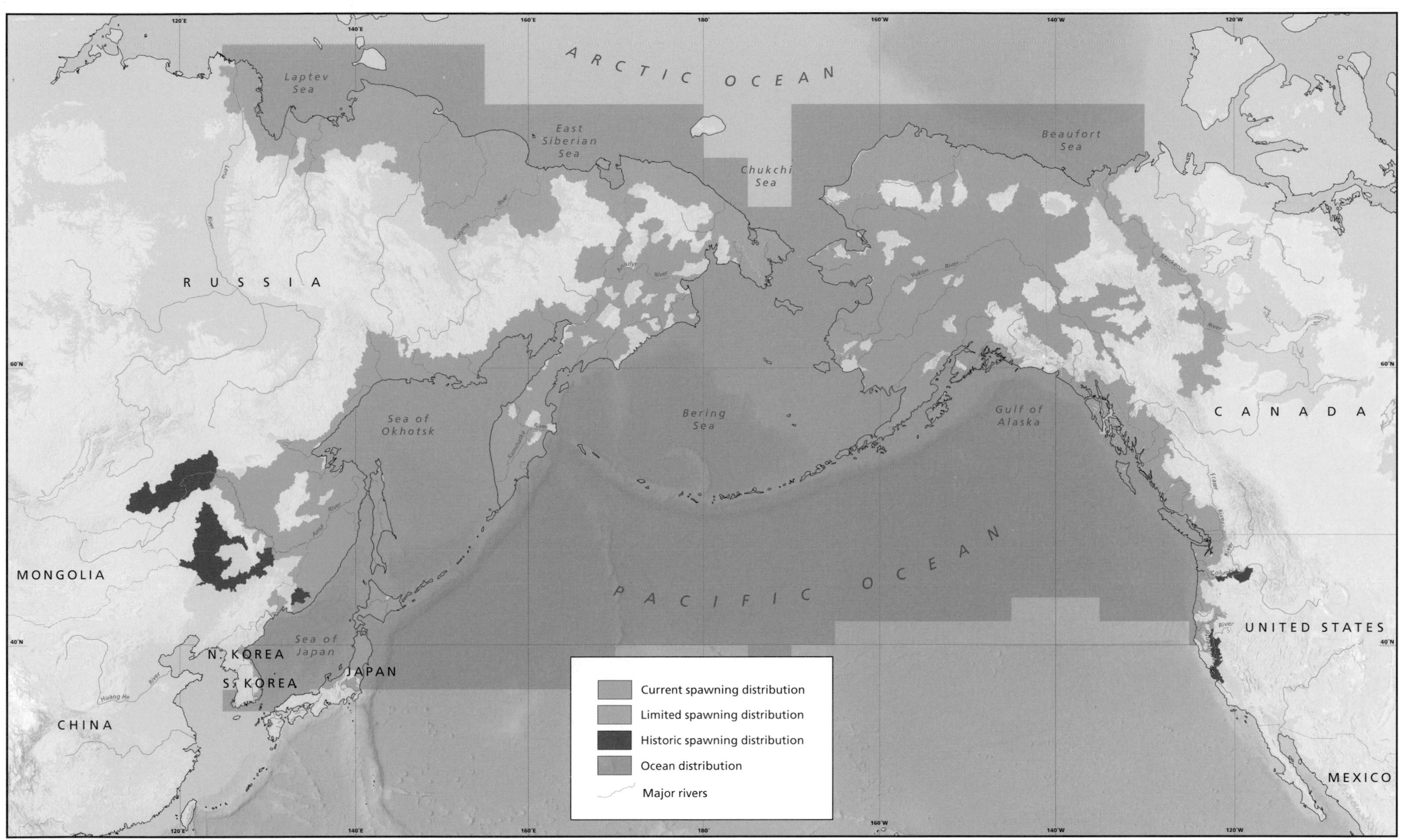


**CHUM DISTRIBUTION** Chum and pink have the highest straying rates among *Oncorhynchus*—up to 10 percent, as evidenced by the breadth of limited spawning distribution. Chum may swim thousands of kilometers up the Mackenzie River, which rarely hosts salmon. Although widely distributed, chum populations are sparse at southern latitudes, which contributes to inland extirpations along the Amur, the Columbia, and the Sacramento-San Joaquin basins. Overharvest, urbanization, dams, and water diversions have eradicated the few populations that have migrated deep into these rivers' interiors. (This map is based on two-season-run data.)

# Chum Risk of Extinction

**Chum are a major source of food in** Japan and Russia—particularly their roe, which is highly prized. In North America, however, chum are among the least valued salmon. In the 1970s Russians implemented formal annual stock forecasting to manage chum and other fisheries more responsibly. Populations have been extirpated throughout the Pacific.

Chum fry

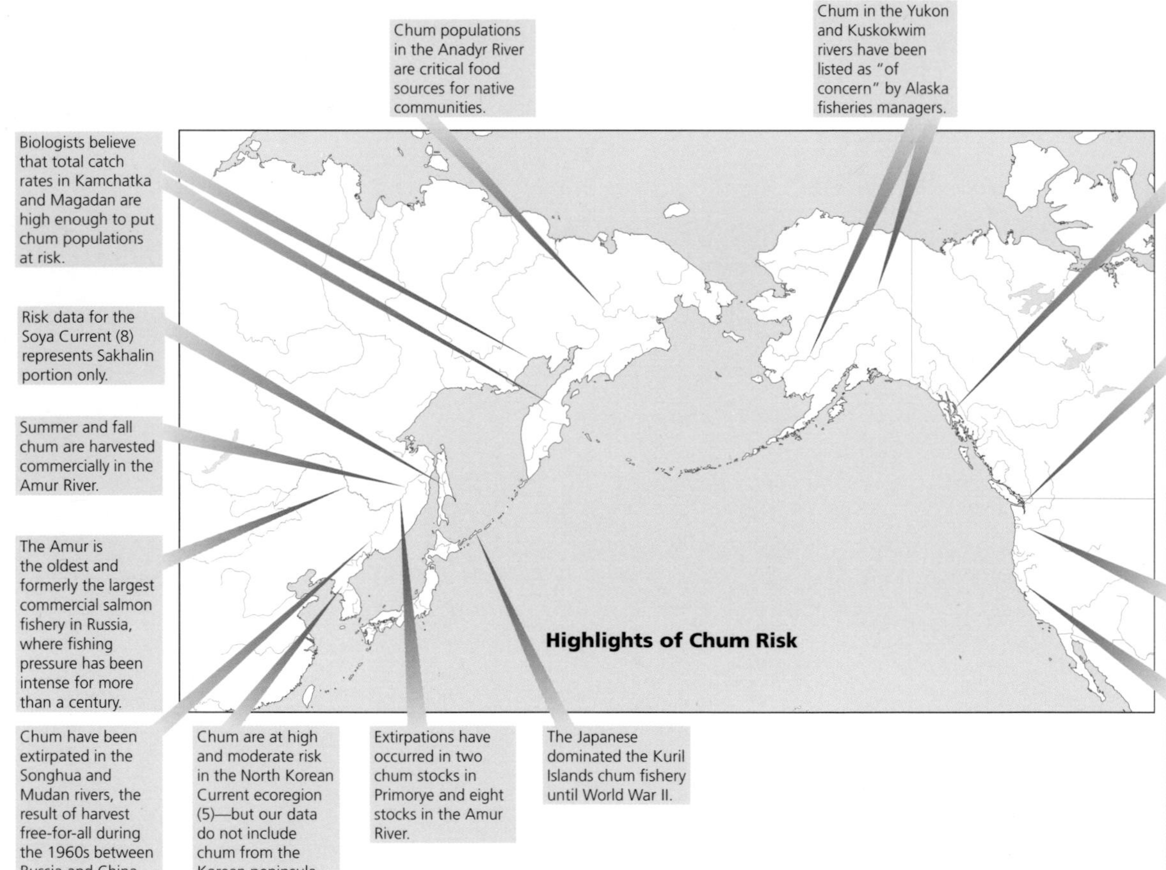


**Highlights of Chum Risk**

**Percentage of Relative Risk/Ecoregion**

| | ECOREGION | %HIGH | %MOD | %LOW | #EXT |
|---|---|---|---|---|---|
| 5 7 | Primorsky/N. Korean Current* | 62 | 38 | 0 | 0 |
| 8 | Soya Current** | 0 | 0 | 100 | 0 |
| 9 | Okhotsk-Oyashio Mixing | 0 | 0 | 100 | 0 |
| 10 | Liman Current | 47 | 53 | 0 | 2 |
| 11 | Southeast Sakhalin Current | 0 | 0 | 100 | 0 |
| 12 | Amur River*** | 0 | 0 | 100 | 0 |
| 12 | Amur River**** | 29 | 32 | 39 | 8 |
| 13 | Southwest Sakhalin Current | 0 | 0 | 100 | 0 |
| 14 | East Sakhalin Current | 0 | 0 | 100 | 0 |
| 15 | Shantar Sea | 0 | 0 | 100 | 0 |
| 16 | Uda River | 0 | 50 | 50 | 0 |
| 17 | Sea of Okhotsk Current | 0 | 100 | 0 | 0 |
| 25 | Anadyr River | 0 | 0 | 100 | 0 |
| 27 | Anadyr Current | 0 | 0 | 100 | 0 |
| 28 | Velikaya River | 0 | 0 | 100 | 0 |
| 29 | Penzhina River | 0 | 100 | 0 | 0 |
| 30 | Penzhina Intracoastal | 0 | 100 | 0 | 0 |
| 31 | Shelikhov Gulf | 0 | 100 | 0 | 0 |
| 32 | Bering Slope-Kamchatka Currents | 3 | 97 | 0 | 0 |
| 33 | Kamchatka River | 0 | 100 | 0 | 0 |
| 34 | Western Kamchatka Current | 0 | 100 | 0 | 0 |
| 50 51 52 53 | Southeast Alaska***** | 0 | 7 | 93 | 1 |
| 52 | Nass River****** | 31 | 0 | 69 | 0 |
| 54 | Skeena River | 18 | 0 | 82 | 0 |
| 55 | Hecate Strait-Q.C. Sound | 17 | 1 | 82 | 0 |
| 57 | Fraser River | 3 | 0 | 97 | 0 |
| 58 | Puget Sound-Georgia Basin | 11 | 2 | 87 | 22 |
| 59 | Vancouver Island Coastal Current | 12 | 1 | 87 | 1 |
| 60 | Seasonal Upwelling Cline | 41 | 18 | 41 | 3 |
| 61 | Columbia River | 33 | 67 | 0 | 2 |

*All of ecoregion 7 and the Russia portion of ecoregion 5.

**The Russia portion of ecoregion 8.

***The Sakhalin portion of ecoregion 12.

****The mainland Russia portion of ecoregion 12.

*****The southeast Alaska portions of ecoregions 50, 51, 52, and 53.

******The British Columbia portion of ecoregion 52.

*******The ecoregion 64 Sacramento-San Joaquin Rivers (in California) features full extinction, though historical number of stocks is unknown.

see foldout Ecoregion map

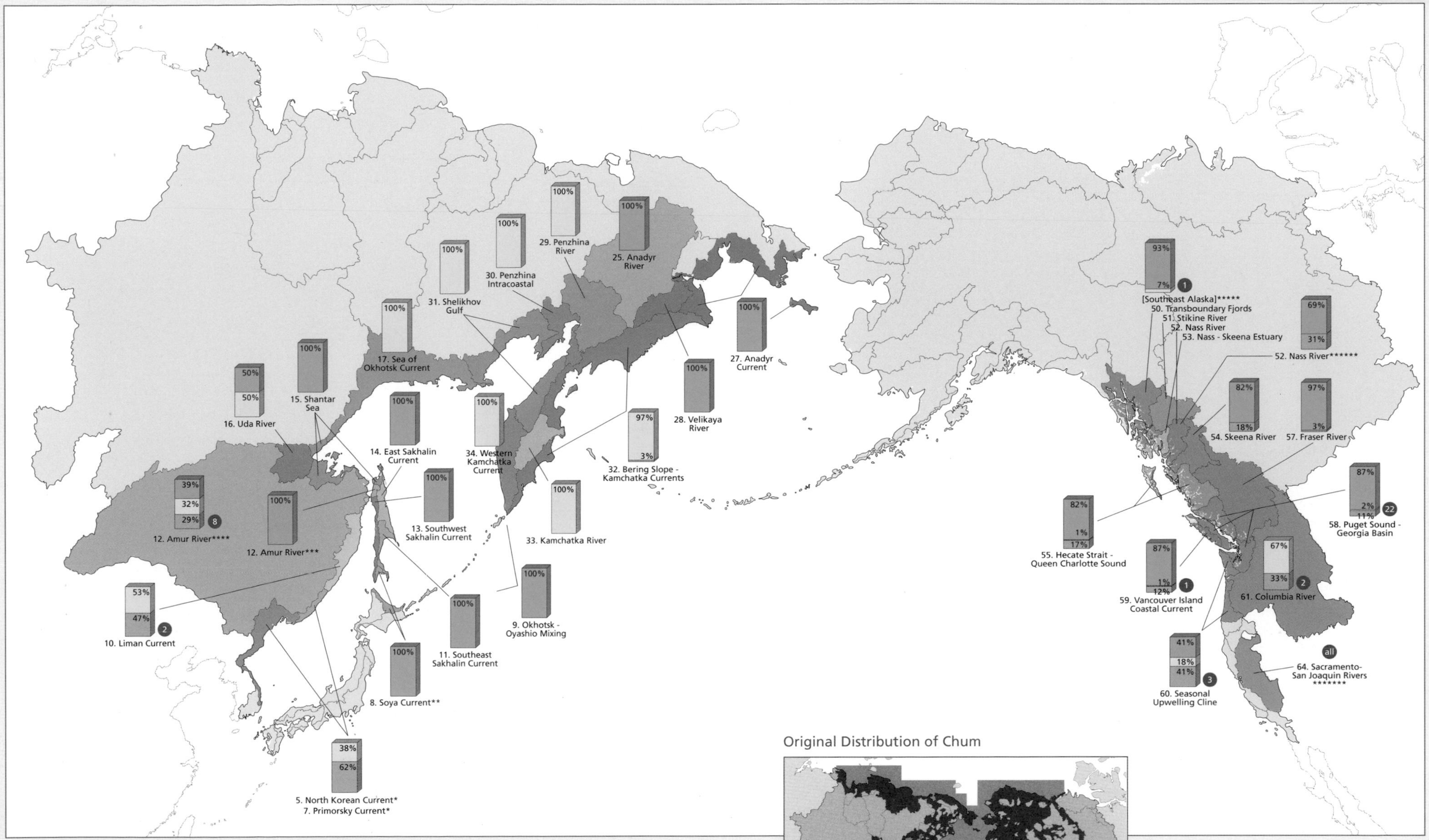

100%
29. Penzhina River
100%
25. Anadyr River
100%
30. Penzhina Intracoastal
100%
31. Shelikhov Gulf
100%
17. Sea of Okhotsk Current
100%
27. Anadyr Current
100%
15. Shantar Sea
50%
50%
16. Uda River
100%
28. Velikaya River
100%
14. East Sakhalin Current
100%
34. Western Kamchatka Current
97%
3%
32. Bering Slope - Kamchatka Currents
100%
13. Southwest Sakhalin Current
100%
33. Kamchatka River
39%
32%
29%
8
12. Amur River****
100%
12. Amur River***
53%
47%
2
10. Liman Current
100%
9. Okhotsk - Oyashio Mixing
100%
11. Southeast Sakhalin Current
100%
8. Soya Current**
38%
62%
5. North Korean Current*
7. Primorsky Current*

93%
7%
1
[Southeast Alaska]*****
50. Transboundary Fjords
51. Stikine River
52. Nass River
53. Nass - Skeena Estuary
69%
31%
52. Nass River******
82%
18%
54. Skeena River
97%
3%
57. Fraser River
87%
2%
11%
22
58. Puget Sound - Georgia Basin
82%
1%
17%
55. Hecate Strait - Queen Charlotte Sound
87%
1%
12%
1
59. Vancouver Island Coastal Current
67%
33%
2
61. Columbia River
41%
18%
41%
3
60. Seasonal Upwelling Cline
all
64. Sacramento-San Joaquin Rivers *******

Original Distribution of Chum

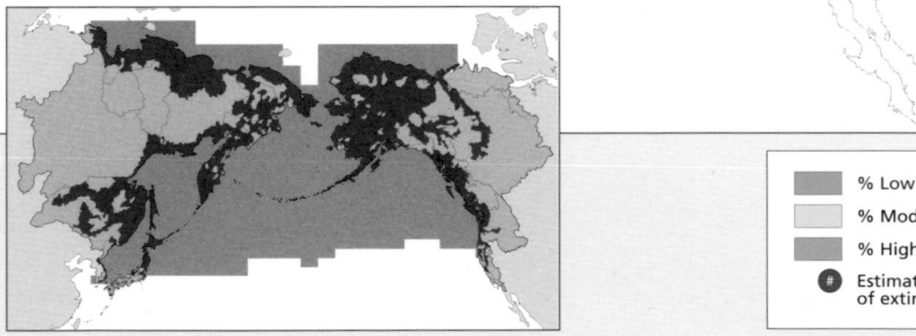

% Low risk
% Moderate risk
% High risk
Estimated number of extinct stocks

**CHUM RISK OF EXTINCTION** Generally chum populations are either at low or high risk throughout their British Columbia and southeast Alaska range, a pattern not seen in other species of *Oncorhynchus*. Because chum populations are thinner in southerly latitudes, threats can have dramatic impacts. In the Russian Far East, biologists are concerned about runs in Kamchatka, Magadan, Khabarovsk, and Primorye. Like pink, chum prefer colder waters, which may result in increased risk and more extirpations at the southern edge of their range and in arid inlands. (This map is based on two-season data.)

# Pink Distribution

**Pink salmon are the most abundant of** *Oncorhynchus spp.* They (along with masu) are the smallest; pink are also the most limited upriver migrators, and, with regard to life history diversity, the most homogenous and predictable. Pink are more populous in the western Pacific, perhaps because ocean waters tend to be cooler.

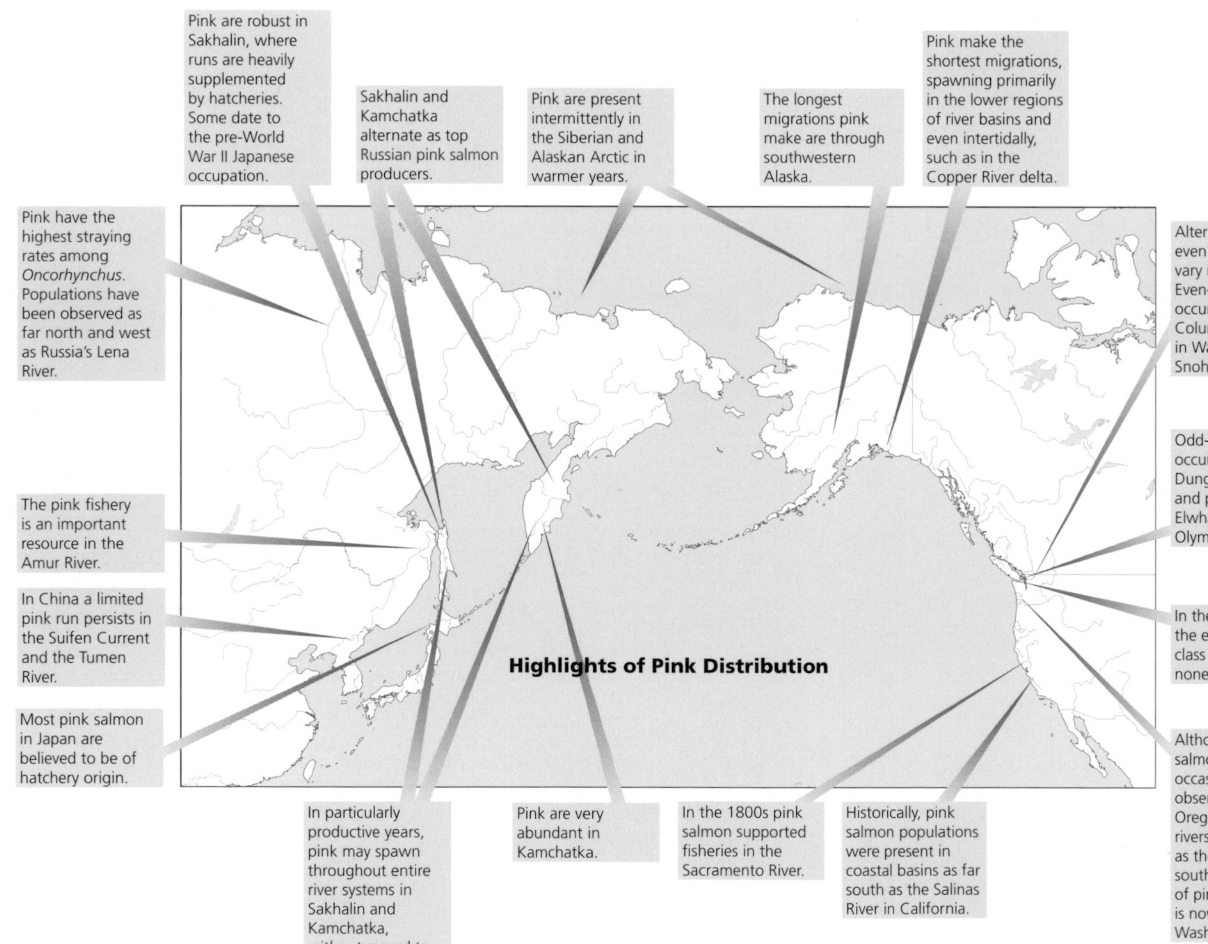


## PINK SALMON

**SCIENTIFIC NAME**

*Oncorhynchus gorbuscha*

**ABUNDANCE**

Pink salmon is the most abundant species of North Pacific salmon, particularly at higher latitudes.

**SIZE**

On average, pink salmon weigh about 1.5–2.5 kg and measure around 75 cm in length.

**LIFE HISTORY**

Among *Oncorhynchus,* pink salmon are least dependent on the freshwater environment and begin ocean migration immediately upon emerging from the redd. Pink mature after two summers at sea, whereupon they begin their migration to natal waters. With a predictable two-year life cycle, these fish are present in even- or odd-year cohorts, which are reproductively isolated, even among those sharing watershed spawning grounds. Because these populations are genetically distinct, biologists can distinguish "even-year" and "odd-year" classes.

**CULTURAL ROLE**

Traditionally, pink salmon were consumed in fishing season but were overlooked during the winter because they did not preserve well. (The species has the lowest oil content of any of the Pacific salmon species.) Demand for pink salmon increased dramatically, however, with the advent of canning technology during the mid- to late 19th century—but more recently has declined with the advent of farmed fish. Current markets are very weak for pink salmon.

Today pink harvest represents nearly half the biomass of commercial North Pacific *Oncorhynchus* catch and more than half the total catch in numbers of fish.

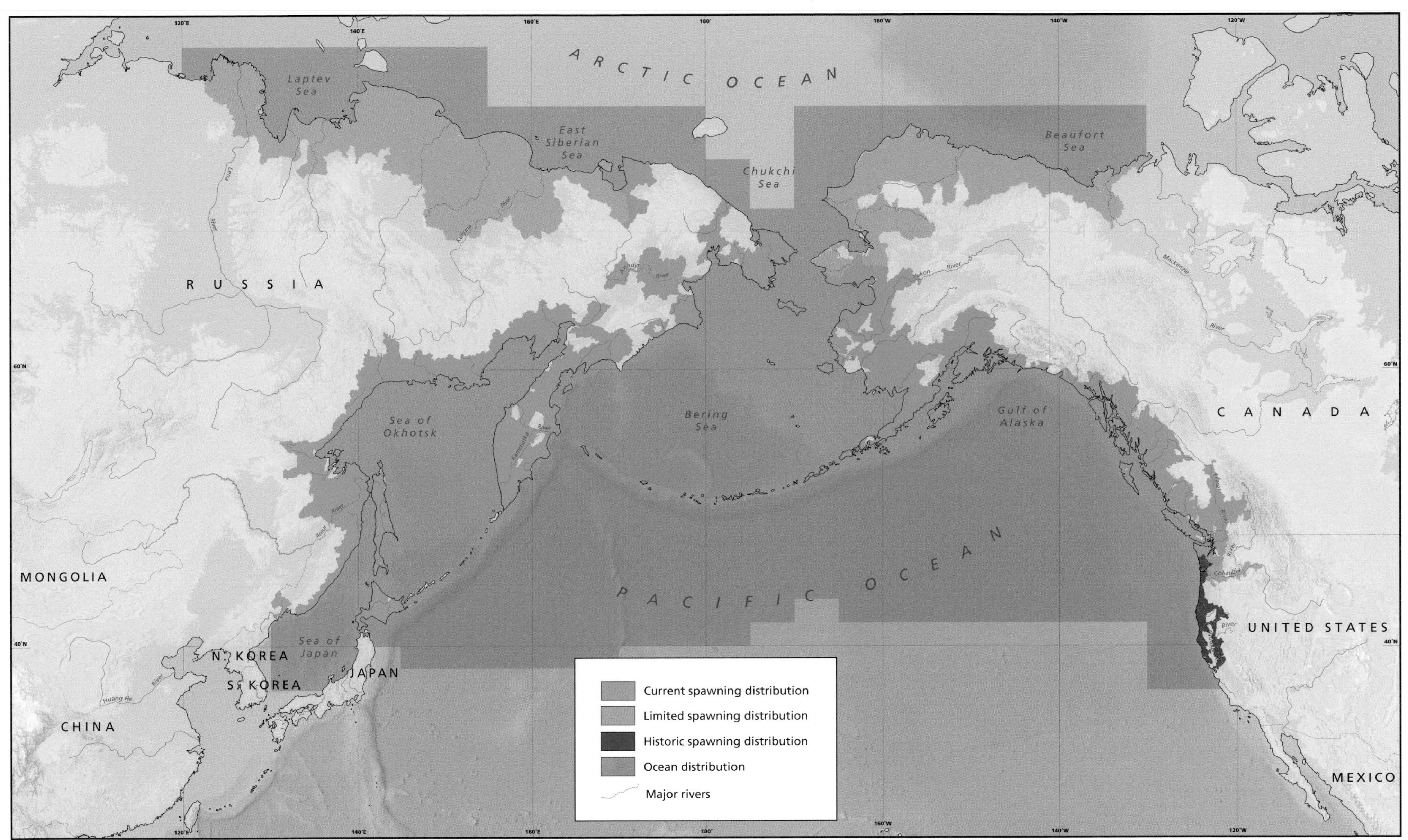


**PINK DISTRIBUTION** As with chum, pink demonstrate the highest straying rates among migrating salmon—up to 10 percent—as evidenced by their expansive limited distributional range. Also like chum, pink prefer colder waters, which may explain straying to the Arctic Ocean during warmer years. Pink have been extirpated south of the Columbia River basin, with limited spawning into the basin. Our map provides whole-basin data, aggregated for even- and odd-year classes (see Life History box, facing page), watershed boundaries, and the best available information on upriver extent.

# Pink Risk of Extinction

**Pink salmon, a high-volume commercial** species, exhibit the lowest risk of extinction among *Oncorhynchus* across the North Pacific, except at the southern edge of their range. Notable is the finding that of all North Pacific salmon, pink is the only species that does not exhibit a belt of moderate extinction risk across the Russian Far East.

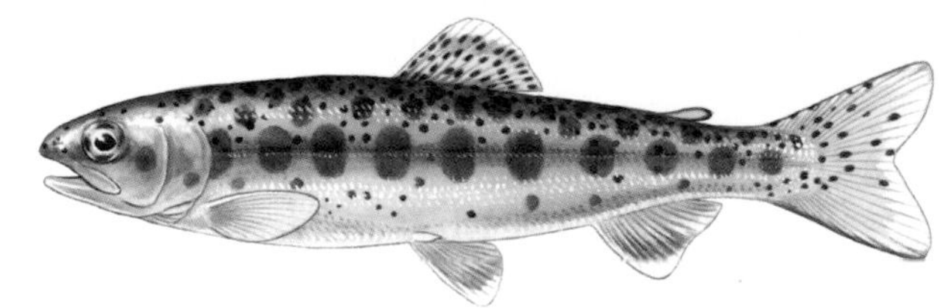

Pink fry

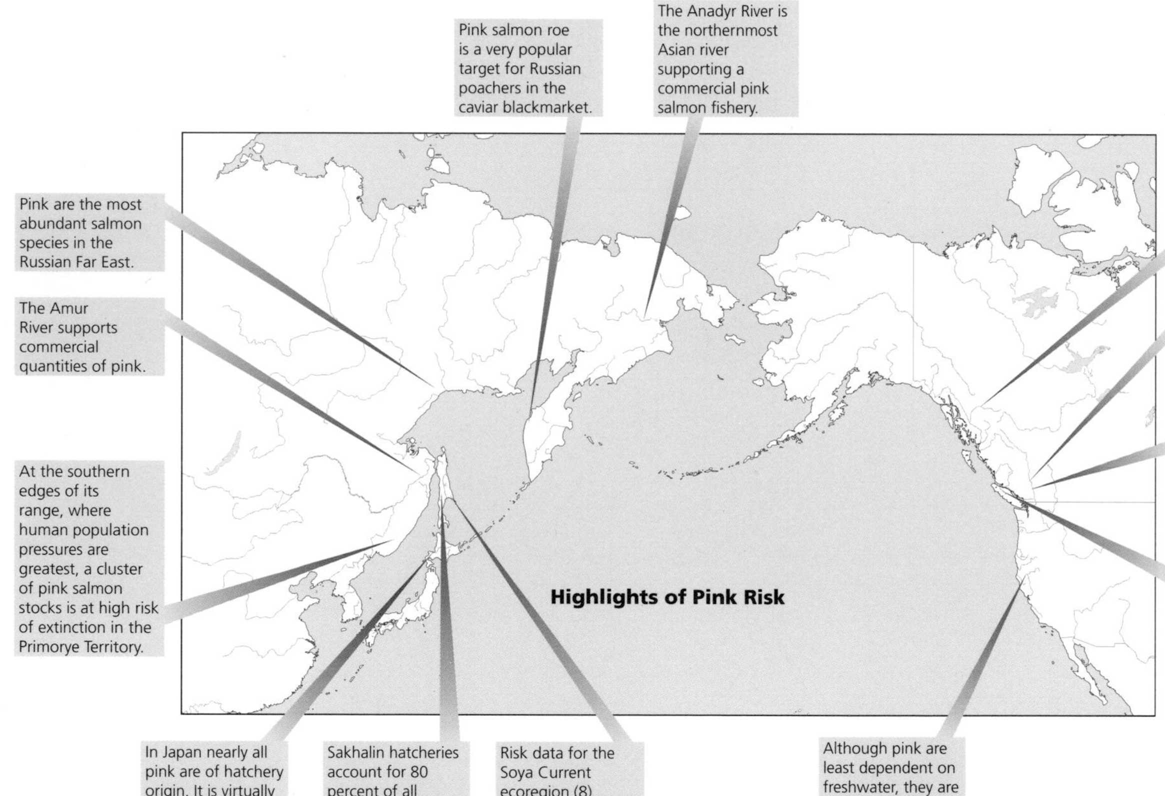


**Percentage of Relative Risk/Ecoregion**

| | ECOREGION | %HIGH | %MOD | %LOW | #EXT |
|---|---|---|---|---|---|
| 5<br>7 | Primorsky/North Korean Current* | 100 | 0 | 0 | 0 |
| 8 | Soya Current** | 0 | 0 | 100 | 0 |
| 9 | Okhotsk-Oyashio Mixing | 0 | 0 | 100 | 0 |
| 10 | Liman Current | 0 | 60 | 40 | 0 |
| 11 | Southeast Sakhalin Current | 0 | 0 | 100 | 0 |
| 12 | Amur River*** | 0 | 0 | 100 | 0 |
| 12 | Amur River**** | 0 | 0 | 100 | 0 |
| 13 | Southwest Sakhalin Current | 3 | 0 | 97 | 0 |
| 14 | East Sakhalin Current | 0 | 0 | 100 | 0 |
| 15 | Shantar Sea | 0 | 0 | 100 | 0 |
| 16 | Uda River | 0 | 50 | 50 | 0 |
| 17 | Sea of Okhotsk Current | 0 | 0 | 100 | 0 |
| 25 | Anadyr River | 0 | 0 | 100 | 0 |
| 27 | Anadyr Current | 0 | 0 | 100 | 0 |
| 28 | Velikaya River | 0 | 0 | 100 | 0 |
| 29 | Penzhina River | 0 | 0 | 100 | 0 |
| 30 | Penzhina Intracoastal | 0 | 0 | 100 | 0 |
| 31 | Shelikhov Gulf | 0 | 0 | 100 | 0 |
| 32 | Bering Slope-Kamchatka Currents | 0 | 0 | 100 | 0 |
| 33 | Kamchatka River | 0 | 0 | 100 | 0 |
| 34 | Western Kamchatka Current | 0 | 0 | 100 | 0 |
| 50<br>51<br>52<br>53 | Southeast Alaska***** | .13 | .40 | 99.47 | 0 |
| 52 | Nass River****** | 9 | 0 | 91 | 0 |
| 54 | Skeena River | 3 | 1 | 96 | 0 |
| 55 | Hecate Strait-Q.C. Sound | 8 | 1 | 91 | 0 |
| 57 | Fraser River | 6 | 1 | 93 | 3 |
| 58 | Puget Sound-Georgia Basin | 11 | 3 | 86 | 11 |
| 59 | Vancouver Island Coastal Current | 34 | 4 | 62 | 2 |

*All of ecoregion 7 and the Russia portion of ecoregion 5.

**The Russia portion of ecoregion 8.

***The Sakhalin portion of ecoregion 12.

****The mainland Russia portion of ecoregion 12.

*****The southeast Alaska portions of ecoregions 50, 51, 52, and 53.

******The British Columbia portion of ecoregion 52.

*******The lower portion of WOCI ecoregion 60 Seasonal Upwelling Cline feature full extinction, though historical number of stocks is unknown.

********The portions of WOCI in ecoregions 62 Klamath River, 63 Strong Upwelling Year Round, 64 Sacramento-San Joaquin Rivers feature full extinction, though historical number of stocks is unknown.

see foldout Ecoregion map

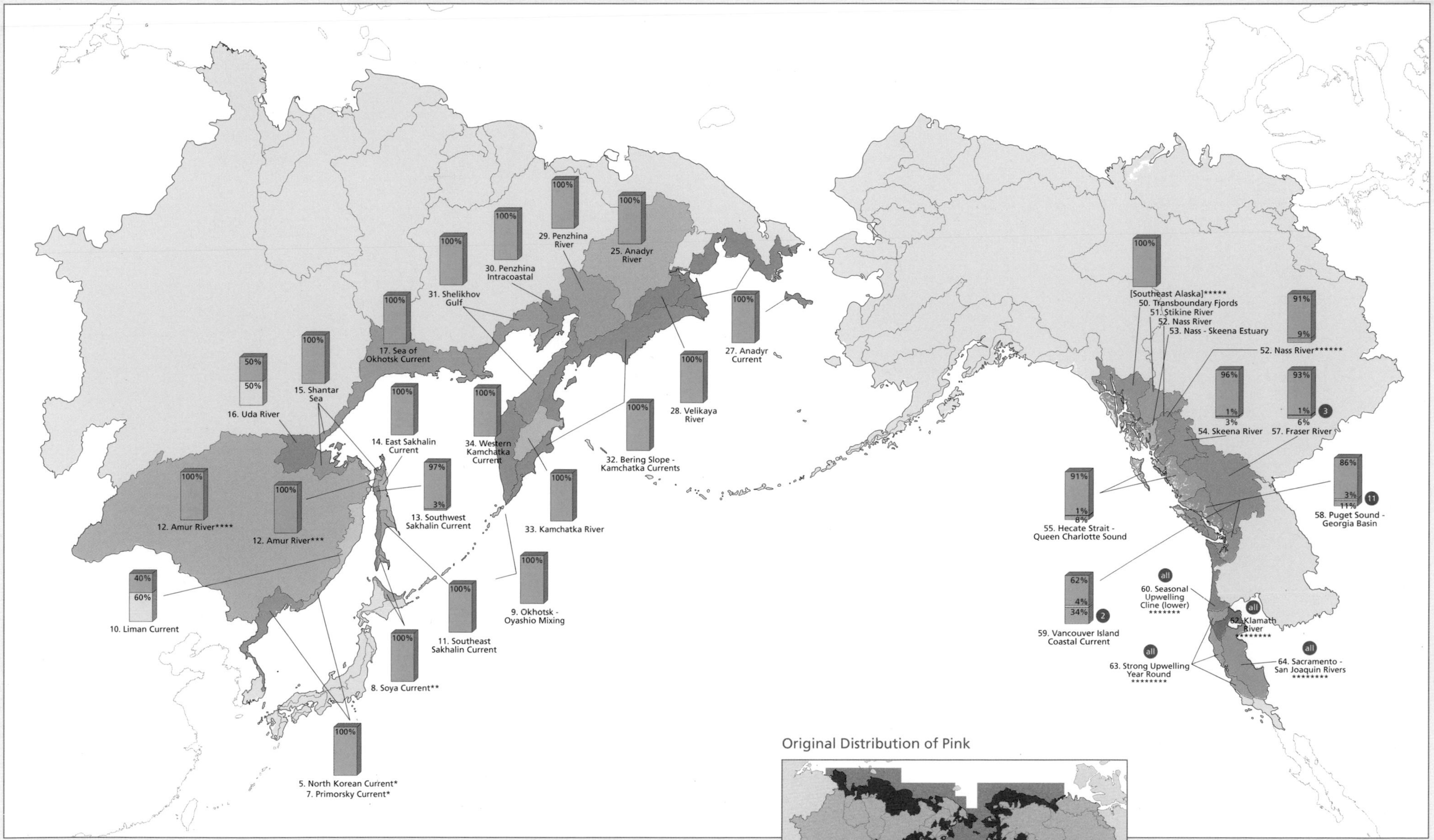

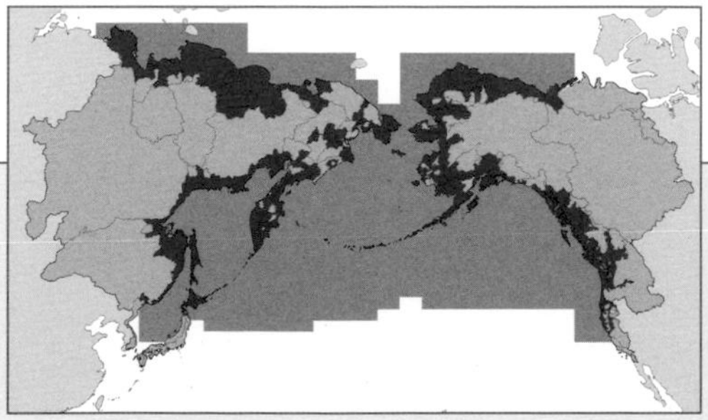

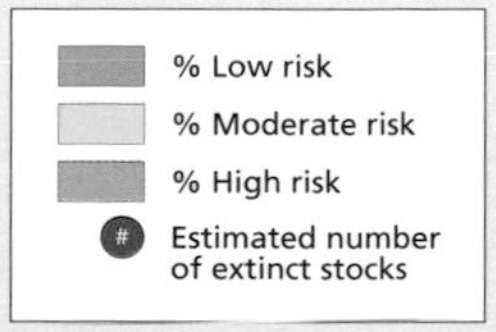

**PINK RISK OF EXTINCTION** Western Pacific biologists deem pink salmon the most robust among *Oncorhynchus* species, at low risk in all but four ecoregions—the Uda River (16), the North Korean (5) and the Primorsky Currents (7), and the heavily populated Liman Current (10). In the eastern Pacific, pink are also regarded as robust, although risk increases toward the southern extent of the range—particularly in the Vancouver Island Coastal Current (59), where one-third of populations are at high risk of extinction. Such a pattern of risk may demonstrate that pink are vulnerable to threats. (This map aggregates data for even- and odd-year classes.)

# Sockeye Distribution

**SOCKEYE SALMON IS THE PREMIUM PACIFIC** salmon species in the international marketplace, selling for more per pound than all Pacific salmon but chinook. Because of their value, sockeye are among the best studied of all species of Pacific salmon. River-rearing sockeye are common in Kamchatka but not in the eastern Pacific, where most sockeye rear in lake habitats.

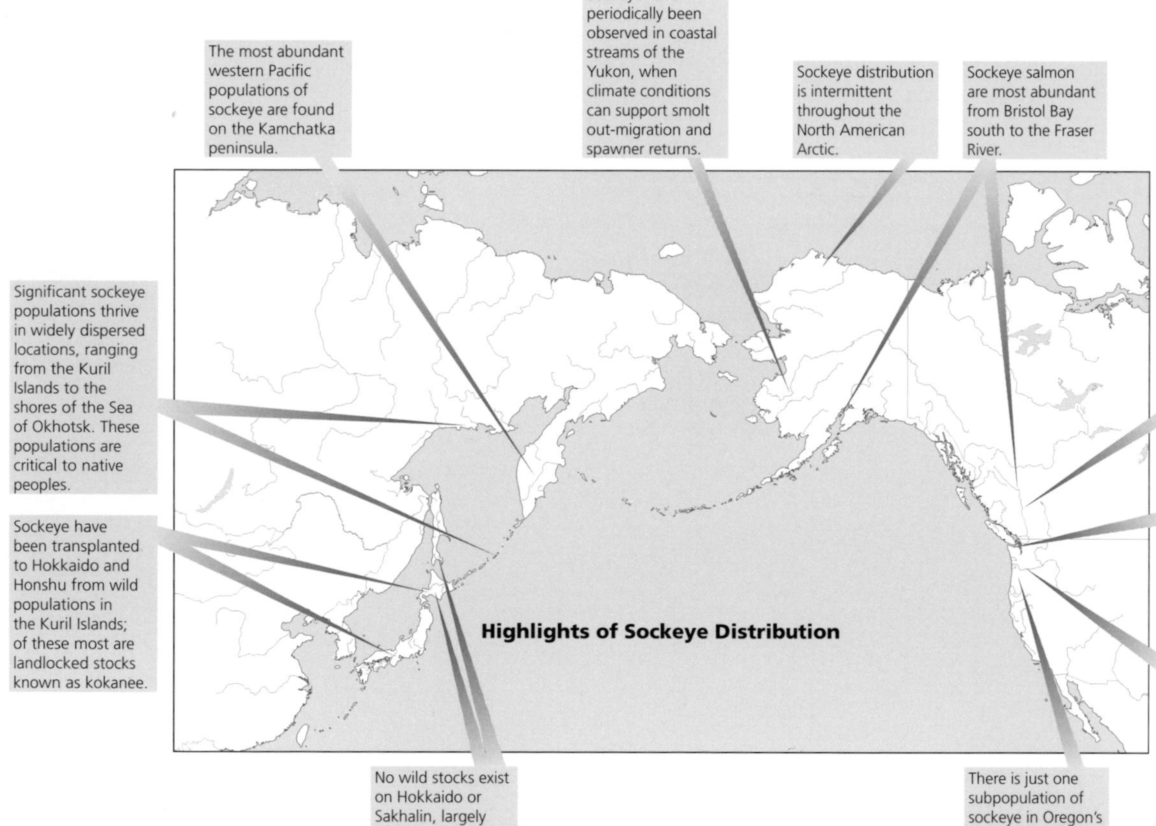


## SOCKEYE SALMON

**SCIENTIFIC NAME**
*Oncorhynchus nerka*

**ABUNDANCE**
In the western Pacific, sockeye salmon are most abundant in Kamchatka. In the eastern Pacific, they are most abundant from Bristol Bay south to the Fraser River. Among *Oncorhynchus*, sockeye salmon are third in abundance across the North Pacific.

**SIZE**
Among Pacific salmon, sockeye are third in terms of average weight, weighing 1.5–3.5 kilograms. Sockeye measure around 80 centimeters in length. In contrast to pink and chum, sockeye do not show a latitudinal gradient in average weights; the largest fish are from Alaska's Chignik River fishery.

**LIFE HISTORY**
Spawning and rearing habitat needs for sockeye salmon are complex and vary significantly among populations across the North Pacific. Sockeye may spawn in lakes, beaches, or river gravel; they may spend anywhere from a few weeks to three years in freshwater. They may feed in the ocean 1–4 years prior to spawning. Landlocked sockeye, kokanee, are not anadromous and have not been included in this study.

**CULTURAL ROLE**
Although the canned salmon industry developed first around the chinook harvest on the Columbia River, sockeye soon became the favored species for canning. Its concentrated run timing made it efficient to catch and process at remote sites, while its deep red meat, good flavor, and firm flesh made it appealing to consumers.

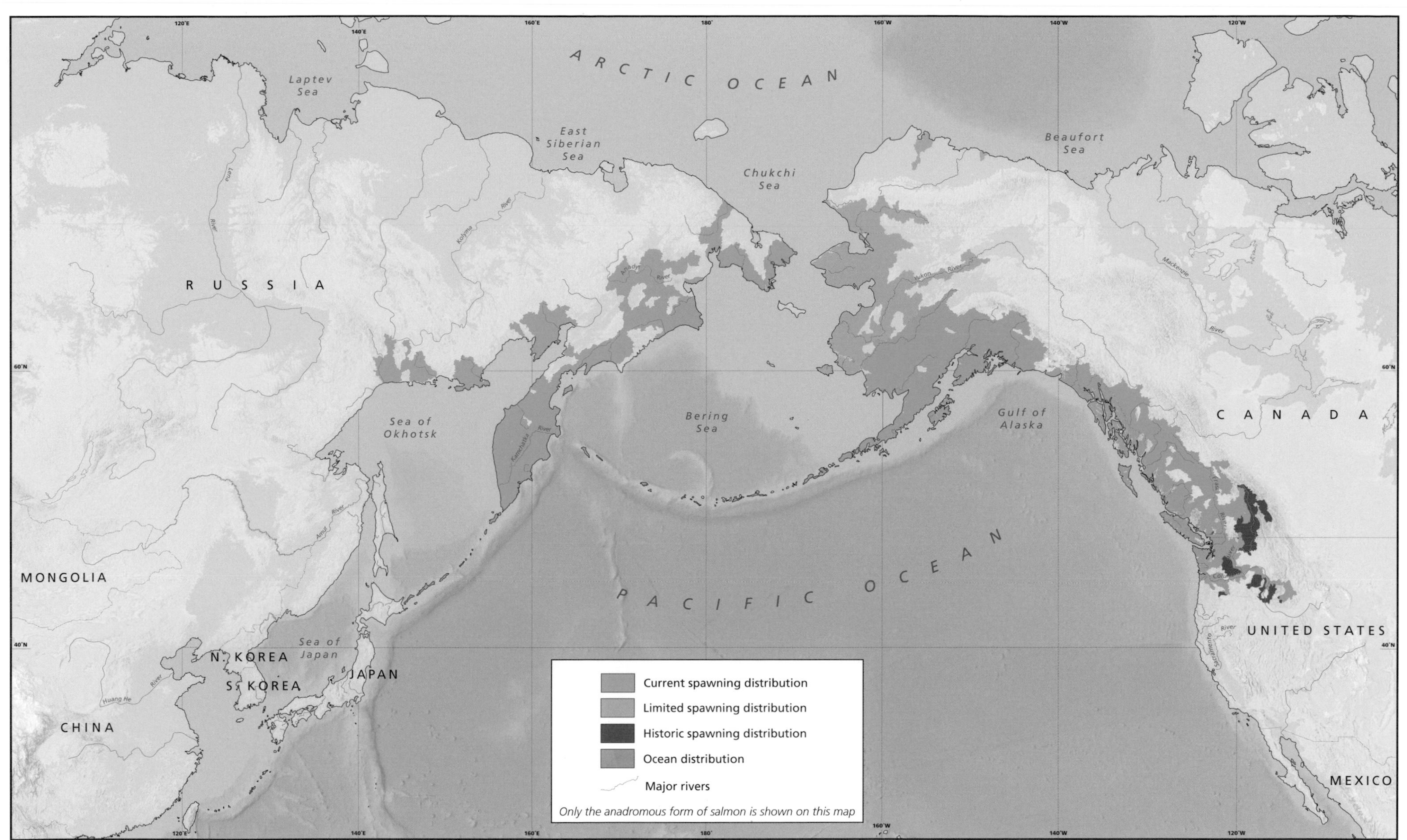


**SOCKEYE DISTRIBUTION** Sockeye depend largely on lake habitats for spawning and are thus more irregularly distributed across the North Pacific than are other species of *Oncorhynchus*. Lake habitat suitable for sockeye may be found in the geologically active Kamchatka peninsula and glaciated regions from the Kuskokwim River to the Puget Sound and interior Columbia River basin. Sockeye have one of the narrowest latitudinal ocean ranges within *Oncorhynchus*, limited by their sensitivity to temperature and dependence on glacial landscapes (with abundant lakes). Dams in the Columbia River basin have extirpated most inland populations (see page 109).

# Sockeye Risk of Extinction

**The use of lakes by sockeye is unique** within the genus *Oncorhynchus*. Lake spawners are vulnerable to localized threats. River spawners are more opportunistic and scout out new spawning grounds. Although lake-spawning populations are genetically very distinct from one another, river spawning sockeye populations are genetically much more homogenous.

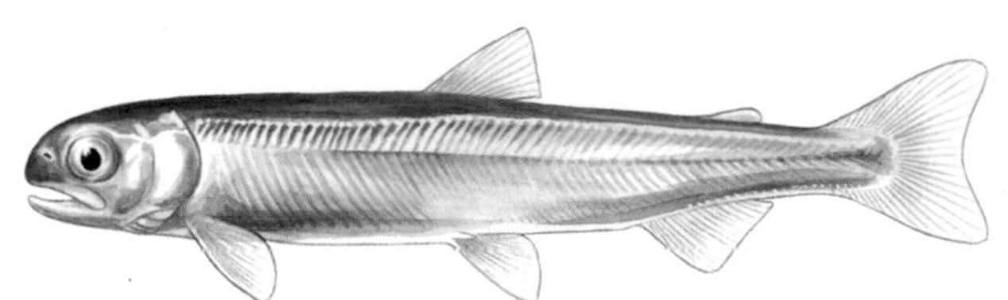

Sockeye lake juvenile

**Percentage of Relative Risk/Ecoregion**

| | ECOREGION | %HIGH | %MOD | %LOW | #EXT |
|---|---|---|---|---|---|
| 9 | Okhotsk-Oyashio Mixing | 0 | 0 | 100 | 0 |
| 17 | Sea of Okhotsk Current | 0 | 100 | 0 | 0 |
| 25 | Anadyr River | 0 | 0 | 100 | 0 |
| 27 | Anadyr Current | 0 | 0 | 100 | 0 |
| 28 | Velikaya River | 0 | 0 | 100 | 0 |
| 30 | Penzhina Intracoastal | 0 | 100 | 0 | 0 |
| 31 | Shelikhov Gulf | 0 | 100 | 0 | 0 |
| 32 | Bering Slope-Kamchatka Currents | 4 | 95 | 1 | 0 |
| 33 | Kamchatka River | 0 | 0 | 100 | 0 |
| 34 | Western Kamchatka Current | 17 | 81 | 2 | 0 |
| 50 51 52 53 | Southeast Alaska* | 0 | 0 | 100 | 1 |
| 52 | Nass River** | 14 | 0 | 86 | 0 |
| 53 | Nass-Skeena Estuary*** | 33 | 33 | 33 | 0 |
| 54 | Skeena River | 20 | 0 | 80 | 0 |
| 55 | Hecate Strait-Q.C. Sound | 18 | 0 | 82 | 1 |
| 56 | Outer Graham Island | 33 | 33 | 33 | 0 |
| 57 | Fraser River | 6 | 1 | 93 | 11 |
| 58 | Puget Sound-Georgia Basin | 18 | 2 | 80 | 2 |
| 59 | Vancouver Island Coastal Current | 2 | 0 | 98 | 2 |
| 60 | Seasonal Upwelling Cline | 0 | 100 | 0 | 0 |
| 61 | Columbia River | 50 | 25 | 25 | 11 |

*The southeast Alaska portions of ecoregions 50, 51, 52, and 53.

**The British Columbia portion of ecoregion 52.

***The British Columbia portion of ecoregion 53.

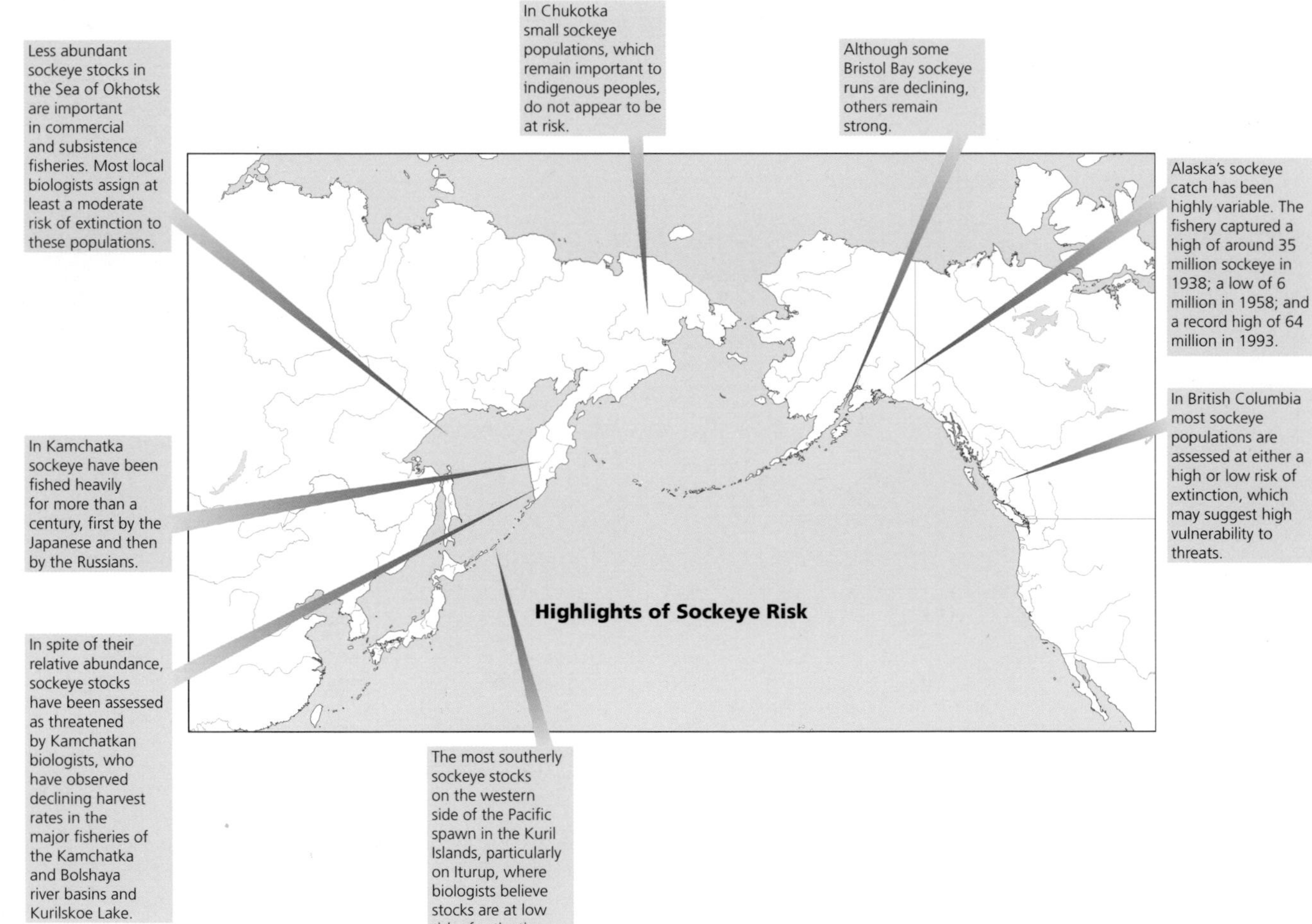


**Highlights of Sockeye Risk**

see foldout Ecoregion map

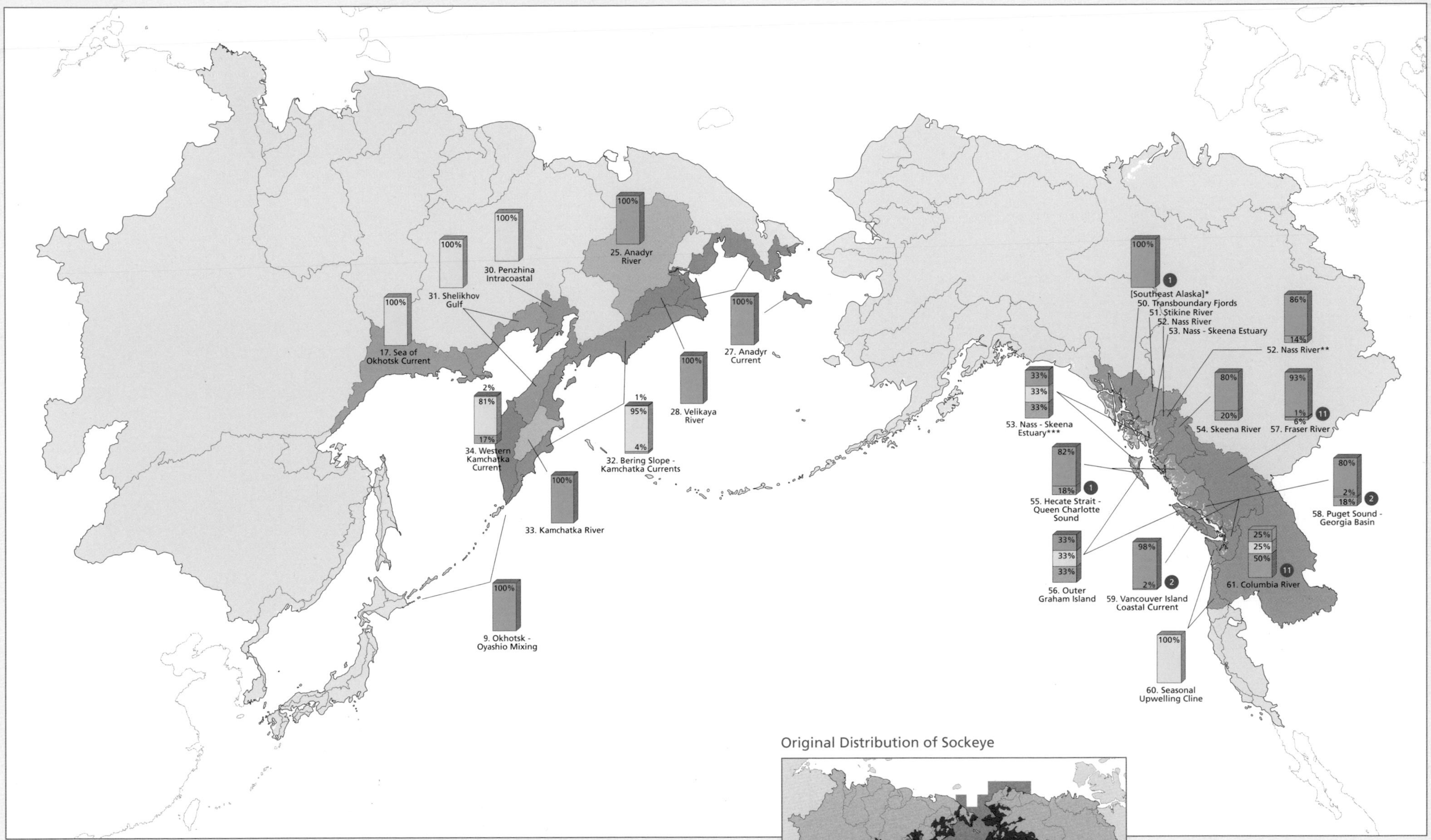

9. Okhotsk - Oyashio Mixing 100%
17. Sea of Okhotsk Current 100%
25. Anadyr River 100%
27. Anadyr Current 100%
28. Velikaya River 100%
30. Penzhina Intracoastal 100%
31. Shelikhov Gulf 100%
32. Bering Slope - Kamchatka Currents 1% 95% 4%
33. Kamchatka River 100%
34. Western Kamchatka Current 2% 81% 17%
[Southeast Alaska]* 100%
50. Transboundary Fjords
51. Stikine River
52. Nass River
53. Nass - Skeena Estuary
52. Nass River** 86% 14%
53. Nass - Skeena Estuary*** 33% 33% 33%
54. Skeena River 80% 20%
55. Hecate Strait - Queen Charlotte Sound 82% 18%
56. Outer Graham Island 33% 33% 33%
57. Fraser River 93% 1% 6%
58. Puget Sound - Georgia Basin 80% 2% 18%
59. Vancouver Island Coastal Current 98% 2%
60. Seasonal Upwelling Cline 100%
61. Columbia River 25% 25% 50%

Original Distribution of Sockeye

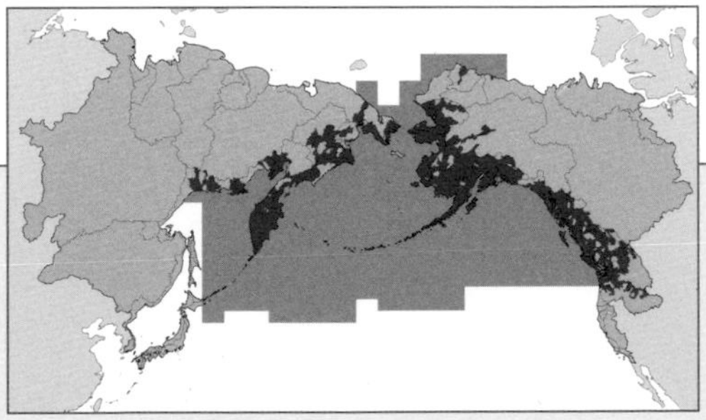

% Low risk
% Moderate risk
% High risk
Estimated number of extinct stocks

**SOCKEYE RISK OF EXTINCTION** Russian biologists have assigned low or moderate risk of extinction to most sockeye populations, except in the Western Kamchatka Current (ecoregion 34) and the Bering Slope-Kamchatka Currents (32), where some populations are at high risk. Sockeye are not faring as well in North America. With the exception of southeast Alaska, all eastern Pacific ecoregions exhibit some degree of moderate or high risk of extinction; of these, three ecoregions—Nass-Skeena Estuary (53), Outer Graham Island (56), and Columbia River (61, see page 109)—contain one-third or more sockeye populations assessed at high risk.

# Chinook Distribution

**Chinook is the largest species of** *Oncorhynchus.* Their formidable size belies an ability to spawn in small tributaries and headwaters, which are particularly vulnerable to the effects of natural resource extraction, development, and agriculture. In North America chinook once spawned outside of WOCI, as far east as Montana and as far south as Nevada.

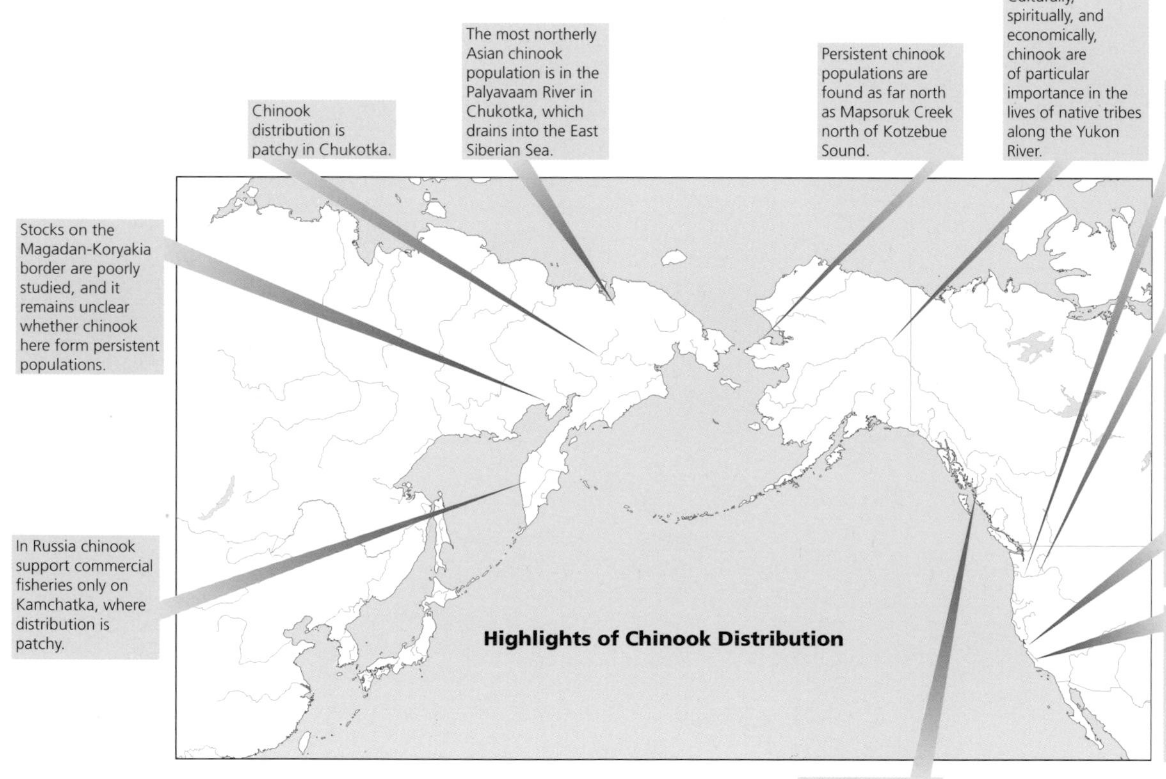


## CHINOOK SALMON

**SCIENTIFIC NAME**

*Oncorhynchus tshawytscha*

**ABUNDANCE**

The least abundant of the five North American species overall, chinook populations are more numerous on the eastern side of the Pacific than they are in Asia.

**SIZE**

Chinook salmon are the largest of the Pacific salmon, measuring up to 150 cm and weighing 5–11 kg.

**LIFE HISTORY**

Chinook live from 3 to 6 years. Their life histories vary greatly both within river basins and across the range of the species. Chinook spawn in mainstem rivers and have four seasonal spawning runs, each with different time periods rearing in freshwater and in their ocean phase. In the largest river basins, chinook return virtually year-round, though there is usually a strong seasonal peak some time between May and September.

There are stream- and ocean-type chinook. Most chinook at the northern and southern extremes of their distribution are stream-type, rearing for up to two winters in freshwater before leaving for the ocean. Toward the center of their range, ocean-type chinook predominate, leaving home rivers within weeks for marine waters, where they spend an average of 4–5 years. The two types generally use different rearing areas in the ocean, but when they return, their runs may overlap and they may share the same spawning grounds.

Chinook appear to have the lowest straying rate among *Oncorhynchus.*

**CULTURAL ROLE**

Chinook salmon are a staple of native peoples' diets, particularly south of Puget Sound. Chinook salmon are also favored among sportfishers.

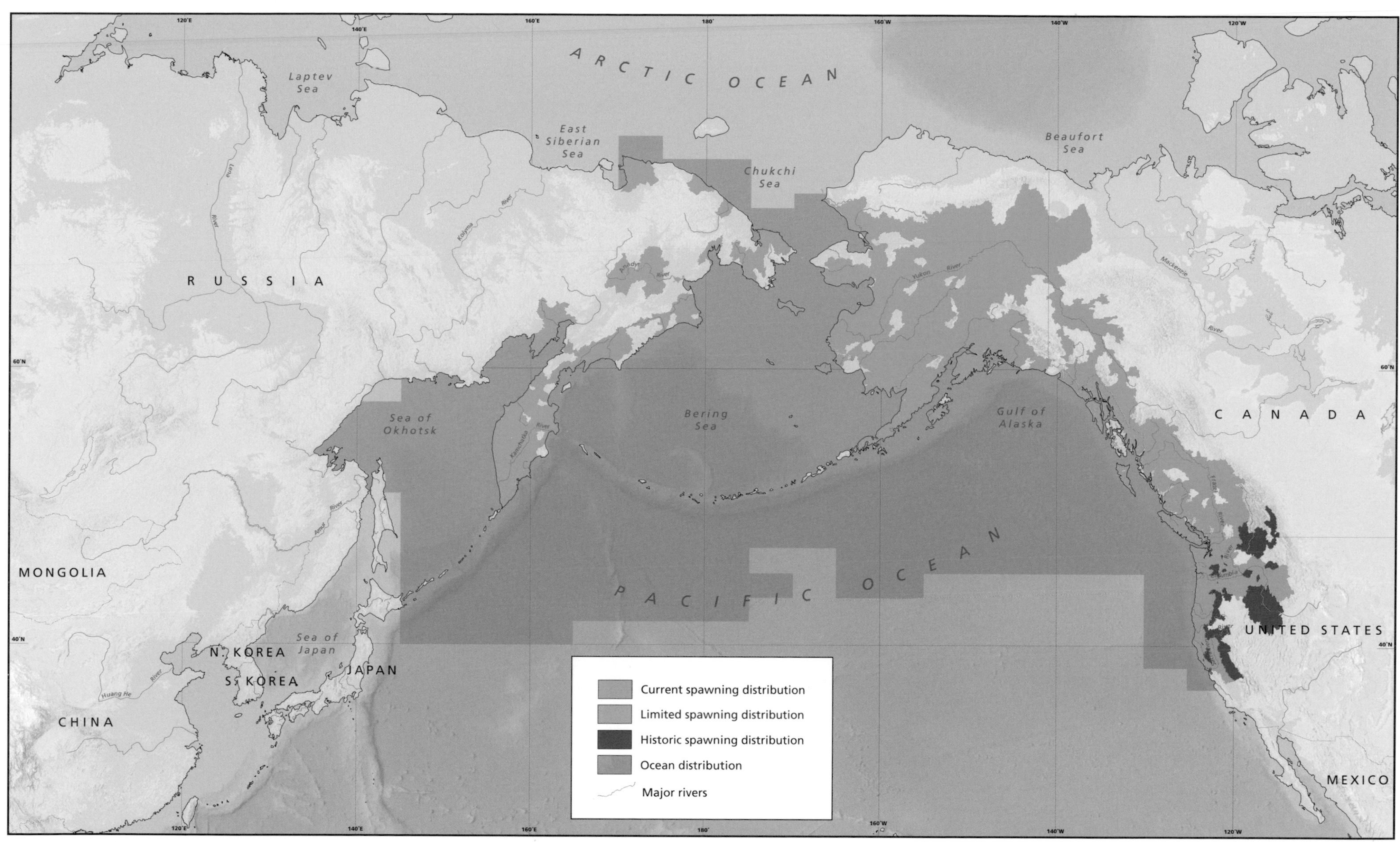


CHINOOK DISTRIBUTION In the western Pacific, Kamchatka is the only region with relatively consistent chinook presence; elsewhere, distribution is patchy. Chinook form the most persistent *Oncorhynchus* populations at southern latitudes in North America. Inland extirpations have occurred east of the Coast Ranges in California and the Cascade Range in Oregon, Washington, and Idaho, particularly along the Snake River. This map represents distribution data for ocean- and stream-type populations (see Life History box, facing page) of juveniles as well as all four seasonal runs of adult chinook.

# Chinook Risk of Extinction

**BECAUSE OF OUR LACK OF RISK DATA** for Alaska, we are missing status information for at least half of the North Pacific chinook populations. Edge-of-range effects appear to be important to chinook, as risk of extinction increases at the southerly extent. In WOCI inland threats of habitat loss and water diversions have resulted in extirpations.

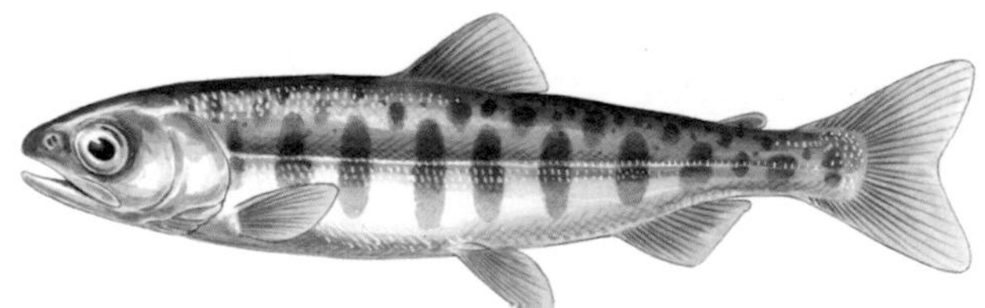

Chinook parr, stream-type

**Percentage of Relative Risk/Ecoregion**

| | ECOREGION | %HIGH | %MOD | %LOW | #EXT |
|---|---|---|---|---|---|
| 25 | Anadyr River | 0 | 0 | 100 | 0 |
| 27 | Anadyr Current | 0 | 0 | 100 | 0 |
| 28 | Velikaya River | 0 | 0 | 100 | 0 |
| 30 | Penzhina Intracoastal | 0 | 100 | 0 | 0 |
| 31 | Shelikhov Gulf | 0 | 100 | 0 | 0 |
| 32 | Bering Slope-Kamchatka Currents | 6 | 94 | 0 | 0 |
| 33 | Kamchatka River | 0 | 100 | 0 | 0 |
| 34 | Western Kamchatka Current | 100 | 0 | 0 | 0 |
| 50<br>51<br>52<br>53 | Southeast Alaska* | 0 | 3 | 97 | 0 |
| 52 | Nass River** | 0 | 0 | 100 | 0 |
| 54 | Skeena River | 2 | 0 | 98 | 0 |
| 55 | Hecate Strait-Q.C. Sound | 18 | 0 | 82 | 0 |
| 57 | Fraser River | 7 | 0 | 93 | 8 |
| 58 | Puget Sound-Georgia Basin | 27 | 8 | 65 | 7 |
| 59 | Vancouver Island Coastal Current | 31 | 0 | 69 | 2 |
| 60 | Seasonal Upwelling Cline | 10 | 3 | 87 | 0 |
| 61 | Columbia River | 68 | 14 | 18 | 33 |
| 62 | Klamath River | 7 | 86 | 7 | 6 |
| 63 | Strong Upwelling Year Round | 38 | 54 | 8 | 12 |
| 64 | Sacramento-San Joaquin Rivers | 24 | 29 | 47 | 14 |

*The southeast Alaska portions of ecoregions 50, 51, 52, and 53.

**The British Columbia portion of ecoregion 52.

**Highlights of Chinook Risk**

Stocks in the Kamchatka River ecoregion (33) likely vary in life histories, providing natural insurance against risk. Even with heavy commercial and poaching pressures, these chinook are ranked at moderate risk.

Significant numbers of chinook are taken as bycatch in the sockeye fishery on Kamchatka.

In Penzhina Bay and northward to Chukotka, fishing pressure is limited and stocks fare well. Nonetheless, chinook populations are small and poorly studied.

Chinook stocks are healthy in Alaska, except in the Yukon and Kuskokwim Rivers, where they are listed as stocks of special concern by the Alaska Department of Fish and Game.

Today chinook run three seasons in the Columbia River, with the spring and winter runs federally listed under the Endangered Species Act.

Although chinook are most abundant on the Kamchatka peninsula, where they are generally at moderate risk, most west coast stocks were ranked at high risk of extinction

Impoundments and water diversions along the Snake River in Idaho have resulted in extirpations.

Along the San Joaquin and Sacramento Rivers, chinook historically demonstrated rich life history diversity with a year-round presence. Yet water diversions and mining influences have accelerated declines.

In the San Joaquin and Sacramento Rivers, fall and spring run timing and stock complexity have been significantly compressed by hatchery practices.

see foldout Ecoregion map

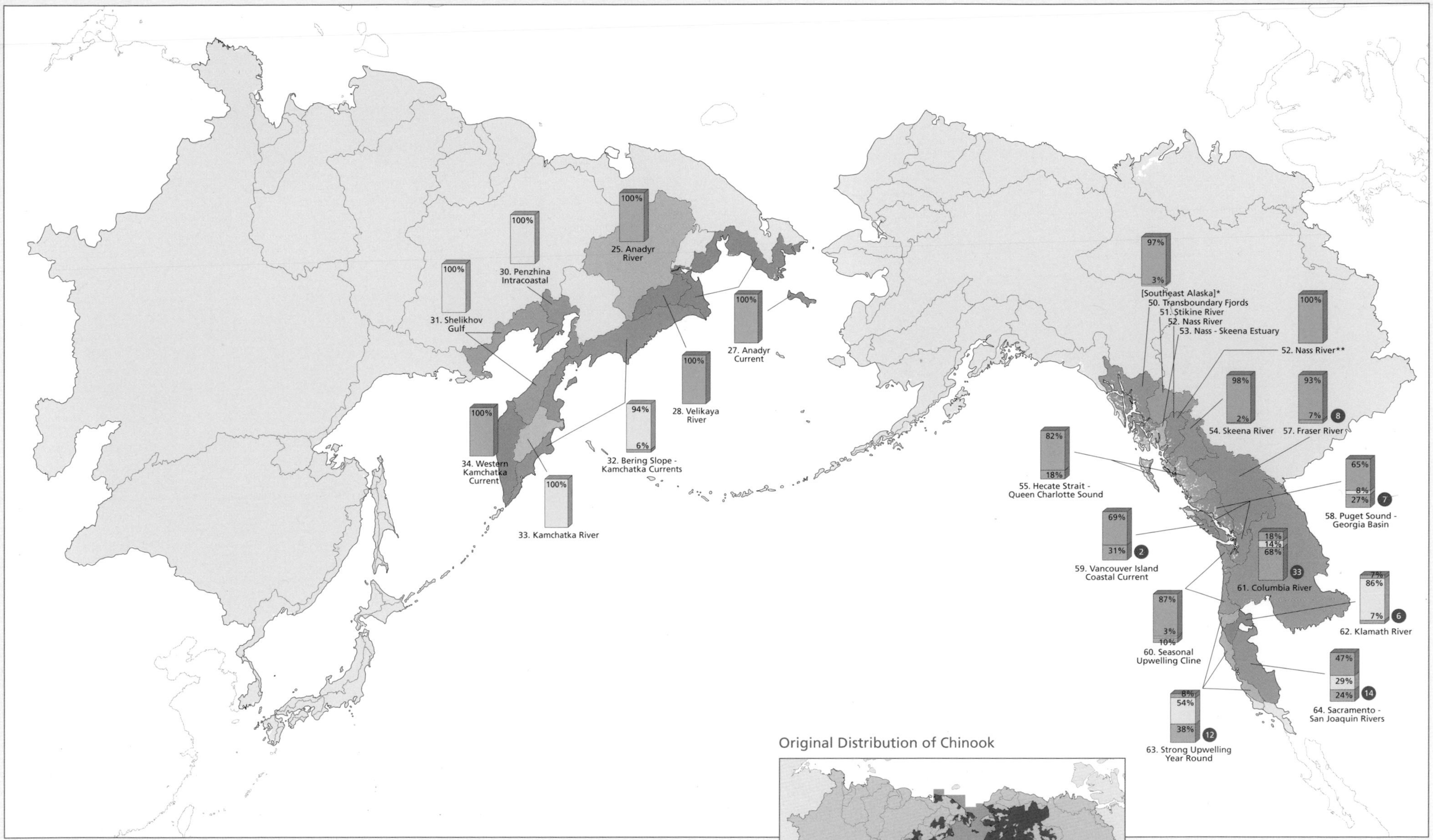

25. Anadyr River 100%
27. Anadyr Current 100%
28. Velikaya River 100%
30. Penzhina Intracoastal 100%
31. Shelikhov Gulf 100%
32. Bering Slope - Kamchatka Currents 94% 6%
33. Kamchatka River 100%
34. Western Kamchatka Current 100%

[Southeast Alaska]* 97% 3%
50. Transboundary Fjords
51. Stikine River
52. Nass River
53. Nass - Skeena Estuary
52. Nass River** 100%
54. Skeena River 98% 2%
55. Hecate Strait - Queen Charlotte Sound 82% 18%
57. Fraser River 93% 7% 8
58. Puget Sound - Georgia Basin 65% 8% 27% 7
59. Vancouver Island Coastal Current 69% 31% 2
60. Seasonal Upwelling Cline 87% 3% 10%
61. Columbia River 18% 14% 68% 33
62. Klamath River 7% 86% 7% 6
63. Strong Upwelling Year Round 8% 54% 38% 12
64. Sacramento - San Joaquin Rivers 47% 29% 24% 14

Original Distribution of Chinook

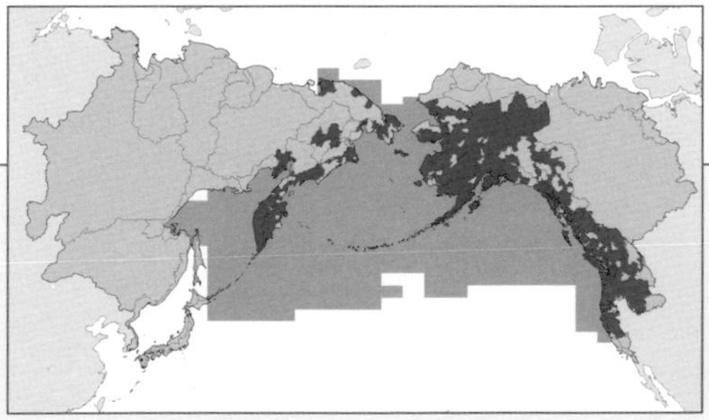

% Low risk
% Moderate risk
% High risk
# Estimated number of extinct stocks

**CHINOOK RISK OF EXTINCTION** Except within WOCI, chinook populations within any given ecoregion exhibit similar degrees of risk. The Western Kamchatka Current (34) is the only ecoregion classified entirely at high risk of extinction. Moving east, risk decreases to moderate in eastern and northern Kamchatka and to low in Chukotka, where populations appear to be healthy. In addition southeast Alaska and British Columbia chinook are assessed at low risk. But as with Russian populations at the edges of their range, WOCI chinook demonstrate high risk of extinction at southerly latitudes—particularly in the Columbia River ecoregion (61), with 33 estimated extinctions.

# Coho Distribution

**Coho populations are small and widely** distributed. They use coastal streams and off-channel wintering habitats, which are highly vulnerable to the effects of agriculture and logging. Entering in fall and early winter, coho usually are the last *Oncorhynchus spp.* to migrate upriver. In Russia coho populations have not been well documented or targeted.

## COHO SALMON

**SCIENTIFIC NAME**

*Oncorhynchus kisutch*

**ABUNDANCE**

Although coho salmon are broadly distributed across the North Pacific, they form networks of small populations. Coho represent approximately 10 percent of the North Pacific commercial catch. Marine distribution is generally restricted to shelf waters.

**SIZE**

Coho typically weigh 2.5–5 kg when they return to their natal rivers. Mature adults measure around 95 cm in length.

**LIFE HISTORY**

Coho, which live 2–4 years, have simpler life histories than sockeye and chinook. Coho typically spawn in coastal streams, selecting sites with groundwater seepage. In contrast to chum and pink salmon, coho spend at least one winter (in some populations two) in freshwater. Coho make short feeding migrations compared to most other anadromous Pacific salmon. They are often associated with slow current, pool, and side channel habitat in rivers.

Early maturing male coho (jacks) leave the riverine system after one winter but return in the fall instead of overwintering in the ocean. Numbers of returning jacks provide a good indicator of feeding conditions and are used to forecast the following year's spawning run.

**CULTURAL ROLE**

Early in the 20th century, niche markets developed around coho. High-quality, line-caught fish were sold fresh or frozen for the European smoked fish market. More recently, coho have become a popular sportfishing target.

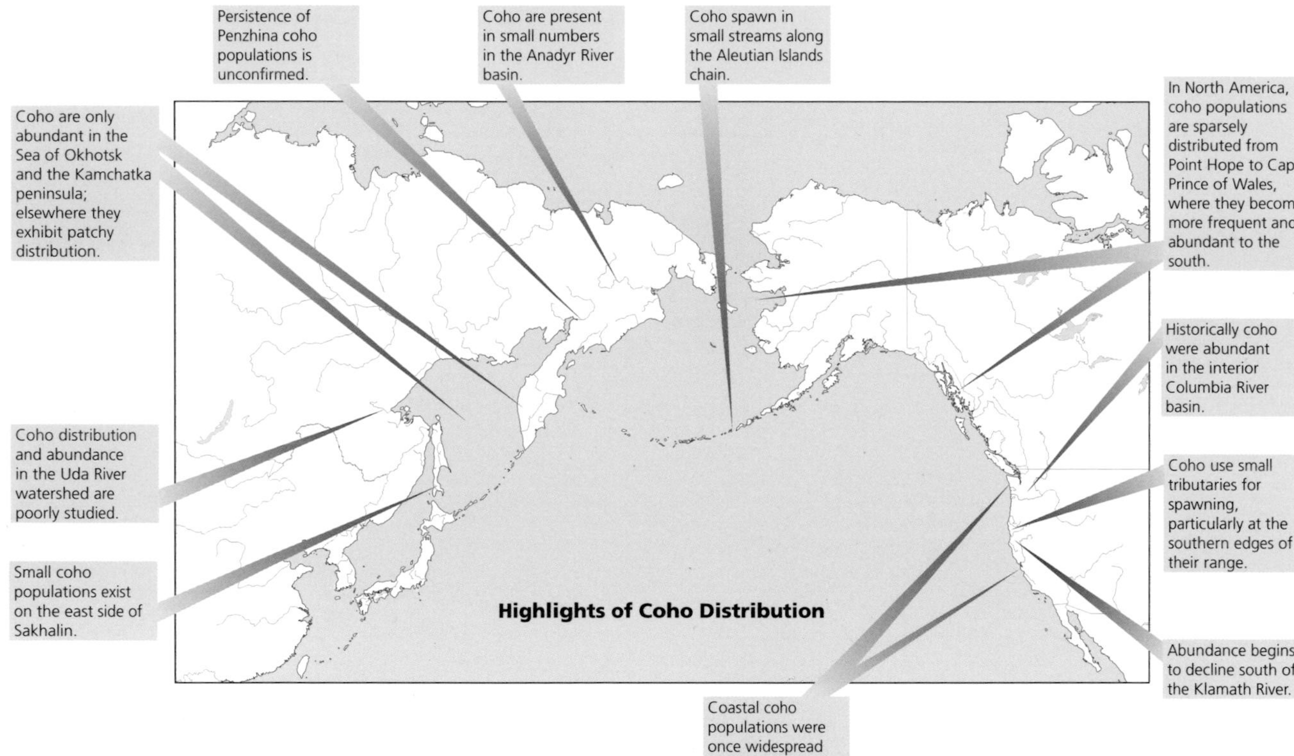

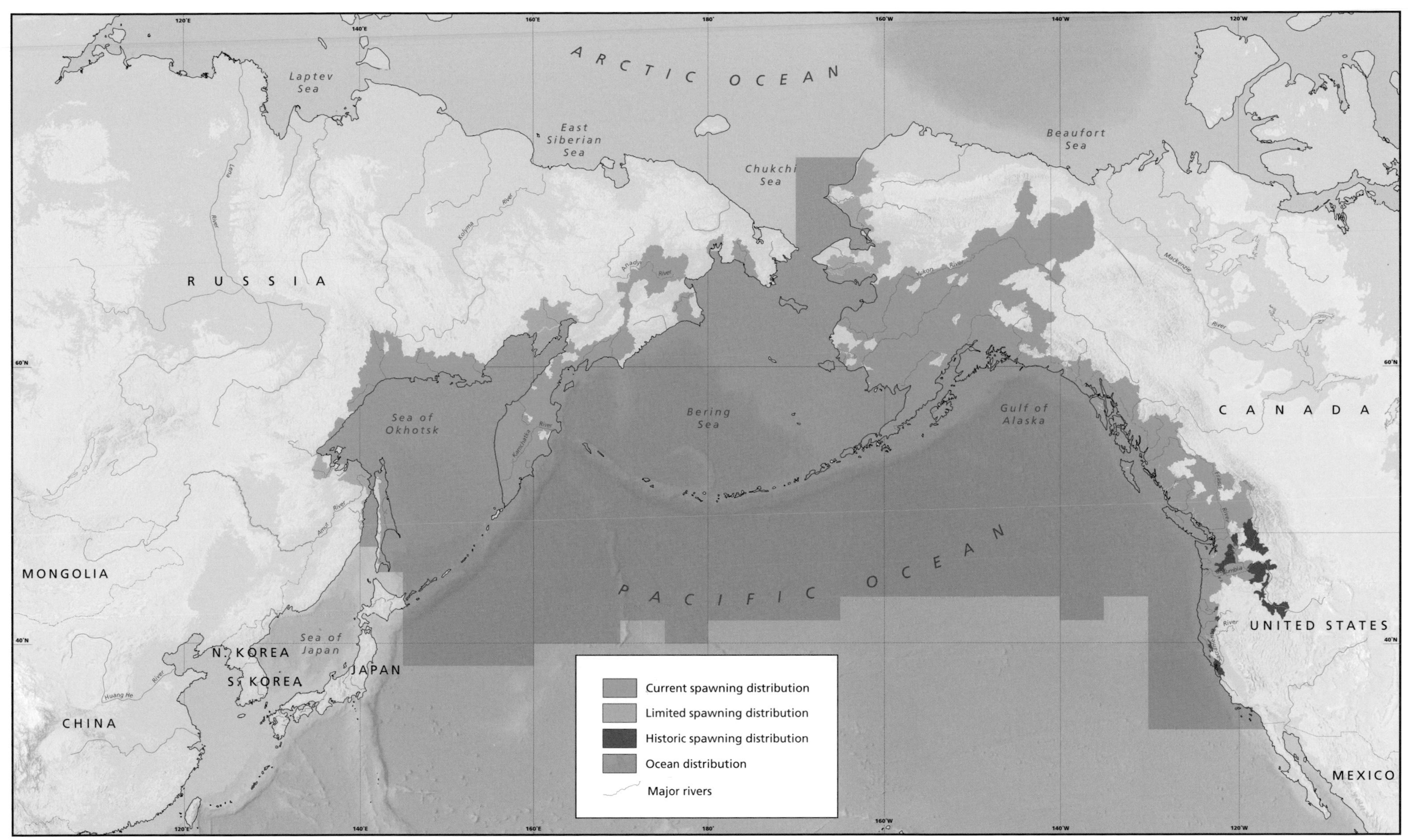


COHO DISTRIBUTION In the western Pacific, coho distribution clusters around the shores of the Sea of Okhotsk, covering Sakhalin and Kamchatka and much of the Magadan coastline. Distribution is continuous from the Koryakia coastline to the Anadyr River basin. In Alaska coho venture deep inland via the Yukon and Kuskokwim Rivers. As the latitude decreases, coho make shorter migrations. Historically, coho spawned hundreds of kilometers inland, in the Snake and Columbia Rivers. Populations there have been extirpated as they have at the southern extent of their former range in the Sacramento-San Joaquin Rivers basin.

# Coho Risk of Extinction

**Because they migrate upstream so late in** the season when waters tend to freeze up, coho are not harvested efficiently by larger fishing boats, which are commonly used in Russia and Japan. Rather, coho are more effectively targeted by smaller boats using short-line troll gear—used largely in North America, where coho are a significant commercial species.

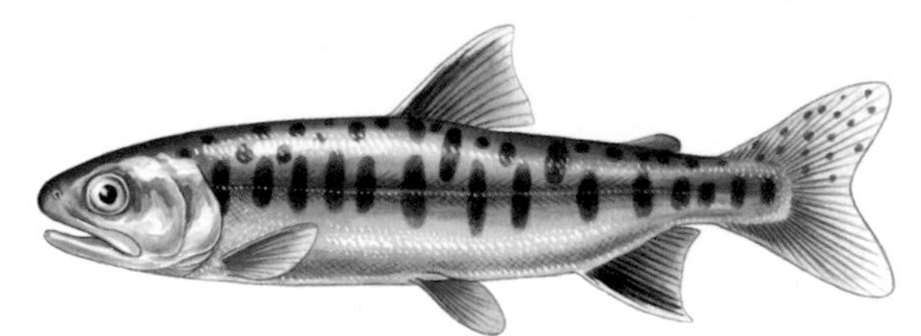

Coho parr

**Percentage of Relative Risk/Ecoregion**

| | ECOREGION | %HIGH | %MOD | %LOW | #EXT |
|---|---|---|---|---|---|
| 9 | Okhotsk-Oyashio Mixing | 0 | 0 | 100 | 0 |
| 11 | Southeast Sakhalin Current | 0 | 0 | 100 | 0 |
| 12 | Amur River* | 0 | 0 | 100 | 0 |
| 14 | East Sakhalin Current | 0 | 0 | 100 | 0 |
| 15 | Shantar Sea | 0 | 0 | 100 | 0 |
| 17 | Sea of Okhotsk Current | 0 | 100 | 0 | 0 |
| 25 | Anadyr River | 0 | 0 | 100 | 0 |
| 27 | Anadyr Current | 0 | 0 | 100 | 0 |
| 28 | Velikaya River | 0 | 0 | 100 | 0 |
| 30 | Penzhina Intracoastal | 0 | 100 | 0 | 0 |
| 31 | Shelikhov Gulf | 0 | 100 | 0 | 0 |
| 32 | Bering Slope-Kamchatka Currents | 3 | 97 | 0 | 0 |
| 33 | Kamchatka River | 0 | 100 | 0 | 0 |
| 34 | Western Kamchatka Current | 0 | 100 | 0 | 0 |
| 50<br>51<br>52<br>53 | Southeast Alaska** | 1 | 1 | 98 | 0 |
| 52 | Nass River *** | 8 | 0 | 92 | 0 |
| 54 | Skeena River | 20 | 0 | 80 | 2 |
| 55 | Hecate Strait-Q.C. Sound | 23 | 0 | 77 | 0 |
| 57 | Fraser River | 7 | 1 | 92 | 1 |
| 58 | Puget Sound-Georgia Basin | 20 | 5 | 75 | 26 |
| 59 | Vancouver Island Coastal Current | 19 | 2 | 79 | 0 |
| 60 | Seasonal Upwelling Cline | 12 | 53 | 35 | 0 |
| 61 | Columbia River | 83 | 17 | 0 | 13 |
| 62 | Klamath River | 38 | 62 | 0 | 2 |
| 63 | Strong Upwelling Year Round | 53 | 43 | 4 | 5 |
| 64 | Sacramento-San Joaquin Rivers | 100 | 0 | 0 | 4 |

*The Sakhalin portion of ecoregion 12.

**The southeast Alaska portions of ecoregions 50, 51, 52, and 53.

***The British Columbia portion of ecoregion 52.

In Magadan, Koryakia, and Kamchatka, coho are at higher risk of extinction than they are elsewhere in the Russian Far East.

Abundant coho stocks in the Bering Slope-Kamchatka Currents ecoregion (32) are heavily targeted by poachers for roe.

Because chum populations proliferate in the Anadyr River, coho fishing is not an important priority.

For overwintering coho require side-channels or sloughs, which have been heavily altered in WOCI and British Columbia by farming, logging, and urbanization.

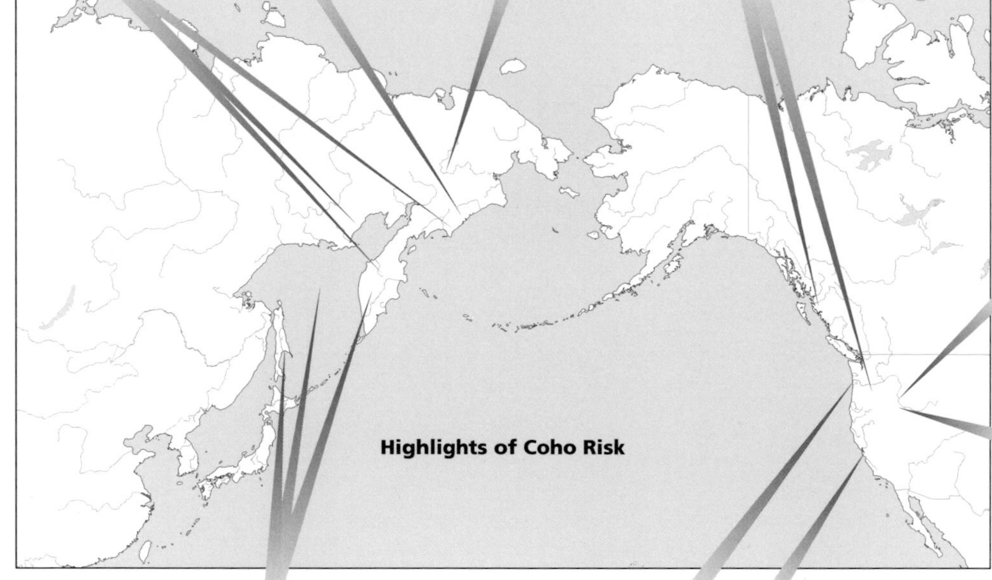



Coho have been largely extirpated from much of their former range in the Columbia River system.

Because coho use tributaries for spawning, damming and coastal development in WOCI have devastated populations.

Coho are abundant enough to have commercial significance on Kamchatka and Sakhalin, and within a few river systems on the Sea of Okhotsk.

Along the Oregon-Washington coast, some coho populations were harvested at rates of 70 to 80 percent in the latter part of the 20th century.

In the Sacramento-San Joaquin Rivers ecoregion (64) only one-third of coho populations remain, and these are all ranked at high risk of extinction.

see foldout Ecoregion map

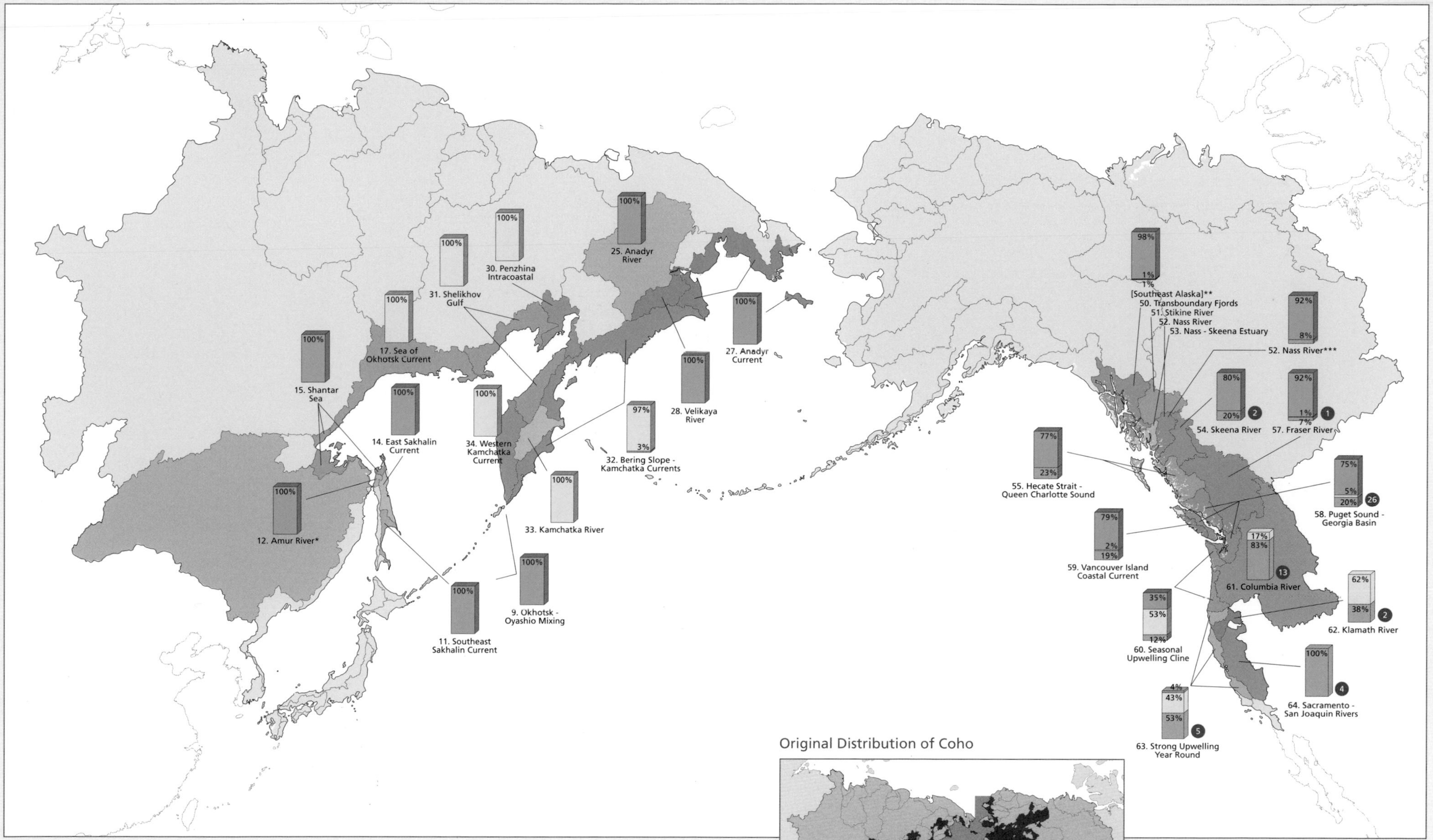

12. Amur River*
14. East Sakhalin Current
15. Shantar Sea
17. Sea of Okhotsk Current
31. Shelikhov Gulf
30. Penzhina Intracoastal
25. Anadyr River
27. Anadyr Current
28. Velikaya River
32. Bering Slope - Kamchatka Currents
34. Western Kamchatka Current
33. Kamchatka River
9. Okhotsk - Oyashio Mixing
11. Southeast Sakhalin Current
[Southeast Alaska]**
50. Transboundary Fjords
51. Stikine River
52. Nass River
53. Nass - Skeena Estuary
52. Nass River***
54. Skeena River
57. Fraser River
55. Hecate Strait - Queen Charlotte Sound
58. Puget Sound - Georgia Basin
59. Vancouver Island Coastal Current
61. Columbia River
60. Seasonal Upwelling Cline
62. Klamath River
64. Sacramento - San Joaquin Rivers
63. Strong Upwelling Year Round

Original Distribution of Coho

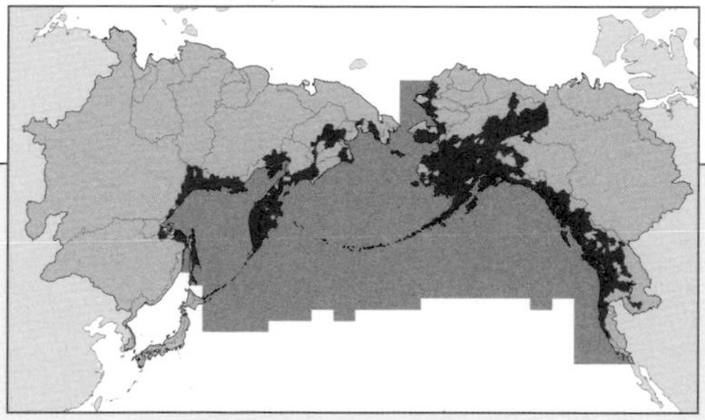

% Low risk
% Moderate risk
% High risk
Estimated number of extinct stocks

**COHO RISK OF EXTINCTION** In Russia coho populations are perceived to be small and healthy, except in Magadan, Kamchatka, and Koryakia, where they are at moderate risk. In the eastern Pacific, every ecoregion includes some coho populations that are at high risk of extinction. Overall risk increases in lower latitudes; for example, in the Puget Sound-Georgia Basin (ecoregion 58) we estimate 26 extinctions. In WOCI risk increases dramatically: more than half the populations are assessed at high risk in the Strong Upwelling Year Round (63); more than three-quarters in the Columbia River (61); and all in the Sacramento-San Joaquin Rivers (64).

# Masu Distribution

**JAPAN IS A MAJOR HARVESTER OF MASU,** but the fishery's commercial importance is impossible to quantify because masu and pink harvest is recorded in aggregate. Historically, Japanese fisheries caught a portion of Russian-origin masu, a practice that likely continues. In Japan in-river recreational fishing for masu adults is illegal; fishing for juveniles is permitted.

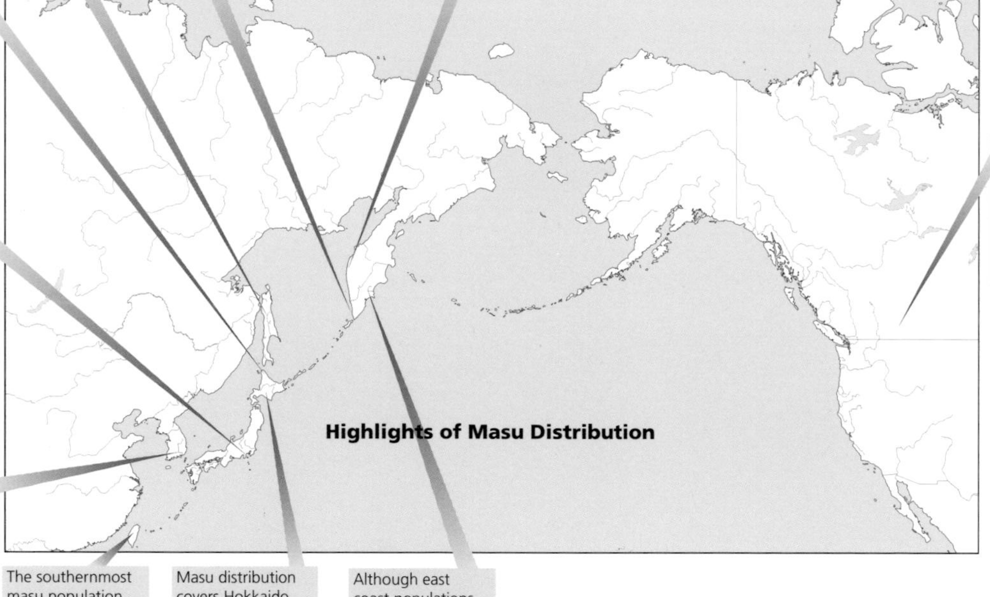


## MASU SALMON

**SCIENTIFIC NAME**

*Oncorhynchus masou*

**ABUNDANCE**

Masu salmon, found only in the Asian Pacific, have the most limited distribution and are the least abundant of the anadromous Pacific salmon. We have not included data on resident *amago* (also referred to as *O. rhodurus*), which are found only in southeastern Japan.

**SIZE**

Masu and pink are the smallest Pacific salmon. Adult masu measure around 50 cm and weigh on average 2–2.5 kg.

**LIFE HISTORY**

Masu, like coho, reside 1–3 winters in freshwater; residence is longer in northerly populations. Masu spend one year in salt water with short ocean migrations, spending most of their time in either the Sea of Japan or the Sea of Okhotsk. There are spring and fall seasonal races of anadromous masu.

Spawning and rearing habitats for masu salmon are similar to those preferred by coho: fast water in the middle to upper reaches of river systems for spawning, and side channels and sloughs in the lower river for rearing. Although the anadromous fish die after spawning, some of the resident males may survive to spawn again.

The most ancestral form of *Oncorhynchus,* masu inspires debate about whether it is one species or two (*amago* (anadromous) and *rhodurus* (resident)). Some treat the southerly landlocked form separately, based on genetically linked coloration patterns. However, the two forms can hybridize and produce viable offspring.

**CULTURAL ROLE**

The extensive freshwater life history stage of masu makes them particularly vulnerable to subsistence fishing and sportfishing pressure. They are harvested as juveniles by sportfishers and subsistence fishers in Russia and Japan. In Japan masu juveniles are the only salmon that can be legally caught in rivers. In Russia juvenile masu are sought after, particularly in rural communities. Most Russians, in fact, do not believe the juveniles are the same species as the robust mature fish that return to spawn.

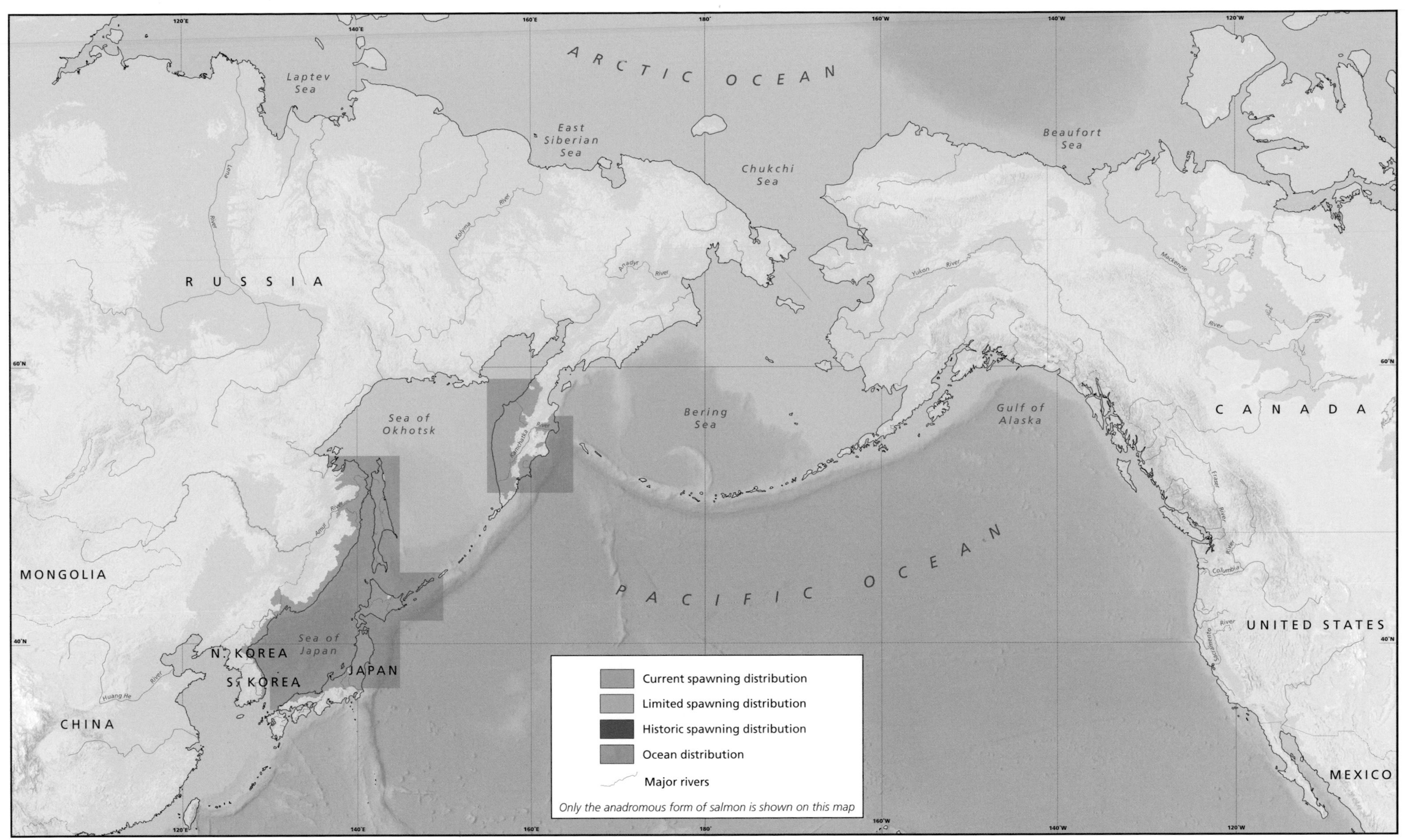


MASU DISTRIBUTION Masu have the most limited distribution among *Oncorhynchus* and are present only in the western Pacific. As the cladogram on page 5 illustrates, masu appeared early and quite separately from other species in the evolutionary history of the genus *Oncorhynchus*. More recently, masu may have diverged further from other salmonids during the last ice age, when lower sea levels likely created a brackish lake of the Sea of Japan (see page 52). The island of Hokkaido is approximately the center of the range for masu salmon, where masu enjoy protection from harvest in 32 masu conservation rivers (see page 44).

# Masu Risk of Extinction

**Beyond the Sea of Japan, masu populations** have not been well monitored or studied. Perception of risk remains controversial. Russia's premier masu expert asserts that the species is at high risk of extinction, particularly as recreational fishermen target juvenile masu. In Japan juveniles are also heavily fished, but status of native populations remains unassessed.

Masu parr

**Percentage of Relative Risk/Ecoregion**

| | ECOREGION | %HIGH | %MOD | %LOW | #EXT |
|---|---|---|---|---|---|
| 5<br>7 | Primorsky/North Korean Current* | 100 | 0 | 0 | 0 |
| 8 | Soya Current** | 0 | 0 | 100 | 0 |
| 9 | Okhotsk-Oyashio Mixing | 0 | 0 | 100 | 0 |
| 10 | Liman Current | 73 | 27 | 0 | 0 |
| 11 | Southeast Sakhalin Current | 0 | 0 | 100 | 0 |
| 12 | Amur River*** | 0 | 0 | 100 | 0 |
| 12 | Amur River**** | 0 | 100 | 0 | 0 |
| 13 | Southwest Sakhalin Current | 0 | 0 | 100 | 0 |
| 14 | East Sakhalin Current | 0 | 0 | 100 | 0 |
| 15 | Shantar Sea | 0 | 33 | 67 | 0 |
| 31 | Shelikhov Gulf | 0 | 0 | 100 | 0 |
| 32 | Bering Slope-Kamchatka Currents | 0 | 0 | 100 | 0 |
| 33 | Kamchatka River | 0 | 0 | 100 | 0 |
| 34 | Western Kamchatka Current | 0 | 0 | 100 | 0 |

*All of ecoregion 7 and the Russia portion of ecoregion 5.

**The Russia portion of ecoregion 8.

***The Sakhalin portion of ecoregion 12.

****The mainland Russia portion of ecoregion 12.

On Kamchatka scientists believe that masu populations are at least at a moderate risk due to small population sizes and poor enforcement of the fishery.

In Sakhalin some biologists deem masu stocks healthy; others believe masu are at risk from heavy harvest pressure from personal-use fisheries, interception in the pink salmon fishery, and recreational harvest of juvenile masu in freshwater.

Commercial masu fishing was prohibited in the Liman Current ecoregion (10) in 1957.

Masu populations, small and suffering from heavy fishing pressure, are at high risk of extinction in southern Primorye.

**Highlights of Masu Risk**

South of the Klevka River, masu populations are not large enough to support commercial fisheries.

On Hokkaido the several wild populations of masu are deemed of special concern.

Data for Soya Current ecoregion (8) represent Sakhalin status only.

see foldout Ecoregion map

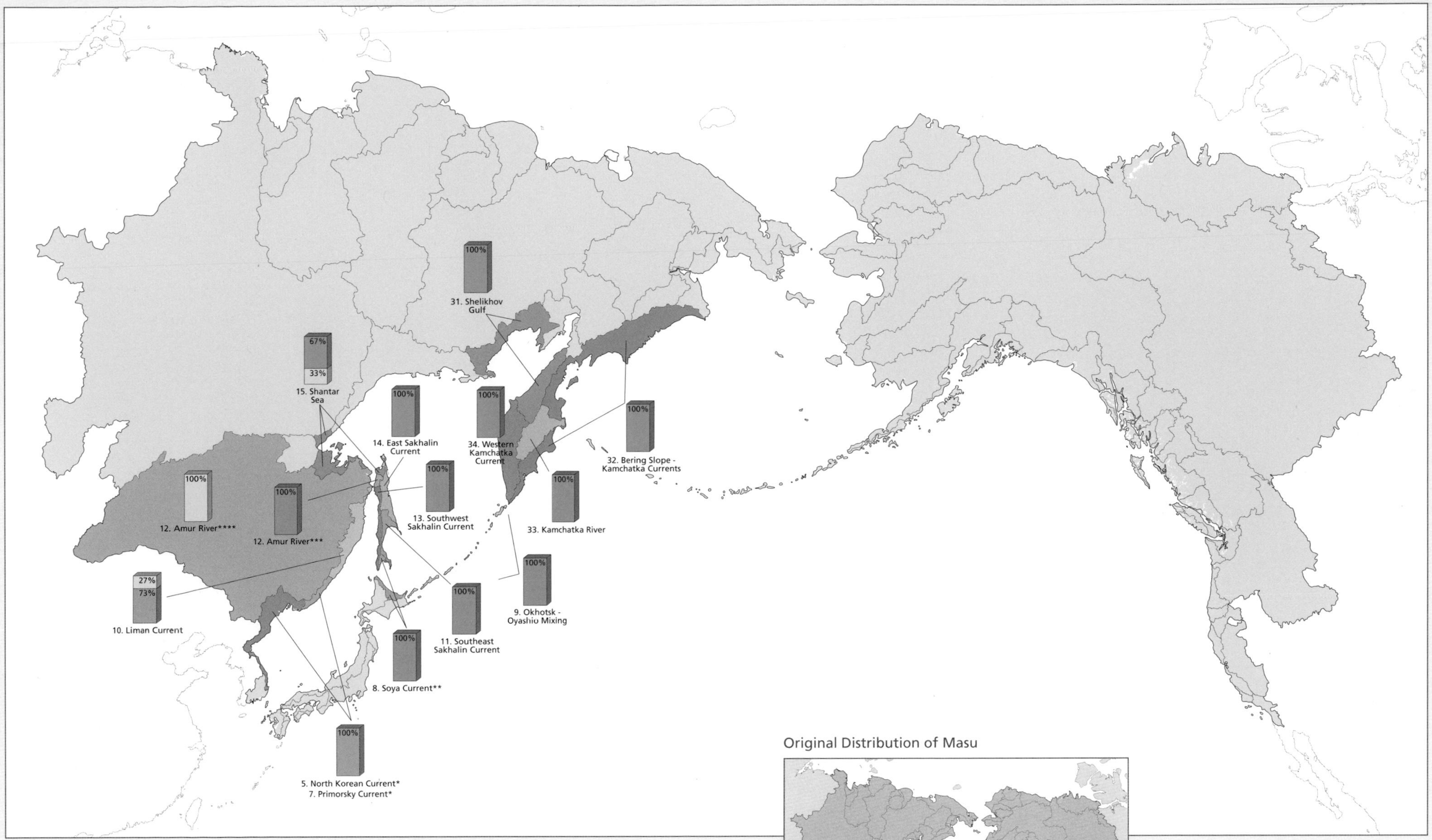

31. Shelikhov Gulf
100%

15. Shantar Sea
67%
33%

14. East Sakhalin Current
100%

34. Western Kamchatka Current
100%

32. Bering Slope - Kamchatka Currents
100%

13. Southwest Sakhalin Current
100%

33. Kamchatka River
100%

12. Amur River****
100%

12. Amur River***
100%

10. Liman Current
27%
73%

9. Okhotsk - Oyashio Mixing
100%

11. Southeast Sakhalin Current
100%

8. Soya Current**
100%

5. North Korean Current*
7. Primorsky Current*
100%

Original Distribution of Masu

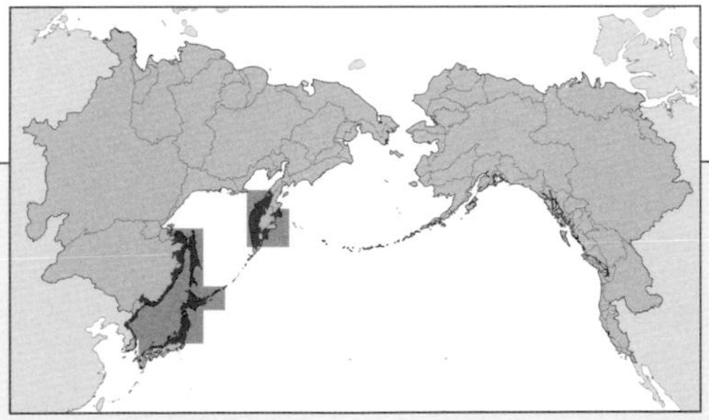

% Low risk
% Moderate risk
% High risk
Estimated number of extinct stocks

**MASU RISK OF EXTINCTION** Because Hokkaido is approximately the center of the masu range, our risk of extinction map paints an incomplete picture of masu status. Assessing masu status is further complicated by the fact that Japanese pink and masu catch are aggregated. Some estimates put masu catch at no more than 2 percent of Japan's total harvest; it is generally believed that Japanese masu catch is decreasing. Note that risk may appear overreported due to the nature of our ecoregions construct: masu are not present in the Magadan portion of the Shelikov Gulf ecoregion (31) or in the Kuril Islands portion of the Soya Current ecoregion (8).

# Steelhead Distribution

**Steelhead have the most southerly** coastal distribution among *Oncorhynchus* in the eastern Pacific; in the west, chum and masu share that distinction. Contemporary studies demonstrate that the anadromous and resident forms interbreed, but we have mapped only the former. Because steelhead are no longer commercially fished, data remain elusive.

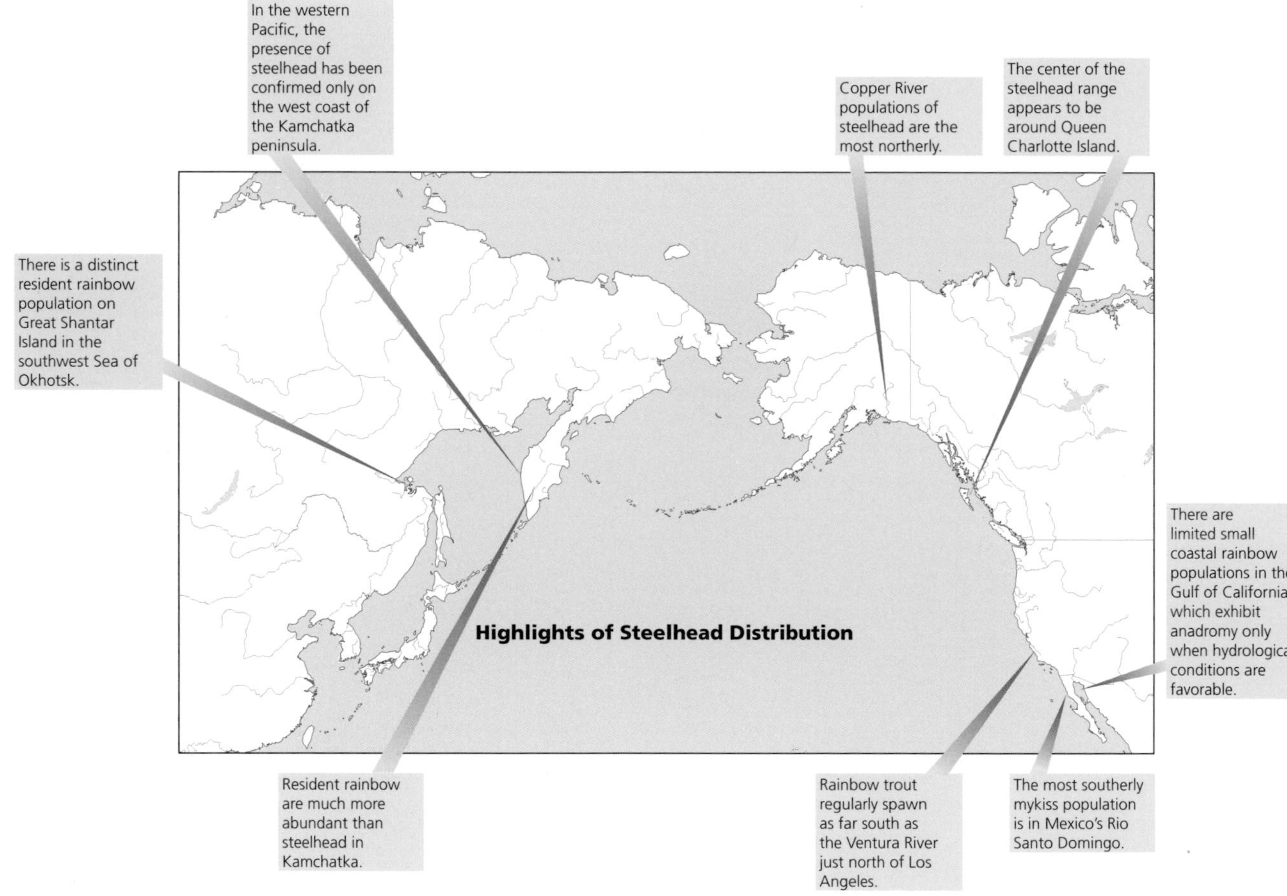


## STEELHEAD

**SCIENTIFIC NAME**

*Oncorhynchus mykiss*

**CLASSIFICATION**

Classification of *O. mykiss* has been hotly debated for years. Rainbow-steelhead and cutthroat trout were moved from the genus *Salmo* to *Oncorhynchus* as a result of analyses from Gerald R. Smith and Ralph Stearley in 1993. Russian biologists refer to steelhead and rainbow trout collectively as "mikizha," and classify them as *Parasalmo* rather than *Oncorhynchus,* largely due to the fact that rainbow-steelhead are repeat spawners and have distinctive morphometric features.

**ABUNDANCE**

*Mykiss* tend to form small populations. In the western Pacific, steelhead have been confirmed only in western Kamchatka. They are much more widespread in the eastern Pacific.

**SIZE**

Steelhead vary greatly in size. They can weigh 2.5–9 kg and measure 60–100 cm as adults.

**LIFE HISTORY**

Unlike other species of *Oncorhynchus* (with the exception of cutthroat and some resident sockeye and masu), *mykiss* do not necessarily die after spawning and therefore live relatively long lives, from 4 to 7 years or more. Steelhead adapt to a variety of environmental conditions and utilize all parts of a river basin, from small tributary streams to mainstem river channels. Life histories vary widely and include diverse combinations of freshwater, estuarine, and saltwater residence. Kamchatka *mykiss* populations appear to be particularly diverse, compared to North American populations.

**CULTURAL ROLE**

Steelhead were harvested commercially in North America until the 1930s and are still harvested commercially by native peoples under original treaties. In Russia *mykiss* were harvested commercially until the 1980s. Both the anadromous and resident forms of *mykiss* are popular with sport fishers in North America and on Kamchatka.

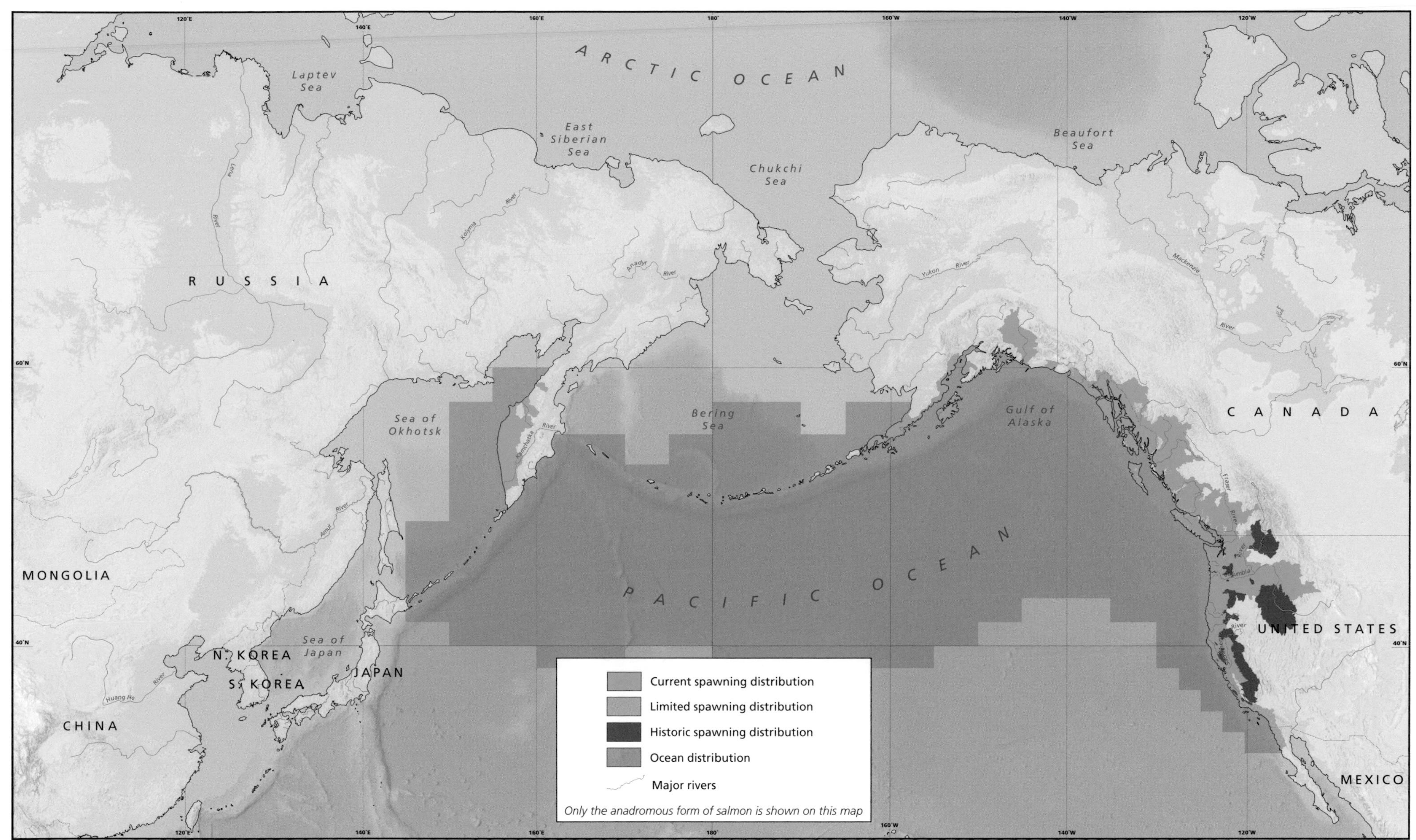


**STEELHEAD DISTRIBUTION** *Mykiss* may thrive in any number of habitats, from lakes to rivers, from shallow tributaries to estuaries. Kamchatka's west coast is the only western Pacific region to host steelhead. Steelhead are robust in southeast Alaska and British Columbia but patchy along the Gulf of Alaska. *Mykiss* have experienced extinctions inland, east of the Coast Ranges and the interior Columbia River basin. Current distribution may reach as far south as Santa Barbara, California. Limited spawning manifests well into Baja California, Mexico. (Note that the distributional footprint for the resident form of *mykiss*—rainbow trout—is more extensive.)

# Steelhead Risk of Extinction

**As with our distribution data, risk of** extinction data include only the anadromous form of *mykiss*. The unusual plasticity of this species and its complex life history diversity would suggest an increased resilience to threats, but that is not the case; steelhead exhibit a pattern of risk across the North Pacific similar to that displayed by other salmonids.

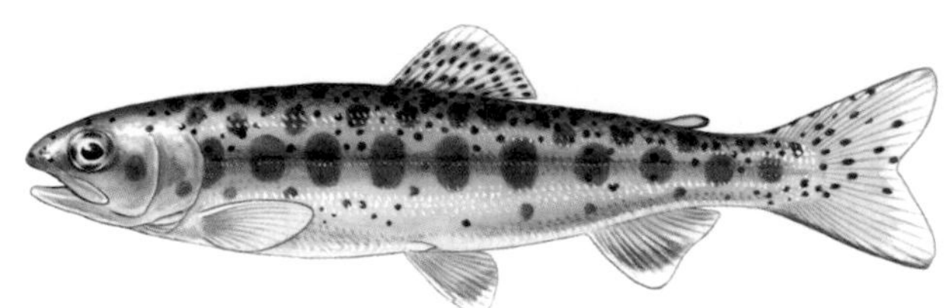

Steelhead parr

**Percentage of Relative Risk/Ecoregion**

| | ECOREGION | %HIGH | %MOD | %LOW | #EXT |
|---|---|---|---|---|---|
| 31 | Shelikhov Gulf | 0 | 75 | 25 | 0 |
| 34 | Western Kamchatka Current | 17 | 39 | 44 | 0 |
| 50 51 52 53 | Southeast Alaska* | 0 | 0 | 100 | 0 |
| 52 | Nass River** | 0 | 0 | 100 | 0 |
| 54 | Skeena River | 0 | 0 | 100 | 0 |
| 55 | Hecate Strait-Q.C. Sound | 0 | 17 | 83 | 0 |
| 57 | Fraser River | 4 | 4 | 92 | 2 |
| 58 | Puget Sound-Georgia Basin | 9 | 6 | 85 | 4 |
| 59 | Vancouver Island Coastal Current | 0 | 14 | 86 | 0 |
| 60 | Seasonal Upwelling Cline | 5 | 10 | 85 | 1 |
| 61 | Columbia River | 48 | 30 | 22 | 14 |
| 62 | Klamath River | 17 | 33 | 50 | 6 |
| 63 | Strong Upwelling Year Round | 54 | 23 | 23 | 0 |
| 64 | Sacramento-San Joaquin Rivers | 82 | 18 | 0 | 14 |
| 65 | Weak Upwelling Cline | 62 | 38 | 0 | 0 |
| 66 | California Undercurrent | 100 | 0 | 0 | 19 |

*The portions of southeast Alaska ecoregions 50, 51, 52, and 53.

**The British Columbia portion of ecoregion 52.

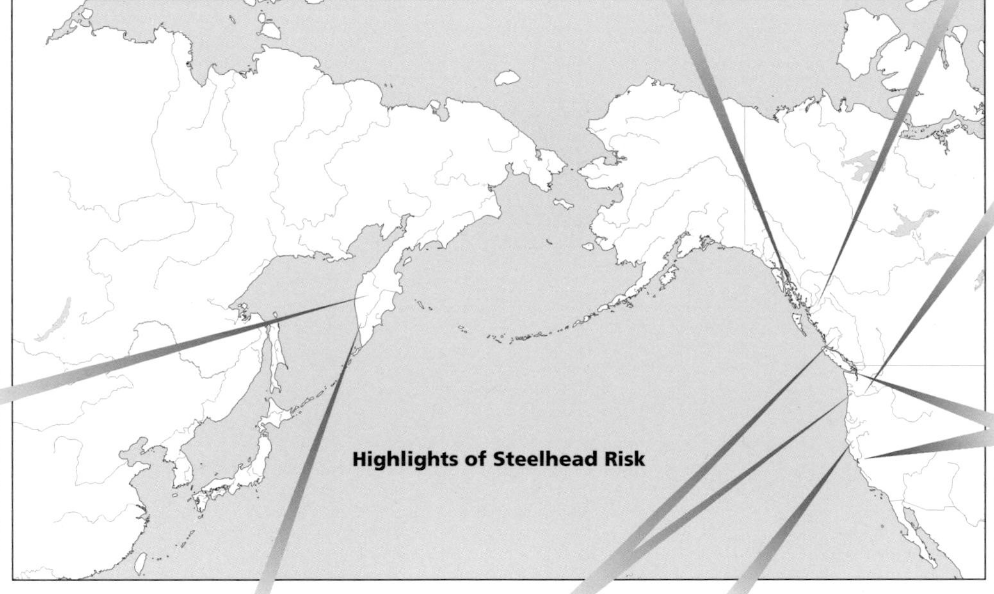

see foldout Ecoregion map

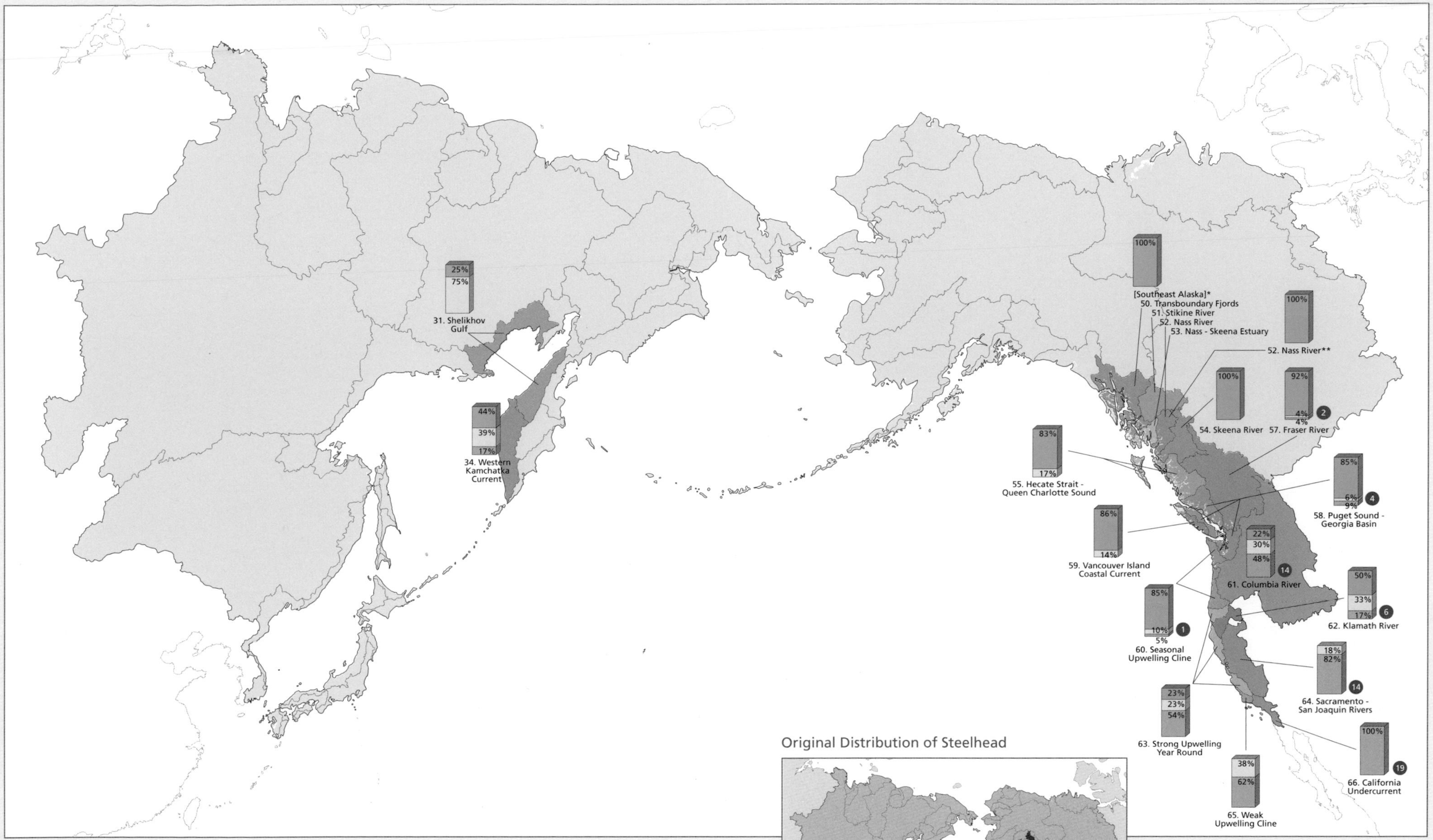

25%
75%
31. Shelikhov Gulf

44%
39%
17%
34. Western Kamchatka Current

100%
[Southeast Alaska]*
50. Transboundary Fjords
51. Stikine River
52. Nass River
53. Nass - Skeena Estuary

100%
52. Nass River**

100%
54. Skeena River

92%
4%
4%
2
57. Fraser River

83%
17%
55. Hecate Strait - Queen Charlotte Sound

85%
6%
9%
4
58. Puget Sound - Georgia Basin

86%
14%
59. Vancouver Island Coastal Current

22%
30%
48%
14
61. Columbia River

85%
10%
5%
1
60. Seasonal Upwelling Cline

50%
33%
17%
6
62. Klamath River

18%
82%
14
64. Sacramento - San Joaquin Rivers

23%
23%
54%
63. Strong Upwelling Year Round

100%
19
66. California Undercurrent

38%
62%
65. Weak Upwelling Cline

Original Distribution of Steelhead

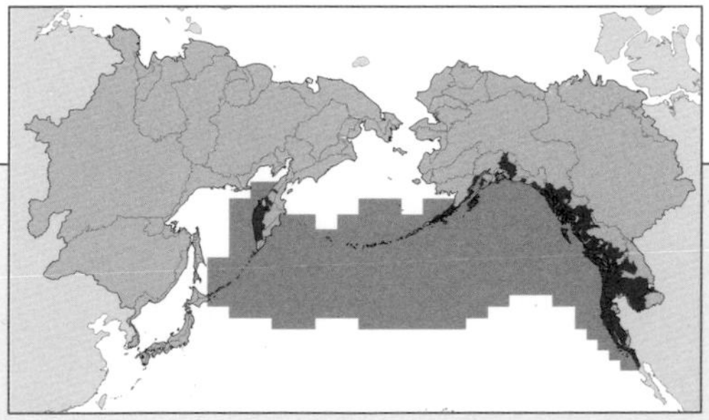

% Low risk
% Moderate risk
% High risk
# Estimated number of extinct stocks

**STEELHEAD RISK OF EXTINCTION** More than half the steelhead populations in Kamchatka are at risk. Note that risk of extinction may appear overreported due to the nature of our ecoregions construct: no steelhead exist in the Magadan portion of the Shelikov Gulf ecoregion (31). *Mykiss* exhibit unusually adaptable life history patterns and demonstrate robustness at the center of their range. Populations are classified entirely at low risk in southeast Alaska and northern British Columbia. Risk greatly increases toward the edges of the range. In four WOCI ecoregions, more than half the steelhead populations are at high risk.

# 5

# Threats to Biodiversity

*Falling into the human footprint*

- Human Population Density
- Logging in Frontier Forests
- Mineral Development
- Oil, Gas, and Pipelines
- Dams
- Climate Change

Legacies of a nearly extirpated population, these year-old Snake River sockeye are raised in tanks and will be pioneers for recolonization in native waters.

The distribution and risk findings that we present in this chapter describe a general latitudinal pattern of increasing risk of extinction for Pacific salmon populations toward the southern margins of their natural range in both Asia and North America. The gradient is consistent despite differences in geology and climate in North America and Asia (described in chapter 3).

For millennia, salmon have adapted to natural variabilities, including changes in sea surface temperature, currents, and nutrient availability. But the advent of the 19TH century industrial economy increased extinction risk across the Pacific Rim as salmon confronted the effects of ecosystem alteration at the hand of humans.

A single smolt might need to survive passage through or over dams; compete for food and habitat with hatchery fish; navigate through siltation caused by logging or effluent from manufacturing plants; wend through shallow waters diverted for agricultural purposes; journey through runoff from city stormwater pipes into a sea netted by trawlers; and find food in waters warmed by climate change to the point that the window of phytoplankton bloom has been shortened or altered entirely. The spawning adult may experience some of the same threats upon return to natal waters, with the added and considerable threat of overharvest.

Global climate change, human population trends, fisheries market pressures, natural resource extraction—these are all among the threats that have direct and indirect impacts on salmon populations. Overharvest is a constant threat to salmon populations throughout the Pacific Rim; it is a broad category that includes commercial over-harvest itself, high-seas bycatch, poaching, as well as personal-use and recreational sport fishing.

Forested hillsides filter, absorb, and divert runoff during heavy rains. Clearcuts eliminate such protections, resulting in flooding, siltation, lack of dissolved oxygen, and gravel disturbance to streams.

The increased use of hatcheries and aquaculture to supplement wild salmon catch is represented across the Pacific Rim. Because the overall effects of these practices on wild fish populations are difficult to quantify, they remain controversial (see pages 34 to 37).

To bypass Oregon's Lower Granite Dam along the Columbia River, wild salmon smolts are barged downstream for outmigration.

Pollution from sewage-treatment plants or factories (point source pollution) can debilitate waterways. Nonpoint source pollution (NSP)—discrete, intermittent releases of pollutants into ground or surface waters, conveyed through rain or snow—is also a major yet unquantifiable threat to rivers and oceans. NSP may include such diverse sources as lawn fertilizers, animal feces, and runoff from roads. According to the American Fisheries Society, NSP is "probably the most pervasive and ubiquitous water quality problem in North America."

Threats are also specific to regions. Poaching is the major threat to salmon in the Russian Far East. Habitat loss, invasive species, oil and gas development in Russia, and hatcheries pose the greatest risk to wild salmon biodiversity in

Japan. In Alaska and British Columbia, logging represents a major threat, but mining and habitat loss loom as well; in British Columbia in particular, the moratorium on offshore oil and gas development may be lifted imminently, leading to possibilities of spills and leaks, as well as the disruption of natural ecosystems. In Washington, Oregon, California, and Idaho (WOCI) major threats include dams, logging, habitat loss, urbanization, water diversion, and human population growth.

**PREDATION**

At least 137 species rely on salmon; humans are just one. Among other predators, mammals, birds, and fish all depend on healthy salmon populations. For example, hake, mackerel, walleye, pollock, channel catfish, northern pikeminnow, striped bass, herons, common mergansers, cormorants, and bears are but a handful of the predators salmon encounter from stream to ocean.

Human impacts on natural ecosystems have rendered salmon more vulnerable to predation. Water diversions for agriculture have created low flows and higher temperatures, exposing salmon to predators, disease, and crowding, and increasing stress and metabolism. Turbulence from dams and industrial-sized water conveyances increases stress, disorients fish, and reduces their ability to evade attack. Reservoirs and dams create predator foraging areas. Furthermore, marine mammals, such as harbor seals and California sea lions, prey upon flatfish, lampreys, and other benthic species for food, but they are increasingly turning to salmon when other prey is scarce and when salmon congregate in small areas as a result of low water flow and poor water quality and habitat.

**REDUCING AND MITIGATING THREATS**

Because salmon migrate, they confront a spectrum of threats—beginning in natal streams, to estuaries, up to the Bering Sea, and back—that resident fish and other localized wild species may never encounter. As the human footprint advances northward, anthropogenic threats to *Oncorhynchus* will expand in their complexities, and extirpations at the southern extents of the species' range will continue to occur.

If recovery efforts aim to remedy isolated threats, they will remain costly and ineffective. Instead, recovery efforts must consider the full reach of the ecosystem and all the factors that affect salmon habitat.

Limits of ocean and freshwater carrying capacity, especially in impaired ecosystems, render salmon increasingly vulnerable to prey.

## SALMON POACHING IN THE RUSSIAN FAR EAST

Poaching is the biggest threat to salmon in the Russian Far East and claims hundreds of tons of fish annually. A walk along most Russian salmon streams during spawning season will likely lead to a camp where pools of maggots testify to discarded fish and the poachers' presence. One kilogram of salmon roe, an amount that can be harvested from about 10 healthy pink salmon, can sell for US$35 on the black market. One or two small buckets of roe can be the equivalent of a monthly income, and a lone poacher can collect several buckets in a day. A few weeks of work can support a rural family for a year. Because the financial stakes are so high and the risks so low, poaching and fish smuggling is a very lucrative and sometimes deadly business. Two recent high-profile assassinations—of Border Service General Vitaly Gamov and Magadan Governor Valentin Tsvetkov—have direct links to the fishing industry.

Poaching may come in many forms, from lone villagers making ends meet to large bands of organized and armed professionals. A severe lack of resources and widespread corruption in oversight agencies make poaching extremely difficult to stop. Compounding matters is the lack of a legal chain of custody in the Russian Far East, which means illegally harvested products become "legalized" by the time they hit store shelves; it also means that we have no real way to measure the true economic and ecological impacts of poaching. ■

In 2003 Russia's Far East federal district reported 135 tons of illegally caught caviar. Above, poachers collect buckets of salmon roe.

# Human Population Density

**Four of the world's ten most populous** countries rim the North Pacific: China, with 1.3 billion people; Japan, 128 million; Russia, 146 million (including 7 million in the Russian Far East); and the United States, nearly 300 million, (including 50 million people in WOCI and Alaska). The Koreas and Hong Kong contribute another 70 million people to the region; British Columbia adds around 4 million to the eastern Pacific rim. Population density increases toward southerly latitudes and is more continuous in the western Pacific. Differences aside, the effects of several-hundred-million people on North Pacific watersheds and seas are significant.

Population modeling posits that human population growth will increase the number of threatened species—including salmon—by 14 percent by 2050. Urbanization and landscape alteration for homes, agriculture, and natural resource extraction (e.g., timber harvest, oil and gas development, mining) degrade essential salmon spawning and rearing habitats. As a result, there is a high correlation between human population density and the risk of salmon extinction in the more southerly latitudes of the Pacific Rim. Compounding the risk are edge-of-range effects, which render stocks particularly vulnerable at the outer extent of their distributional range.

In large, complex river basins like the Amur, the whole of the damage wrought by human population can be greater than the sum of its parts; habitat loss, overfishing, pollution from manufacturing, natural resource extraction for energy, and lack of sewage treatment in major cities all contribute to salmon declines.

In Japan flood control projects, a priority in protecting densely populated cities vulnerable to flash flooding, have reengineered wild rivers. Above the 50TH parallel (bisecting Sakhalin and Vancouver Island), human populations are sparse, and salmon are comparatively buffered from urbanization. But impacts persist as these regions serve as wood, mineral, and energy sources.

An indirect result of human population growth is increased institutional fragmentation. Ecosystem stewardship is divided among varied interests, including local, regional, state, and federal entities, as well as tribal and private institutions that may have conflicting missions or goals that do not prioritize salmon conservation. When river basin and ecosystem management becomes fragmented, restoration efforts cannot be implemented effectively.

## TALE OF TWO RIVERS

The Amur River basin was once a major producer of commercially harvested chum and pink salmon. Today more than 80 million people live in this region; 90 percent are Chinese. The Amur and the Ussuri Rivers, which feed the Amur basin, serve as the border between China and Russia for about 1,800 kilometers. At the Russian city of Khabarovsk, the China-Russia border follows the Amur and its major tributary, the Ussuri River, which is fed by Lake Khanka to the south (also shared by China and Russia). The Ussuri offers a dramatic and unusual illustration of the effects of human population on a river: to the west, densely populated Chinese villages crowd the water's edge; the eastern banks, however, are sparsely inhabited. Polluted runoff from Chinese industry along the banks affects water quality for those who depend on the river, Chinese and Russians alike. As resources are limited and poorly protected, competition for salmon and sturgeon catch can lead to gross overharvest. The pressures on the river have led to salmon extirpations upstream. ■

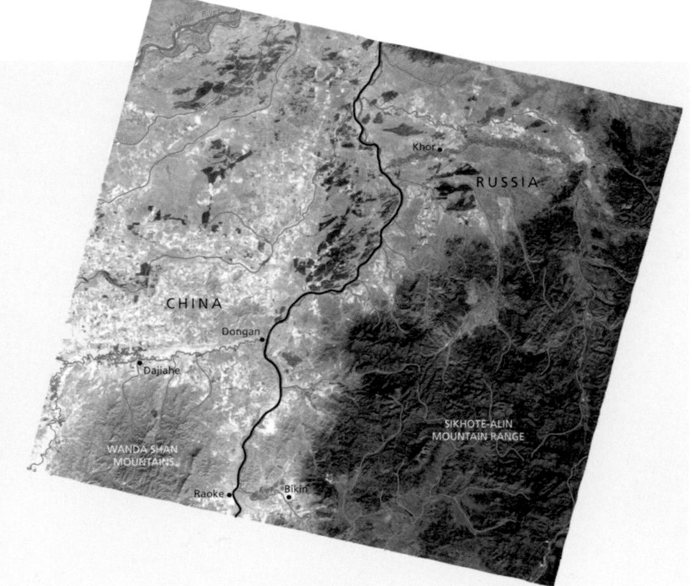


**STRADDLING TWO COUNTRIES** The satellite image of the Ussuri River on the right depicts light and heat intensity (as measures of deforestation). From this, we note how the river endures intense landscape change on its Chinese bank, yet little intrusion from the Russian bank, particularly along the Sikhote-Alin mountain range.

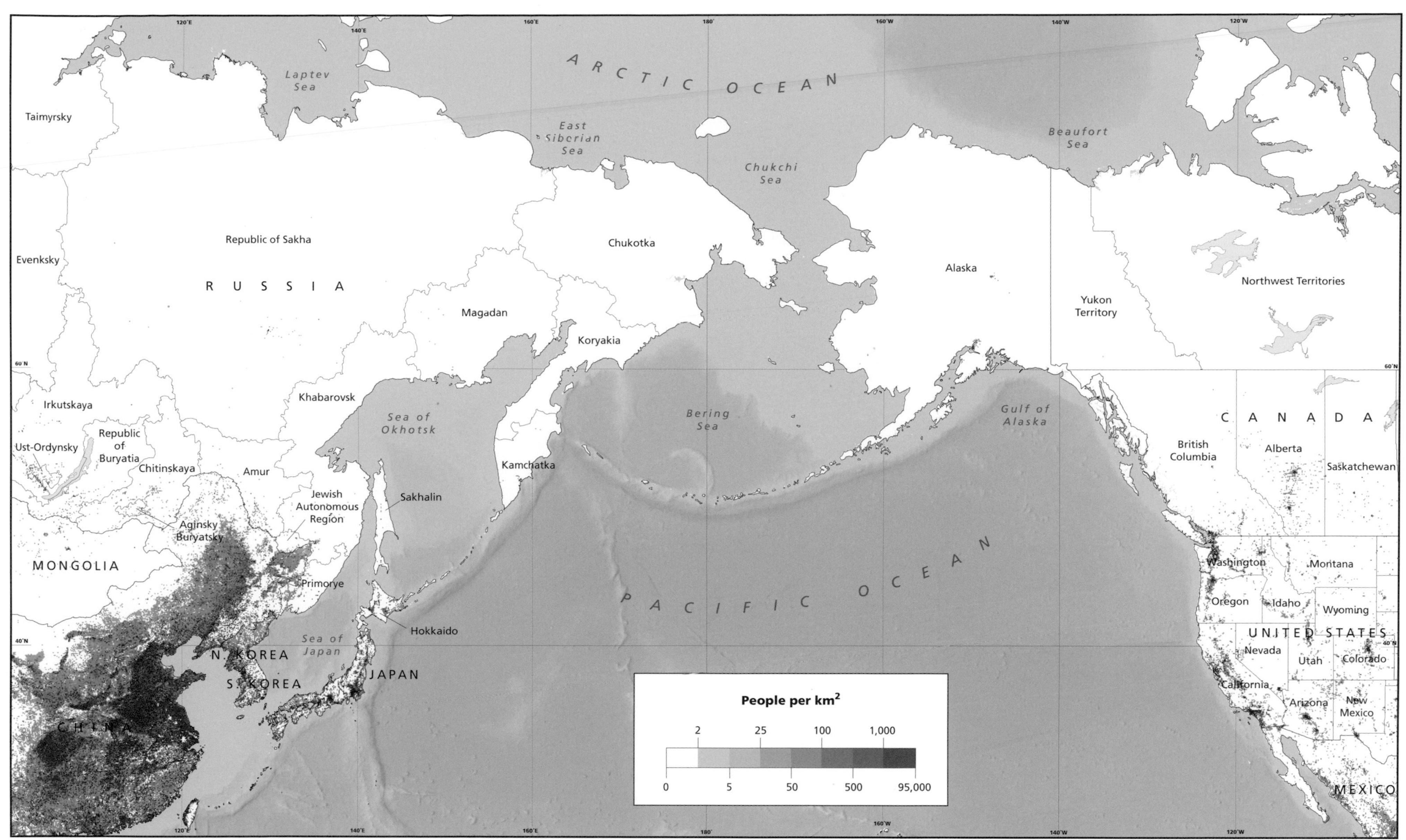


AMBIENT HUMAN POPULATION DENSITY Most population data is census-driven—information that may not reflect where people spend their time. Our LandScan 2002 dataset maps average ambient human population density over a 24-hour period and demonstrates not just where people live but also where they work, eat, drive, and generally how they use the landscapes around them. Using several inputs—including roads, terrain slope, land cover, nighttime lights, as well as conventional census counts to determine patterns of human movement—this dataset more accurately represents where people use the land most, such as river valleys and coastal regions.

# Logging in Frontier Forests

**EIGHT THOUSAND YEARS AGO FORESTS IN** the Pacific Rim covered half the landmass. Today around one-third of original forest cover remains, according to the World Resources Institute (WRI). The WRI calls these regions "frontier" forests, which may be defined as "large, ecologically intact, and relatively undisturbed" self-sustaining ecosystems.

The rest has been logged, often repeatedly, and the pace of deforestation has increased exponentially since the advent of industrial-scale logging in the 19TH century. Two major long-term studies, in Carnation Creek in British Columbia and the Alsea watershed in Oregon, explore the complexities of logging in forested salmon streams. Within a coastal stream habitat, salmon depend on an assortment of, and interactions among, natural factors: clean gravel, a variety of pools, side channels, and alcoves, and natural riparian vegetation that regulate the movement of sediments, provide organic matter and insects, and moderate water temperatures through shade. Without these natural buffers, habitat values diminish, resulting in, for example, erosion, siltation, desiccation, debris, disease, and pest infestation. Logging roads bisecting habitat also contribute to the degradation of freshwater resources, increasing siltation and runoff.

Second-growth forests make up the vast portion of forested landscape below the 60th parallel. These areas include clearcuts that devastated habitats for fish and wildlife. Between 1971 and 2002, for instance, 29 percent of all the forestland in Washington's Olympic Peninsula was clearcut—more than one million acres.

Pockets of remaining frontier forests dot the landscape throughout the Pacific Rim, but these fragmented parcels may not function adequately ecologically. On the western side of the Pacific Rim, frontier forest logging occurs primarily in the extensive larch and spruce-fir forests of the southern Far East, which remain largely unprotected. A 2002 United Nations Environment Programme study of closed canopy forests (virgin, old-growth, naturally regenerated woodlands) found that Russia has the lowest level of protections in the world—just 2 percent.

In the United States and Canada, virtually all frontier forests remaining in salmon spawning grounds are moderately or highly threatened. Alaska's virgin forests are targeted by logging interests, from the Chugach to the Tongass, as are vast swaths of frontier forest in Canada's Northern Interior and Coast forest regions. In WOCI stretches of pristine forests are threatened at the headwaters of chinook, coho, sockeye, and steelhead distribution.

## THE EFFECTS OF CLIMATE CHANGE ON INLAND FOREST HABITAT

Global climate change represents an as-yet-immeasurable threat to forests and stream habitat. A working group of the Intergovernmental Panel on Climate Change predicted that with a doubling of $CO_2$, Asia's boreal forests would be reduced by 50 percent, largely in the Russian Far East. In North America forest cover would shift, increasing in some areas and decreasing in others, altering habitat and modifying biodiversity altogether. A 2003 study in *Nature* found that many species of fish, birds, plants, and insects have already been averaging a four-mile shift toward the poles each decade—just with an average increase of around 17°C (1°F) during the last century.

In temperate and alpine zones rising temperatures would hasten glacial melt, changing the hydrological pattern, increasing water flow downriver, and affecting the productivity of salmon redds and juvenile development. Early thaw on temperate lakes would change salmon run seasonality, which could disrupt development. Higher lake temperatures would affect levels of dissolved oxygen and could increase incidence of disease and productivity of insects and invasive species.

Higher stream temperatures would modify productivity and possibly render some streams uninhabitable for salmon. (For brief stints, chinook and steelhead can tolerate the highest freshwater temperatures, up to 24°C; chum, the lowest, at around 19.8°C.) Increased temperatures would also speed the process of evapotranspiration and reduce levels of surface water, increasingly exposing salmon to predators; in turn, forests might rely more on groundwater for recharge, which could have the cumulative effect of lowering the water table and diminishing ecosystem productivity. ■

Draining into Russia's Sea of Okhotsk, the Inya River watershed is fed by a series of glacial lakes, including the 15-meter-deep Etergen Lake. In temperate forests, climate change could result in increased evapotranspiration and precipitation, affecting salmon productivity.

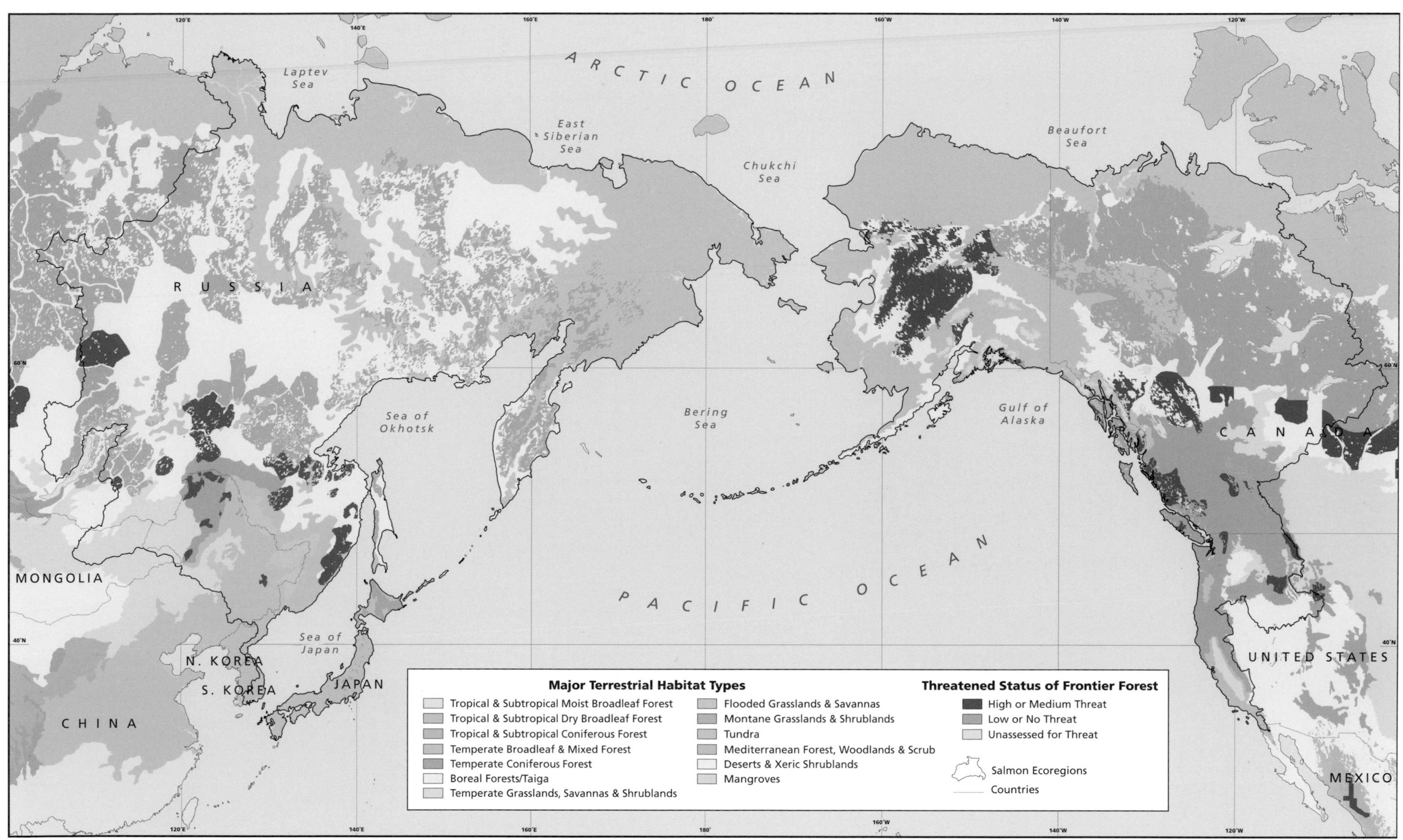


**LOGGING IN FRONTIER FORESTS** Few frontier forests (above, in red, orange, and yellow) remain in the southern half of the North Pacific; those that do are fragmented, relatively small, and at heightened risk for logging. Salmon in these regions are exposed to increased risk of extinction, as threats of urbanization and edge-of-range effects render populations most vulnerable. Canada hosts a comparatively high number of forests at elevated risk, as does southern Russia. Western Alaska contains among the largest swaths of at-risk frontier forest, where many salmon rivers have their headwaters.

# Mineral Development

**Although people have mined for natural** resources for thousands of years, hydraulic mining—the process of sifting through soil and gravel with pressurized water—was invented in 1853 in the United States. It propelled the industry forward and, within a few years, the ecological impacts were calamitous as tailings piled up on riverbeds.

Today mining can take many forms: surface mining (including open-pit mining and quarrying), in-stream mining, underground mining, placer mining (land based or floating), marine mining (including seawater, continental shelves, ocean beaches, and the seafloor), and solution mining (using brine and sulfur).

Mining uses massive amounts of water and energy to run machinery and transport materials and can erode soil and leach toxins into groundwater and surface water. For instance, in-stream gravel mining, a major industry in California, skims off the top layers of streambeds, destroying spawning areas and flooding downstream areas with siltation and pollutants.

The Environmental Protection Agency reported that the mining industry produced 46 percent of all toxins released by all industries in the United States in 2001. In the western United States, 40 percent of watershed headwaters are contaminated by mining waste. Among the leachates produced by mining each year are nearly a half-million kilograms of arsenic, acid, and heavy metals, including lead, cadmium, and mercury. In 2002 in Washington, Idaho, and California alone, metal mining disposed of or released more than 21 million pounds of toxics.

In the west, the effects of mining are of particular concern to salmon biologists from northern Khabarovsk territory to Chukotka, and all along the shores of the Sea of Okhotsk and the western Bering Sea. In these places, regulations are weaker and more difficult to enforce, and reclamation is not consistently required.

Abandoned mines continue to cause environmental damage after they are closed. In the United States, and largely in the western states, there are more than a half-million abandoned mines that are unmonitored. For example, even though Iron Mountain Mine in Shasta County, California, ceased its century-long operations in 1963, acid mine drainage from underground workings, tailings, waste, and open pits continues to pollute local waterways. A natural resources damage assessment conducted on-site estimated that, between 1981 and 1986, toxic releases that escaped from the mine and into the Sacramento River destroyed around 20 million fall-run chinook salmon.

## THE WATER EXTRACTION INDUSTRY

In the fall of 2002 the largest salmon die-off ever recorded in WOCI occurred on a 60-kilometer (36-mile) stretch of the Klamath River, killing one-quarter of the river's fall-run chinook. Each year, an average of 37 million cubic meters (30,000 acre-feet) of water is diverted for farming from the Klamath River to the Rogue River basin on the Oregon-California border. In the spring of 2002 the U.S. Bureau of Reclamation increased water diversions from the Klamath River to satisfy the needs of nearly 900 million square meters (220,000 acres) of farmland. Biologists attributed the die-off to these diversions; as flow and oxygen decreased, temperatures rose, pathogens multiplied, and salmon populations were concentrated in unusually shallow pools and channels.

Water diversions for farming and hydropower are among the most significant threats to salmon populations in WOCI—particularly in the San Joaquin and Sacramento Rivers and the tributaries that feed into them. These diversions modify natural stream flow and temperature patterns. California's Central Valley is one of the most altered water systems in the North Pacific. Naturalist Livingston Stone reported that "the year 1878 was the year of immense gathering of salmon in the McCloud [at the northernmost point of the Central Valley]. Indeed they were more numerous than I have ever known them to be . . . I have never seen anything like it anywhere, not even on the tributaries of the Columbia." Today the fall, winter, and spring chinook runs in the McCloud River, flooded during construction of the Shasta Lake reservoir, have been extirpated. All Central Valley salmon runs are listed as threatened or endangered or are candidates for listing. ■

Agricultural interests divert up to one-quarter of the volume of the Sacramento River each year to flood rice fields, such as those above in Willows, California. The $500 million California rice industry is second only to Thailand in premium rice exports.

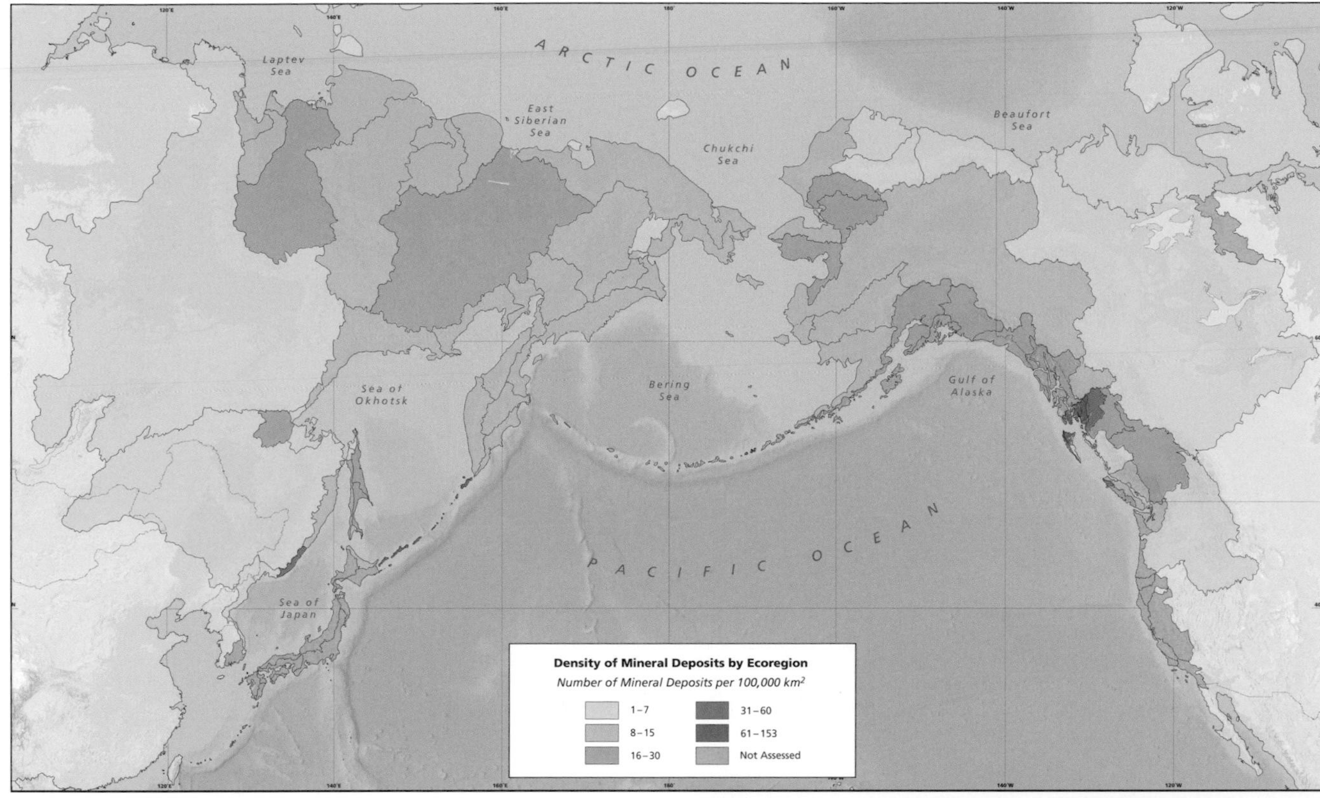


MINERAL DEVELOPMENT Comprehensive international datasets on mines are nonexistent. Location, size, and status of mines are debatable and difficult to verify. Instead we used a dataset of mineral deposits for the Russian Far East, Alaska, and Canada developed by the U.S. Geological Survey and then interpreted the density of deposits using our ecoregion template. At this time no such data exists for the continental United States or Asia.

In sum, mineral deposits are richest at the center of the salmon range, along southeast Alaska and the British Columbia coastline, and diminish outward. Pressures on salmon rivers from population, urbanization, and habitat loss manifest just south of this region.

From 1940 to 1952, a massive gold dredging operation near Stanley, Idaho, quite literally turned the Yankee River upside down. The salmon runs of this tributary deep in the Columbia River basin were destroyed.

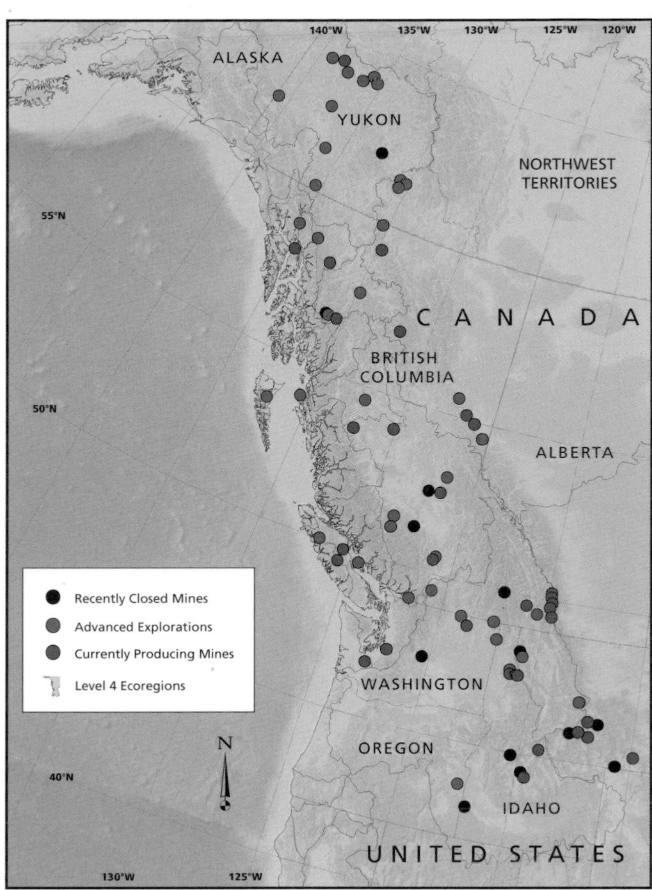


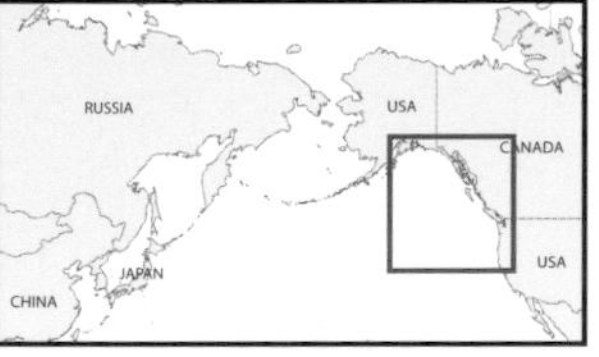


MINES: THE MICRO VIEW
As the map to the left demonstrates, British Columbia contains the highest concentrations of mineral deposits of any North Pacific jurisdiction.

Today British Columbia is actively mined and prospected. And, as the map above shows, many mines are located near major rivers. Past mining efforts can also pose threats to salmon because they are not maintained or overseen as diligently as are active sites.

NOTE: This map is presented in Lambert Azimuthal projection.

# Oil, Gas, and Pipelines

**Estimating oil and gas reserves is an** uncertain science at best: some experts say demand will outstrip supply as early as 2010; others extend the forecast a decade or two, or four. What is certain is that oil and gas are finite resources, and salmon—which can experience sharp declines at the hand of oil and gas development—can be an infinite resource, if ecosystems are adequately protected.

In the 1940s new technologies created the capability to drill for oil and gas off the outer continental shelf (OCS) off North America. Leasing, drilling, and exploration advanced in the 1950s and 1960s. But in the 1970s, following major accidents that caused significant ecological damage, the United States and Canada soon issued moratoria on OCS exploration. However, these areas were not granted permanent protection from drilling. Leases are up for renewal and the moratoria may be lifted at any time.

Yet oil and gas development continues at an ever-increasing pace elsewhere due to the world's growing energy needs. Today five Middle Eastern nations within the Organization of Petroleum Exporting Countries (OPEC)—Saudi Arabia, Iraq, United Arab Emirates, Kuwait, and Iran—control 75 percent of the planet's one trillion barrels of oil reserves. In order to develop energy independence, countries outside of OPEC—particularly the United States, which consumes about one-third of the world's oil, and China, which has the world's largest population of nearly 1.3 billion people—are advancing exploration.

The environmental threats of onshore and offshore oil and gas exploration, extraction, processing, and transportation are profound. Salmon populations confronting these threats may be severely damaged or extirpated. Oil and gas development causes noise pollution during seismic exploration; water and air pollution during drilling, extraction, and processing; and habitat fragmentation and destruction. Accidents are part of the process. The 10.8-million gallon (40.8-million liter) *Exxon Valdez* oil spill, in 1989, was North America's biggest, but it remains only the 37th largest in world history. Smaller spills occur daily; in 1998 a total of 60,000 gallons (227,000 liters) of oil spilled in 511 discrete incidents from platforms into U.S. waters.

New fuel technologies loom, some with unknown implications. Major oil and gas corporations are exploring methane hydrates, gas trapped in ocean floor sediments, which are estimated to contain twice the carbon of all the world's fossil fuels combined. Drilling could cause ocean-floor landslides; large-scale releases could trigger major and rapid climate change.

In addition to hosting seven species of *Oncorhynchus,* Kamchatka is home to rich natural gas reserves, which are undergoing exploration.

Thirteen-hundred kilometers long and 120 centimeters wide, the Alaska pipeline has transported 11 billion barrels of oil since 1977.

Per well, oil drilling generates an average of 680,000 liters of drilling muds, including heavy metals such as lead and mercury.

Sakhalin I and II are the most active and expansive oil and gas developments in the Russian Far East, but dozens of smaller projects are also underway in the Sea of Okhotsk and the Bering Sea. Data on oil and gas reserves across the Pacific Rim are often proprietary and therefore quite difficult to map.

Note that in the map to the right, the region in yellow denotes salmon presence in the path of the pipeline.

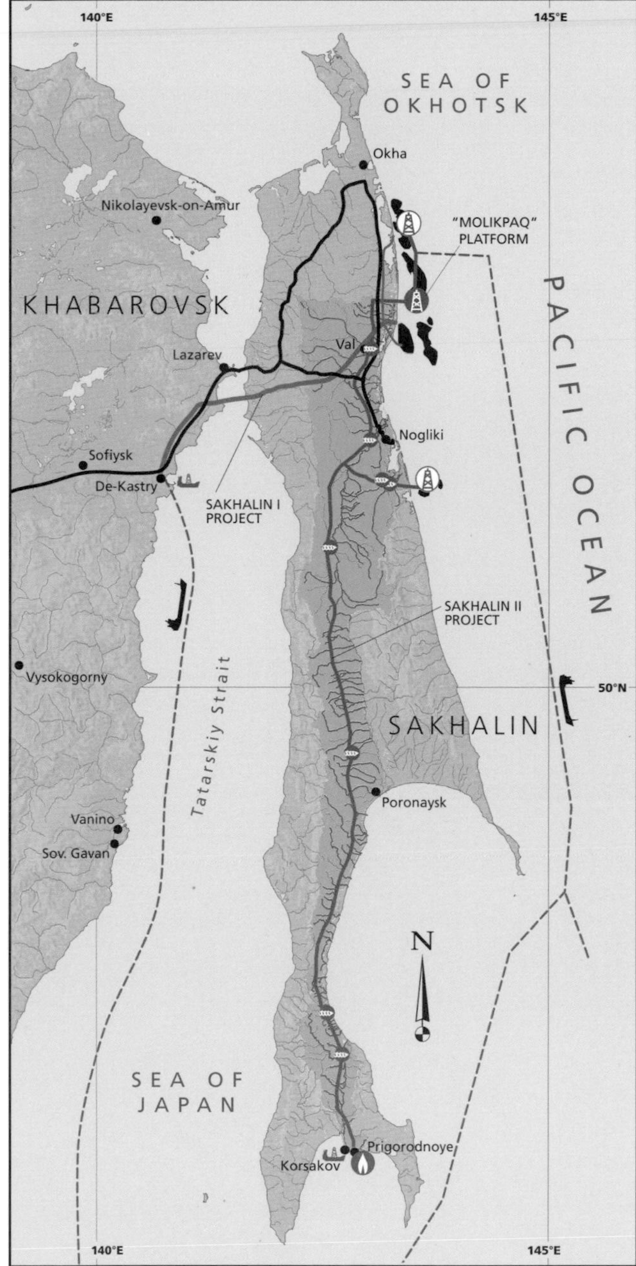


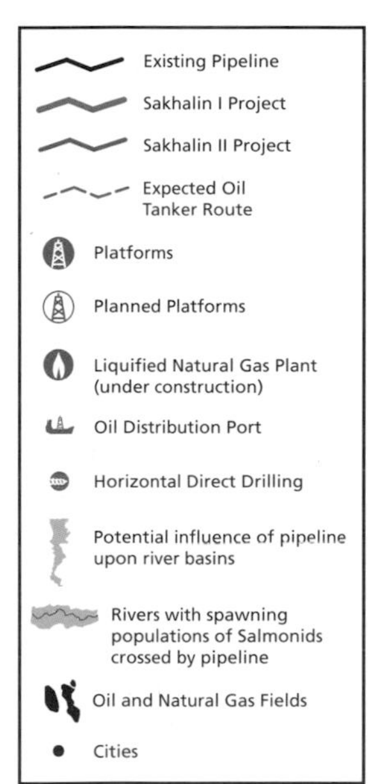


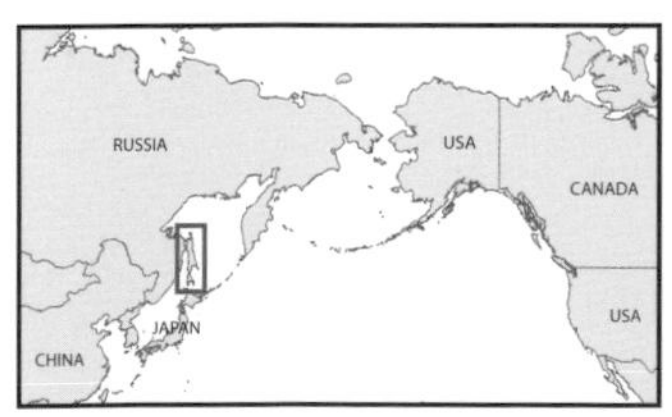


## SAKHALIN PIPELINE

The continental shelf off the Russian Federation is experiencing heightened scrutiny by oil and gas interests, particularly in the Sea of Okhotsk and the Bering Sea—places of extreme importance to Pacific salmon populations. Nearly four dozen sites, including some off the western coast of Kamchatka, are scheduled for exploration off Russia's shore by 2005. Sakhalin is presently experiencing activity on the largest scale.

The continental shelf off Sakhalin and in the Sea of Okhotsk are among the world's richest and as-yet-untapped stores of oil and natural gas. Estimated at around 7.9 billion metric tons, Sakhalin deposits are more than twice the total of U.S. and Chinese deposits combined. Development is underway. ExxonMobil and Royal Dutch/Shell are pushing ahead with two major extraction efforts; five more are in the making. Sakhalin I will include a 225-kilometer oil pipeline bridging Sakhalin and the mainland and offshore oil platforms. At US$12 billion, it is the largest foreign investment in Russia, but it pales in the shadow of much-anticipated Sakhalin II. This project will include two adjacent 800-kilometer pipelines that bisect Sakhalin from north to south.

In Sakhalin, where salmon are relatively healthy, one-third of the economy relies on fishing. The island is near the center of the range for masu as well as anadromous Sakhalin taimen, the most ancient surviving salmonid species. Pipelines for Sakhalin II will cross more than 1,100 salmon rivers and traverse up to four dozen major seismic faults. Pipeline construction will result in siltation from trenching, increased poaching access, and heightened risk of spills and leaks in one of the world's richest salmon regions. ■

A freighter transports fuel in Russian waters. Ecological threats posed by oil and gas development extend beyond exploration and drilling to the post-production transportation.

# Dams

**The World Commission on Dams,** established by the World Bank and IUCN-The World Conservation Union in 1998, reports the existence of at least 45,000 large dams (at least 15 meters high or with reservoir volume greater than 3 million cubic meters). Half the world's extant dams were built primarily for irrigation and contribute to 15 percent of world food production. Hydropower, the other major use of dams, currently provides around 20 percent of the world's energy supply.

In North America dams are concentrated near population centers and increase in density southward. Alaska has around 160 dams; British Columbia, around 2,500 dams. Across the United States, the U.S. Army Corps of Engineers keeps track of 76,000 dams of varying sizes. The major dam-building era in U.S. history spanned the half century following the Great Depression, but the 1973 Endangered Species Act essentially rendered large dam building financially infeasible.

If North America has run the gamut of major dam building, Japan may just now be at the apex. According to the International Rivers Network, virtually all of Japan's rivers have been dammed, largely for flood control. More than 2,600 dams exist today, and 350 more are planned. But since 1988, the growing anti-dam movement—which mobilized in opposition to the Nagara River Estuary Dam, sited on Japan's last wild river—has succeeded in blocking 80 projects.

Water-rich Russia contains 10 percent of the world's total mean annual river stream flow, and hydroelectric interests are looking here for future projects. Russia has just 90 large dams. New dams have had significant impacts on salmon populations in two major Amur River tributaries: the Zeya River Dam, one of the largest storage dams in the world, and the in-progress Bureya River Dam. The Chinese are looking to the Amur River to fulfill massive energy needs.

As technology has advanced, the benefits of dams to humans are increasingly being weighed against social and economic challenges. As many as 80 million people have been displaced by reservoirs; countless more have absorbed cost overruns in dam construction. The atmosphere is at risk as well: scientists estimate that over a 100-year period, reservoir releases of carbon dioxide and methane together contribute to around 7 percent of the global warming impact.

However ambiguous the cost-benefit analysis may be for humans, dams have been disastrous for salmon. In the Columbia River basin, for example, dams have devastated salmon runs, and recovery costs of around US$600 million annually have done little to mitigate or reverse losses.

At the Bonneville Dam on the Columbia River, a counter tallies salmon migrating to natal spawning grounds or hatcheries.

Upstream, wild salmon smolts are barged by the thousands to bypass the Lower Granite Dam on the Columbia River.

Fish screens prevent salmon from entering irrigation canals. Before lowering screens into the river, workers check them for salmon.

## MAJOR DAMS OF THE COLUMBIA RIVER

The Columbia River basin, the fourth-largest river basin in North America, is the most hydroelectrically developed river system in the world. In 1933 the Rock Island Dam, just east of Seattle, was the first major dam constructed in the Columbia River basin. Among the areas inundated was Kettle Falls, the native fishing ground of the Confederated Tribes of the Colville Reservation, which saw harvest numbers plummet from 1,333 fish in 1929 to 159 in 1934. It was a sign of things to come. The last mainstem dam erected was the Mica Dam in British Columbia, in 1973. In the interim, the Columbia River and its main tributary, the Snake River, have become home to an interdependent network of 150 hydroelectric projects, 250 reservoirs, and 18 mainstem dams.

This level of industrialization has resulted in widespread extirpations of salmon populations. In the Columbia River ecoregion an estimated 11 sockeye, 13 coho, 33 chinook, 2 chum, and 14 steelhead populations have been extirpated from their historic range as upriver extents were cut off with the construction of dams.

Dozens of Columbia River basin dams have had an impact on sockeye salmon populations. In 1941, when the Grand Coulee Dam opened without fish passage facilities, 1,084 kilometers (674 miles) of the Columbia River were immediately cut off from sockeye populations heading to native spawning grounds. The John Day Dam tells a similar story: it opened in 1968 just before the fish ladder was completed; estimates of the total sockeye kill amounted to more than 200,000. Today 21 of the Columbia River basin dams within the sockeye distributional range have no fish passage capacity. ■

Moving salmon around dams by barge is stressful to smolts. Here a scientist counts mortalities following barging on the Columbia River.

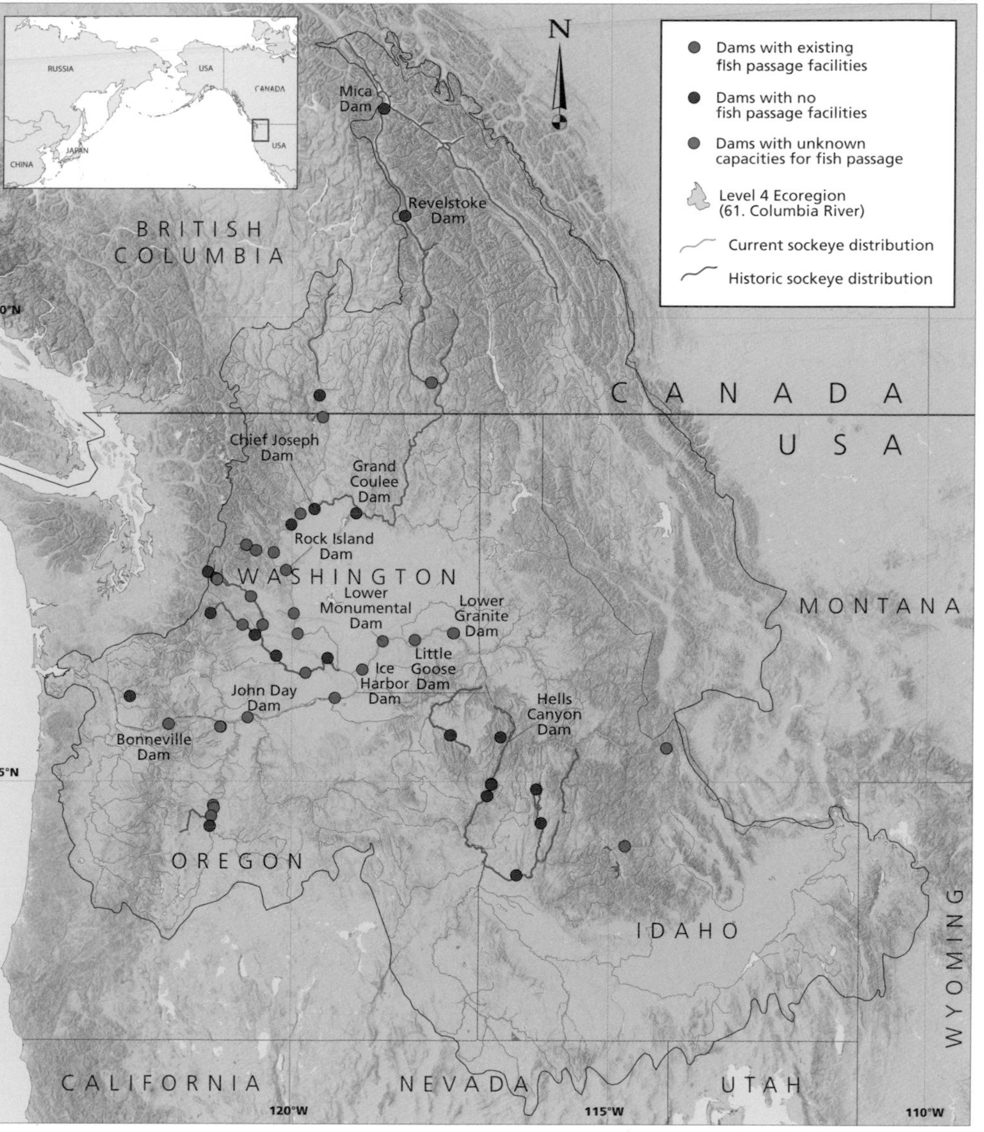


**DAMS** Using best available data, we identify in the map above all dams that sockeye salmon encounter in the Columbia River basin, including major mainstem dams. Among all species of *Oncorhynchus*, sockeye have been hit hardest by Columbia River basin dams, with an estimated 11 populations extirpated; 3 of the 4 sockeye populations that we have identified in this ecoregion are at risk of extinction.

NOTE: This map is presented in Lambert Azimuthal projection.

# Climate Change

**Global climate change will affect—** is already affecting—salmon populations throughout their entire range; the extent of the changes remains unknown.

The map on the facing page, a springboard for discussion, was built on the conservative assumption put forth by the Intergovernmental Panel on Climate Change: if carbon dioxide emissions remain constant at 1994 levels, by the end of the 21ST century, the atmospheric accumulation of greenhouse gases will be twice what they were in preindustrial times. Within this basic scenario, the Pacific Rim coastline will look vastly different. At one end of the spectrum, increased snowmelt and precipitation in more mountainous regions at higher latitudes will increase stream runoff and shift seasonal runoff patterns; at the other end of the spectrum, in regions with desert features, heightened evapotranspiration will reduce stream flows. Therefore, in the late winter and early spring, for example, the Arctic coast of the Russian Far East, southeast Alaska, British Columbia, and WOCI may experience accelerated snowmelt and higher stream levels; and, with few exceptions, the western coastline of the Pacific Rim will experience reduced runoff.

This model represents average levels of stream runoff in a given year and is a generalized picture; more poignant is what climate change will do to stream levels throughout the course of one year. When peak runoff occurs earlier in the year, seasonality may shift by a number of weeks, and this will have a profound impact on salmon runs. Salmon egg incubation periods may shorten, water velocity will increase, and average water temperatures may decrease. But as the season progresses, some areas may experience an extended period of summer drought as low stream flows overlap with periods of high evapotranspiration. With low water levels and high temperatures, some rivers may become inhospitable to salmon. Shifts in fall and winter flood time and intensity may serve as obstacles to spawning salmon. Salmon are particularly sensitive to habitat modifications; even slight changes due to global warming can affect populations.

Scenarios range from mild to extreme. Catastrophic climate change, the subject of a recent Pentagon study, examines scenarios influenced by threshold changes and break points. For example, increased ice and snowmelt has the potential to flood the ocean with freshwater, which could significantly shift the hydrological balance. Cumulative changes in the ocean heat transport system could radically alter drainages and potentially trigger another ice age.

## THE SCIENCE OF CLIMATE CHANGE

The science of climate change is constructed around the certainty that greenhouse gas emissions will continue to accumulate in the atmosphere and cause the earth to warm. Around this constant revolve countless variables, including changes in policy, physics, economics, population, hydrology, geology, precipitation, and types of emission concentrations. Therefore, the effort to assess effects of accumulating greenhouse gases is constantly evolving and can produce very different results. For example, a recent study of climate change impacts on California worked with two models: one found that heatwaves will be four times as frequent and that related deaths will increase by a factor of three; another found that heat waves may be as much as eight times as frequent and that mortalities could increase by a factor of seven.

Models for analysis are constructed largely around two key issues: what will future concentrations and emissions of greenhouse gases be, and how will the earth's climate respond to these gases. Beyond these two central assumptions are a host of unknowns. To build a model, scientists must consider the physics of atmosphere and climate. They test variables for temperature and humidity and conservation laws for mass momentum and energy; they measure the extent to which sea ice reflects sunlight and how currents alter energy transports between the ocean and the atmosphere. Policy choices, emissions scenarios, and human population growth fuel countless inputs. Most of the science around global warming is predicated upon great uncertainty, with the effects on marine systems and cloud cover still largely unknown. Model results produce thousands of scenarios, which may be sensitive to changes in even a single variable. ■

In addition to modeling, the science of global warming is based on data analysis of historic and present climate regimes. Installing a probe pipe in 1980, this geophysicist measured permafrost temperatures in Alaska over 15 years, documenting an increase from 0.5° to 1.5°C.

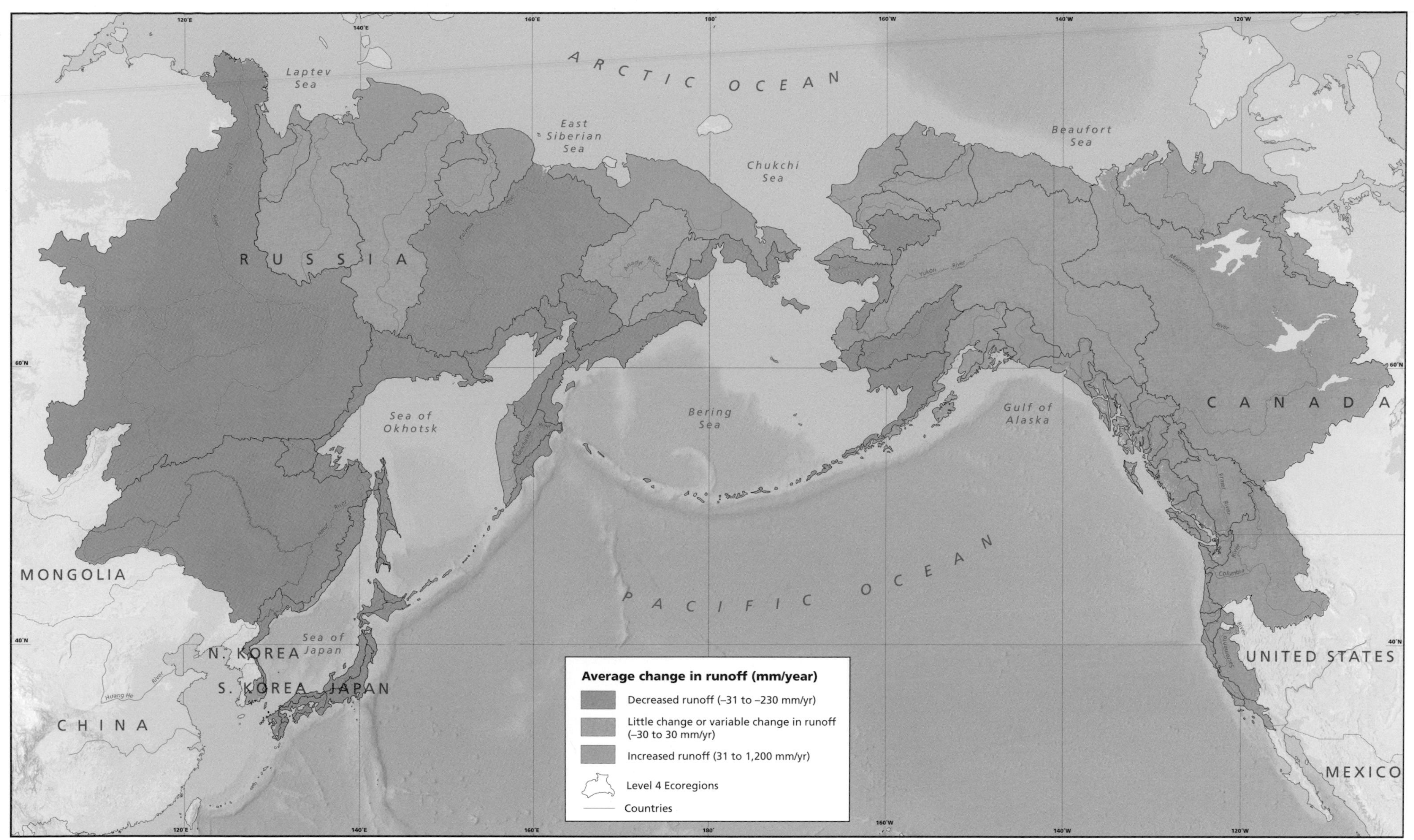


CLIMATE CHANGE The extent and the effects of global climate change remain hypotheses that scientists are exploring through modeling. Assuming doubling of $CO_2$ levels by the close of this century, the hydrological cycle of the North Pacific will have fundamentally changed. At the center of the range for salmon, in the Gulf of Alaska and along the British Columbia coastline, runoff will increase, causing river flows and flooding incidents to increase and temperatures to decrease. The Bering Sea, however, will suffer from decreased freshwater inputs. The western Pacific will become drier and hotter, with the exception of the Arctic coastline.

# 6

# Migration Ahead

*A North Pacific Ecosystem Approach*

Fisherman Bobby Begay cradles a salmon caught upriver from the Dalles Dam, which flooded Celilo Falls, his ancestral fishing grounds, in 1957.

**For millions of years, Pacific salmon** have survived myriad natural challenges, from the expected (e.g., interannual changes in food availability, water quality, sea surface and stream temperatures, spawning habitat, sea ice) to the calamitous (from glaciation and earthquakes to ice ages and volcanic eruptions). As powerful as these obstacles may have been, salmon populations over the millennia have had time to recover, in periods measured not by annual harvest returns or quotas but by the self-paced process of natural selection, marked by dozens or even hundreds of salmon generations that adapted to changing ecological conditions through varied life history strategies.

Pacific salmon in the 21ST century do not have this luxury. While natural threats have remained generally consistent over time, human threats just within the past 150 years have intensified and continue to do so at an accelerated pace.

To measure the extent of these threats and begin to devise effective long-term remedies, we need to understand the moving picture: the trends in salmon abundance, diversity, distribution, and productivity. But before we can do that, we must first gather the still images: the status of salmon within our current North Pacific inventory and data we need to enrich our information state.

We believe that our ecoregional approach offers a useful framework from which to view and study Pacific salmon at a consistent resolution and affords us a new perspective on the vast and vital role of salmon throughout their range. It has enabled us to discern coarse-scale spatial patterns in distribution: that chum have the broadest presence, that masu have the most limited range. Furthermore, we can begin to detect patterns in risk of extinction: that species demonstrate a clear north-south cline throughout their range across the North Pacific; that WOCI ecoregions have the highest concentrations of extinct stocks; that ecoregions in the Sea of Okhotsk and the western Bering Sea have stocks largely at moderate risk; and that the western Pacific has more ecoregions classified entirely in the low-risk category than does the eastern Pacific.

Sockeye swim through the Kennedy River on Vancouver Island in British Columbia. The coastlines in this region are home to the greatest concentration of fish farms in North America.

Fish brokers examine salmon at Tokyo's Tsukiji Market, which sells around 1.5 billion yen (US$14 million) in marine products daily.

Our ecoregional framework also affords us a fuller context in which to study Pacific salmon. *Oncorhynchus* inhabit an enormous and varied territory that crosses political, terrestrial, and geographic boundaries, occupying 4 percent of the world's landmass. The jurisdictional perspective that North Pacific nations have imposed on salmon stocks to increase catch bears little resemblance to the way that salmon use the ecosystem—which is how native peoples have related to salmon for millennia. Within a legislative context, our hard-won, land-based conservation efforts and harvest-directed management efforts may be entirely unrelated to the viability of habitats used by salmon.

Most importantly, our ecoregional framework has allowed us to determine that nearly one-quarter of all North Pacific salmon populations we assessed—a group that represents at best around 10 percent of North Pacific salmon populations—are at moderate or high risk of extinction. These findings reveal just how much more work we have before us.

## IMPROVING OUR KNOWLEDGE BASE

Having catalogued the information that we have, we need to begin to gather the information we lack. The most gaping hole that we presently face in our current knowledge base is data collected at a uniform scale and in a standardized format. Although the distribution and risk of extinction data we have collected for this atlas are roughly representative at the broad North Pacific scale, the information breaks down at finer resolutions where salmon are managed at a local or regional scale. Current salmon monitoring is patchy at best. Agencies operate independently and often without coordination. Scientists work in isolation and use different methods to gather various kinds of data. Data are often unavailable, privileged, or perhaps not comparable with other datasets. As a result, we can't discern whether population changes are due to localized problems, cumulative effects, or larger stressors.

Eggs are stripped from a chinook for the Elwha hatchery, constructed to mitigate losses in wild populations affected by two nearby dams.

Furthermore, salmon monitoring programs have been typically designed to collect data to inform commercially based catch management decisions, not to protect biodiversity for salmon conservation or recovery. Data on salmon are derived from sampling designs ranging from convenience to probabilistic, therefore yielding no comprehensive view of stock status or trends.

We need to establish a sampling effort that yields data at the same resolution across the North Pacific. To characterize status and trends across the natural range of anadromous *Oncorhynchus*, to forge a useful tool for early detection of problems in the sustainable use of salmon and their ecosystems, we need a cooperative international monitoring strategy. It will provide metrics by which we can measure salmon status and improve fisheries management throughout the North Pacific. As the nations of the Pacific Rim confront risk of extinction in the salmon populations that have sustained us for thousands of years, we must pool our efforts and collaborate across borders to understand trends and craft solutions.

Salmon researchers are asking pertinent questions as they design an international salmon monitoring strategy. What are the key drivers affecting species distribution, abundance, diversity, and productivity in oceans and fresh waters? Can we identify as conservation priorities the critical regions of genetic diversity? How much salmon can we harvest for human use without altering a river basin's productivity? How do we evaluate these variables in a warming world?

Drummers issue farewells to the first salmon used during the First Salmon ceremony in Tulalip, Washington.

The answers to these and other weighty questions will shape how we collect data across the vast North Pacific ecosystem, which will help us ensure the continued viability of wild salmon and their ecosystems.

## SCIENCE AND POLICY IN CONTEXT

As we work toward predicting salmon productivity and viability, we need to understand the natural drivers in the North Pacific that affect escapements seasonally and interannually. How do sea surface temperature and primary productivity vary from year to year? To what extent do salmon respond to changes in sea ice melt and phytoplankton blooms, which support the zooplankton on which salmon feed? How

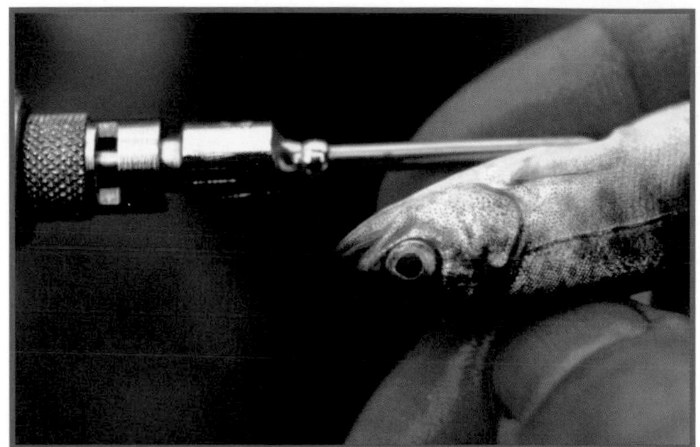

A passive integrated transponder (PIT) tag, a microchip the size of a grain of rice is inserted into the belly of a wild chinook salmon smolt.

do responses to variability affect patterns of salmon ocean rearing and migration? How do Pacific Decadal Oscillation, currents, and climate variability influence seasonal salmon growth and patterns in migration and overall abundance? How do all these factors work together?

We also must understand the extent to which human-made drivers influence salmon. Urbanization concentrates at the southern extremes of the range, where salmon have suffered the most extirpations. Yet in less densely populated regions, the threat of natural resource extraction can be high.

In the face of global climate change, human communities must adapt our uses of natural resources, including salmon. Reliance on fossil fuels has broad implications for salmon habitat: overnight shipping has created an international, year-round salmon market that devalues wild salmon and removes us from our consumer relationships with local salmon populations. Among the key questions we must consider: can we predict the variable productivity of specific areas in the ocean? How will they affect particular salmon populations, species by species? How do we anticipate unpredictable shifts in stream runoff in a warming world?

Specific threats also loom large. Gas exploration may soon transform Hecate Strait and the Sea of Okhotsk; poaching in Russia endangers or extirpates many salmon populations each year; logging in British Columbia and urbanization in WOCI diminish salmon habitats.

The effects of artificial salmon propagation, in the form of hatcheries or fish farms, on wild salmon populations and their freshwater and marine habitats, remain inconclusive and poorly understood. As the world's major consumer of Pacific salmon, Japan is driving a world market that expects biomass to increase. Alaska is responsible for more than 40 percent of the North Pacific salmon catch; Russia and Japan together contribute around one-half. Can these major producers maintain harvest levels, or will the market become increasingly internationalized and dependent on artificial propagation?

Managers of hatchery populations will need to adapt their policies in order to maintain the viability of their fishing industry—particularly given changes in international salmon markets brought about by the farmed fish industry. For example, as they seek to find a balance between hatchery-rearing and natural production, managers will need to consider questions around the relative economic and ecological efficiencies of low-value species such as pink salmon. And they will need to address changes in ocean carrying capacity associated with global warming that may limit the ocean's ability to accommodate increasing numbers of hatchery fish without adversely affecting natural salmon populations.

## TOWARD A NORTH PACIFIC ECOSYSTEM APPROACH

Many farsighted and eloquent treatises have spoken to the need for salmon and salmon ecosystem conservation, and several come to the foreground. In *Salmon Without Rivers* fisheries biologist Jim Lichatowich urged us to change our relationship with salmon from one of dominance to one of tolerant and patient coexistence: if we help salmon save themselves, we will save habitat, culture, and economy as well. In its 1986 report *Return to the River,* the Northwest Power Planning Council's Independent Science Advisory Board predicated healthy salmon runs on the ability to reclaim natural river conditions. Journalist Terry Glavin, in his 2003 "Strategy for the Conservation

Russian fisherman row toward a salmon setnet near Poronaysk, on Sakhalin Island, at the mouth of Terpeniya Bay.

Salmon fight over food and spawning rights, when aggression can be especially intense. Above, chum battle in a stream near Juneau, Alaska.

of Pacific Salmon," pressed Canada's federal government to take an ecosystem approach in maintaining and protecting populations. The National Research Council's "Upstream" report concluded that salmon survival relies on genetic diversity and recommended habitat rehabilitation and conservation at the watershed level to maximize genetic diversity`

We echo these recommendations and place them within our North Pacific ecosystem. We need to reclaim rivers and collaborate at the catchment level, while reshaping our relationship with salmon across jurisdictional boundaries.

In this book, we have improved our understanding of the status of North Pacific salmon using a panoramic perspective, and we encourage further study through this ecoregional lens. We believe that we can protect and restore our native salmon and the food webs they support through a four-step international approach—a North Pacific Ecosystem Approach—to salmon conservation as follows:

### MONITORING

*First, we must collaborate to improve and standardize our monitoring systems.*

We need a comprehensive strategy that will enable us to inventory and monitor the distribution, diversity, abundance, and productivity of wild salmon populations.

Until we can conduct baseline status and trends analysis with data collected at a common scale across the North Pacific, we cannot know if we are succeeding or failing in our efforts to protect and restore wild salmon populations. A comprehensive international monitoring strategy will enable scientists to standardize data collection methods and provide information at a uniform scale so that we can resolve trans-Pacific and regional trends in risk of extinction, threats, and management practices.

### MANAGEMENT

*Second, we must improve our ecological understanding of salmon and their habitats and use it to inform management decisions.*

We need to make fundamental changes in the way we manage fish and shift harvest and hatchery practices from managing for biomass to managing for biodiversity.

We must establish and maintain conservation goals, in every basin, that sustain genetic and life history diversity as well as nutrient loads for entire watersheds by

- advancing genetics research to identify and define the key salmon populations across the Pacific Rim that represent evolutionarily significant building blocks for the species;
- establishing policies that ensure temporal and geographic segregation of hatchery stocks from wild salmon populations;
- implementing escapement-based management that recognizes the critical nutrient-transfer role of spawning salmon in freshwater systems.

### CORE HABITATS

*Third, we must improve protections for core habitats and our understanding of them.*

We need to identify and protect the last remaining and most productive, species-rich salmon ecosystems in each salmon ecoregion. At the same time, we must restore the rest through aggressive local programs.

We must save the large building blocks of genetic and life history diversity within salmon populations by supporting the restoration of endangered stocks while moving aggressively to save the salmon populations and river ecosystems that are not yet severely degraded.

Therefore we propose the creation of

Data are collected differently across the Pacific. Russians, for example, do not have access to the American equipment shown above.

salmon protected areas for wild salmon in every ecoregion—optimally a network of at least several subbasin to basin-level watersheds. This will ensure that we safeguard the full range of salmon biodiversity. This network of salmon protected areas will form the building blocks for salmon conservation and recovery over a period of several hundred years. Specifically we need to

- create a system of observatory rivers where we can improve our understanding of natural patterns and processes;
- refine and codify salmon ecoregion boundaries;
- identify the remaining healthy native salmon strongholds in each salmon ecoregion;
- create a broad network of protected areas for salmon populations within each salmon ecoregion identified through consensus among scientists, conservationists, landowners, and legislators.

A biologist conducting field research snorkels down an Idaho creek looking for wild chinook salmon smolts.

### PARTNERSHIPS

*Lastly, we must form partnerships to enrich our knowledge, resources, and efforts on behalf of wild salmon protection.*

We need to reform and strengthen the human institutions that protect salmon and their habitat—from citizen watershed councils to international commissions and development agencies—so that they will be more accountable and increasingly effective.

Scientists, conservationists, fisheries managers, legislators, entrepreneurs, citizens, and activists must combine efforts and skills on behalf of wild salmon. At the regional level, we need to build institutional capacity in each river basin. At the national level, we must increase the capacity of government agencies, tribal organizations, and regional councils. At the international level, we need to establish information networks so we can learn from each other and adopt treaties that set standards for monitoring, management, conservation, and restoration across borders.

Specifically we need to

- build international alliances and partnerships with salmon conservation organizations that emphasize strategic, place-based efforts to protect intact populations and ecosystems, in order to guarantee future representation of multiple genetic legacies and life history strategies;
- amend the Convention for the Conservation of Anadromous Stocks to create an international body to oversee habitat conservation provisions for wild fish and monitoring of Pacific salmon distribution, diversity, abundance, and productivity;
- support programs that emphasize international research collaborations, including the North Pacific Anadromous Fish Commission (NPAFC), the North Pacific Marine Science Organization (PICES), and the Pacific Salmon Commission, to improve our understanding of the biological complexity of salmon populations and their marine and freshwater ecosystems;
- create infrastructure and facilitate funding for data-sharing across the North Pacific, including the communications needs implicit in such activities;
- invest in responsible management and provide economic incentives that strengthen and support fully functioning salmon ecosystems;
- formulate a sustainability index that gauges progress of local stewardship efforts within salmon-river ecosystems.

Spawning salmon can exhibit aggression in protecting territory. Here, a female sockeye chases another adult away from her redd.

## SHARED RESPONSIBILITY

Across a spectrum of interests and across various landscapes and cultures of the North Pacific, we share salmon. We share in the responsibility we must take for the obstacles we have put in their path. We have in common the rivers, oceans, and land that salmon use in their lives and fortify in their deaths. We have in common occupations, cultural traditions, hobbies, and need for nourishment—all of which tie us to salmon. We have in common communities, traditions, and economies that need healthy and abundant salmon and habitats that support us.

So we must share in the solutions. Our discrete efforts must be synchronized. The body of information must be assessed by all of us, and we must take into account our very different perspectives. The North Pacific Ecosystem Approach we present will enable us to move toward informed policy progress and gain a substantially broader constituency for change as we revise our relationship with natural places in energy, forests, farms, urban development, communities, and international markets. So equipped, we can confront the dominant issues that play out on a localized basis: poaching in the Russian Far East, river "hardening" in Japan, the possibility of increased genetic homogeneity in hatcheries, accelerated logging rates in British Columbia, and urbanization in WOCI.

A subsistence fisherman on the Yukon River, near Tanana, Alaska, checks his fishwheel for the day's catch.

Increasing our knowledge base will prepare us for a future that is full of changing variables and unknowns. The threats presented in the previous chapter paint a static picture, without any hint of the reach of catastrophic human error. Global climate change is perhaps the greatest threat looming in our future. Although projections of rising sea levels, stream temperature changes, and unpredictable weather patterns may be gradual, we recognize that human-inflicted environmental change rarely manifests in modulated steps. The North Pacific climate is a unique and ever-changing balance of energy and matter, including solar sunspot activity, nutrients, polar ice melt, currents, storms, temperature, wind, and sea surface temperature. A slight shift in one factor may have a disproportionate effect on countless others. And while we may anticipate gradual transitions, we need to prepare for sudden climatic changes.

From the intimately local to the broadly international, our path must mirror the salmon's. We must work at every level of a salmon's life history, from the aerated gravel underlying streams and tributaries to the oceans and then back again. Most importantly we need to modify our own life histories, reframe our relationship to the natural world, change how we get our food, water, and energy, and reinvent how we build our houses and cities. We must assess our approach to land and river management and reevaluate our understanding of how landscapes function. If human-imposed threats are mitigated in economically sustainable ways, and if our science is sound and our management is adaptive, we can give salmon a better chance at recovery.

Many rural Alaskans rely on subsistence fishing for survival, today, as they have for thousands of years. Above, a woman hangs split chum to dry in preparation for canning.

With a North Pacific Ecosystem Approach, acknowledging that all pieces are part of the greater whole, we can help salmon by improving our own relationship with the world; we can fortify our watersheds by attending to the landscape; and we can help the economy by heeding the needs of ecology. Salmon exhibit a multitude of life history strategies that enhance their chances for survival; so should we.

# Sourcing

## FRONT MATTER

### Forewords

World Wildlife Fund (WWF). *The Status of Wild Atlantic Salmon: A River by River Assessment.* WWF Norway, May 2001.

## FRONT MATTER MAPS

### The North Pacific

U.S. Department of the Interior, U.S. Geological Survey. Dataset: GTOPO30: Global 30 Arc Second Elevation Data Set, 1996. http://edcdaac.usgs.gov/gtopo30/gtopo30.asp.

### Major Stream Networks

Environmental Systems Research Institute. Digital Chart of the World. [machine readable data file]. ArcView format. Redlands, CA: Environmental Systems Research Institute, [1992].

## CHAPTER 1

Allendorf, Fred, and Fred Utter. "Phylogenic Relationships among Species of *Oncorhynchus*: A Consensus View." *Conservation Biology* 8, no. 3 (1994): 864–67.

Augerot, Xanthippe. "An Environmental History of the Salmon Management Philosophies of the North Pacific: Japan, Russia, Canada, Alaska and the Pacific Northwest United States." PhD diss., Oregon State University, 2000.

Augerot, Xanthippe, Dan Bottom, and Jeff D. Rodgers. "Risk of Extinction of Salmon Stocks around the North Pacific." Unpublished paper. Wild Salmon Center and Oregon Department of Fish and Wildlife, 2004.

Brown, Lloyd A. *The Story of Maps*. New York: Dover Publications, Inc., 1979. First published 1949 by Little, Brown and Co.

Cederholm, C. J. *Pacific Salmon and Wildlife—Ecological Contexts, Relationships, and Implications for Management*. Special Edition Technical Report. Olympia, WA: Washington Department of Fish and Wildlife, 2000.

Chereshnev, Igor' A. "Puti i faktory formirovaniia taksonomicheskogo i tipologicheskogo biologicheskogo raznoobraziia presnovodnoi ikhtiofauny Severo-Vostoka Rossii." [Means and Factors of the Formation of Taxonomic and Typologic Biodiversity of Freshwater Ichthyofauna of Northeastern Russia.] Chap. 4 in *Biologicheskoe raznoobrazie presnovodnoi ikhtiofauny Severo-Vostoka Rossi* [Biodiversity of Freshwater Ichthyofauna of Northeastern Russia]. Vladivostok, Russia: Dal'nauka, 1996.

Gall, G. A. E., and P. A. Crandell. "The Rainbow Trout." *Aquaculture* 100 (1992): 1–10. Quoted in R. Froese and D. Pauly, eds., "*Oncorhynchus mykiss*." *FishBase*. http://fishbase.org/Summary/SpeciesSummary.cfm?ID=239&genusname=Oncorhynchus&speciesname=mykiss.

Groot, Cornelius, and Leo Margolis, eds. *Pacific Salmon Life Histories*. Vancouver: University of British Columbia Press, 1991.

Groot, Cornelius, Leo Margolis, and W. C. Clarke, eds. *Physiological Ecology of Pacific Salmon*. Vancouver: University of British Columbia Press, 1995.

Gustafson, R. G., Robin S. Waples, J. M. Myers, G. J. Bryant, O. W. Johnson, and L. A. Weitkamp. "Pacific Salmon Extinctions: Lost Diversity, Populations, and ESUs." NMFS draft paper, 2003.

Hilborn, Ray. "Apparent Stock-recruitment Relationships in Mixed Stock Fisheries." *Canadian Journal of Fisheries and Aquatic Science* 42 (1985): 718–23.

Huntington, Charles, Willa Nehlsen, and Jon Bowers. "A Survey of Healthy Native Stocks of Anadromous Salmonids in the Pacific Northwest and California." *Fisheries* 21, no. 3 (1996): 6–14.

Kuzishchin, Kirill V., Sergei V. Maksimov, V. Y. Upriamov, V. K. Larin, Natal'ia V. Varnaskaia, and Guido R. Rahr. "K probleme ustoichivogo ispol'zovaniia rybnykh resursov Zapadnoi Kamchatki: opredelenie rechnykh basseinov, prioritetnykh dlia sokhraneniia bioraznoobraziia lososevykh ryb." [Regarding Sustainable Use of Western Kamchatka Fish Resource: Identification of Priority River Basins for Salmonid Biodiversity Conservation.] In *Doklady vtoroi Kamchatskoi Oblastnoi nauchno-prakticheskoi konferentsii, Problemy okhrany i ratsional'nogo ispol'zovaniia bioresursov Kamchatki*. [Papers of the Second Regional Scientific-practical Conference, Issues of Protection and Rational Use of Kamchatka Bioresources], 35–40. Petropavlovsk-Kamchatskii, Kamchatrybvod and Kamchatsky Pechatny Dvor', Russia, October 3–6, 2000.

Levin, Phillip S., and Michael S. Schiewe. "Preserving Salmon Biodiversity." *American Scientist* 89 (2001): 220–27.

Lichatowich, James A. *Salmon Without Rivers: A History of the Pacific Salmon Crisis*. Washington, DC: Island Press, 1999.

Long, John A. *The Rise of Fishes*. Baltimore: The Johns Hopkins University Press, 1995.

McKay, Sheldon J., Robert H. Devlin, and Michael J. Smith. "Phylogeny of Pacific Salmon and Trout Based on Growth Hormone Type-2 and Mitochondrial NADH Dehydrogenase Subunit 3 DNA Sequences." *Canadian Journal of Fisheries and Aquatic Sciences* 53 (1996): 1165–76.

Montgomery, David R. *King of Fish: The Thousand-Year Run of Salmon*. Boulder: Westview Press, 2003.

Narver, David W. "Review of Salmon Management in British Columbia: What Has the Past Taught Us?" In *Sustainable Fisheries Management: Pacific Salmon*, edited by E. Eric Knudsen, Cleveland R. Steward, Donald D. MacDonald, Jack E. Williams, and Dudley W. Reiser, 67–74. New York: Lewis Publishers, 2000.

Naylor, Rosamond L., Josh Eagle, and Whitney L. Smith. "Salmon Aquaculture in the Pacific Northwest: A Global Industry with Local Impacts." *Environment* 45, no. 8 (October 2003): 18–39.

Oregon Rivers Council, Inc. *The Economic Imperative of Protecting Riverine Habitat in the Pacific Northwest*. Report 5. Portland, OR: 1992.

Page, L. M., and B. M. Burr. *A Field Guide to Freshwater Fishes of North America North of Mexico*. Boston: Houghton Mifflin Company, 1991. Quoted in R. Froese and D. Pauly, eds., "*Oncorhynchus gorbuscha*." *FishBase*. http://fishbase.org/Summary/SpeciesSummary.cfm?ID=240&genusname=Oncorhynchus&speciesname=gorbuscha; "*Oncorhynchus keta*." http://fishbase.org/Summary/SpeciesSummary.cfm?ID=241&genusname=Oncorhynchus&speciesname=keta; "*Oncorhynchus kisutch*." http://fishbase.org/Summary/SpeciesSummary.cfm?ID=245&genusname=Oncorhynchus&speciesname=kisutch; "*Oncorhynchus nerka*." http://fishbase.org/Summary/SpeciesSummary.cfm?ID=243&genusname=Oncorhynchus&speciesname=nerka; "*Oncorhynchus tshawytscha*." http://fishbase.org/Summary/SpeciesSummary.cfm?ID=244&genusname=Oncorhynchus&speciesname=tshawytscha.

Pavlov, Dmitrii S., Ksenia A. Savvaitova, Kirill V. Kuzishchin, Marina A. Gruzdeva, Sergei D. Pavlov, Boris M. Mednikov, and Sergei V. Maksimov. "The Structure of Local Populations of

Mikizha in Kamchatka." Chap. 5 in *Abstract of the First Regional Scientific-practical Conference, Issues of Protection and Rational Use of KamchatkBioresources,* edited by Dmitrii S. Pavlov and Ksenia A. Savvaitova. Moscow, Russia: Scientific World, 2001.

Ricker, William E. "Hereditary and Environmental Factors Affecting Certain Salmon Populations." In *The Stock Concept in Pacific Salmon*, edited by Raymond C. Simon and Peter A. Larkin, 27–160. Vancouver: Institute of Animal Resources, University of British Columbia, 1972.

———. "Stock and recruitment." *Journal of the Fisheries Research Board of Canada* 11 (1954): 559–623.

Savvaitova, Ksenia A., Kirill V. Kuzishchin, and Sergei V. Maksimov. "Kamchatka Steelhead: Population Trends in Life History Variation." Chap. 14 in *Sustainable Fisheries Management: Pacific Salmon*, edited by E. Eric Knudsen, Cleveland R. Steward, Donald D. MacDonald, Jack E. Williams, and Dudley W. Reiser. Boca Raton: Lewis Publishers, 2000.

Shepard, M. P., C. D. Shepard, and A. W. Argue. "Historic Statistics of Salmon Production around the Pacific Rim." *Canadian Manuscript Report of Fisheries and Aquatic Sciences* 1819 (September 1985): 1–297.

Slaney, Tim L., Kim Hyatt, T. G. Northcote, and R. J. Fielden. "Status of Anadromous Salmon and Trout in British Columbia and Yukon." *Fisheries* 21, no. 10 (1996): 20–35.

Stearley, R. F., and G. R. Smith. "Phylogeny of the Pacific Trouts and Salmons (*Oncorhynchus*) and Genera of the Family Salmonidae." *Transactions of the American Fisheries Society* 122, no. 1 (January 1993): 1–33.

Waples, Robin S., R. G. Gustafson, L. A. Weitkamp, J. M. Myers, O. W. Johnson, P. J. Busby, J. J. Hard, et al. "Characterizing Diversity in Salmon from the Pacific Northwest." *Journal of Fish Biology* 59, suppl. A (2001): 1–41.

Watson, Rupert. *Salmon, Trout and Char of the World*. Shrewsbury, England: Swan Hill Press, 1999.

Welch, David W., Yukimasa Ishida, et al. "Thermal Limits on the Ocean Distribution of Steelhead Trout (*Oncorhynchus mykiss*)." *North Pacific Anadromous Fish Commission* 1 (1998): 396–404.

Willson, Mary F., Scott M. Gende, and Brian H. Marston. "Fishes and the Forest: Expanding Perspectives on Fish-wildlife Interactions." *BioScience* 48 (1998): 455–62.

Willson, Mary F., and Karl C. Halupka. "Anadromous Salmonids as Keystone Species in Vertebrate Communities." *Conservation Biology* 9, no. 3 (1995): 489–97.

Woody, Elizabeth, Jim Lichatowich, Richard Manning, Freeman L. House, and Seth Zuckerman. *Salmon Nation: People, Fish, and Our Common Home*. Edited by Edward C. Wolf and Seth Zuckerman. Portland, OR: Ecotrust, 2003.

## CHAPTER 1 MAPS

**Salmon Ecoregions**
*Peer reviewers:*
*Dan Bottom, Research Fishery Biologist, NOAA Fisheries (National Marine Fisheries Service), Newport, Oregon, United States*

*Jeff Rodgers, Oregon Plan Monitoring Coordinator, Oregon Department of Fish and Wildlife, Corvallis, Oregon, United States*

In 1999 the author hosted the North Pacific Salmon Workshop at Oregon State University, Corvallis, OR, 4–6 May 1999, to determine the extent of what would become Levels 1–4 salmon ecoregions. Participants are listed in Acknowledgments.

**Original Distribution of Genus *Oncorhynchus***
This map is a composite of all distributions in Chapter 4. See *Chapter 4 maps* for sourcing.

**Salmon Diversity**
*Peer reviewers:*
*Robert Behnke, Professor Emeritus, Department of Fishery and Wildlife Biology, Colorado State University, Fort Collins, Colorado, United States*

This map summarizes the number of species by Level 4 ecoregion as a reflection of species distribution. See *Chapter 4 maps* for sourcing.

**Kamchatka Steelhead Life History Types**
*Peer reviewers:*
*Kirill Kuzishchin, Associate Professor, Department of Ichthyology, Moscow State University, Moscow, Russia*

*Dmitrii S. Pavlov, Member of the Russian Academy of Sciences, Moscow State University, Moscow, Russia*

*Ksenia Savvaitova, Biology Faculty, Moscow State University, Moscow, Russia*

Pavlov, Dmitrii S., Ksenia A. Savvaitova, Kirill V. Kuzishchin, Marina A. Gruzdeva, Sergei D. Pavlov, Boris M. Mednikov, and Sergei V. Maksimov. "The Structure of Local Populations of Mikizha in Kamchatka." Chap. 5 in *Abstract of the First Regional Scientific-practical Conference, Issues of Protection and Rational Use of Kamchatka Bioresources,* edited by Dmitrii S. Pavlov and Ksenia A. Savvaitova. Moscow, Russia: Scientific World, 2001.

Savvaitova, Ksenia A., Kirill V. Kuzishchin, and Sergei V. Maximov. "Kamchatka Steelhead: Population Trends in Life History Variation." Chap. 14 in *Sustainable Fisheries Management: Pacific Salmon*, edited by E. Eric Knudsen, Cleveland R. Steward, Donald D. MacDonald, Jack E. Williams, and Dudley W. Reiser. Boca Raton: Lewis Publishers, 2000.

## CHAPTER 2

Agee, James K. *Fire Ecology of Pacific Northwest Forest*. Washington, DC: Island Press, 1993.

Alaska Department of Fish and Game, Sport Fish Division. "Economic Value of Sport Fishing in Alaska." http://www.sf.adfg.state.ak.us/statewide/SFeconomics.cfm.

Allendorf, Fred W. "Ecological and Genetic Effects of Fish Introductions: Synthesis and Recommendations." *Canadian Journal of Fisheries and Aquatic Science* 48 (1991): 178–81.

American Indian Technical Services, Inc. *Anthropological Study of the Hupa, Yurok, and Karok Indian Tribes of Northwestern California*. Final Report for the U.S. Department of the Interior/Bureau of Indian Affairs, Contract No. J50C14205074, January 1982.

Amos, Kevin H., and Andrew Appleby. "Atlantic Salmon: Commercial Production of Atlantic Salmon, Escapes and Recoveries in Washington State." In *Atlantic Salmon in Washington State: A Fish Management Perspective*. Washington Department of Fish and Wildlife, September 1999. http://wdfw.wa.gov/fish/atlantic/toc.htm. (Updated factsheet, August 2001. http://wdfw.wa.gov/factshts/atlanticsalmon.htm.)

Anderson, James L. "The Growth of Salmon Aquaculture and the Emerging New World Order of the Salmon Industry." In *Global Trends: Fisheries Management*, edited by E. K. Pikitch, D. D. Huppert, and M. P. Sissenwine, 176. Bethesda: American Fisheries Society, 1997.

Appleby, Andrew (Washington Department of Fish and Wildlife Aquaculture Coordinator). Personal communication.

Associated Press. "Concerns Raised Over Souped-up Salmon: Tinkering with Genes Sparks Worry over Altered Fish's Impact," August 22, 2000.

———. "Knowles Declares Western Alaska Fishing a Disaster." *Alaska Legislature.com*, August 26, 2002. http://www.aklegislature.com/stories/082602/knowles.shtml.

Augerot, Xanthippe. "An Environmental History of the Salmon Management Philosophies of the North Pacific: Japan, Russia, Canada, Alaska and the Pacific Northwest United States." PhD diss., Oregon State University, 2000.

Augerot, Xanthippe, and Guido R. Rahr. "Protected Areas for Native Salmon: A Strategy for Protecting Salmonid Biodiversity across the Northern Pacific Rim." White paper, Wild Salmon Center, Portland, OR, 2003.

Baumhoff, Martin A. "Ecological Determinants of Aboriginal California Populations." *University of California Publications in American Archaeology and Ethnology* 49, no. 2 (1963): 155–235.

Beamish, Dick (Senior Scientist, Pacific Biological Station, Fisheries and Oceans Canada, Nanaimo, BC). Personal communication, 2004.

Berge, Aslak. "The World's 30 Largest Salmon Farmers." *IntraFish*, April 13, 2001. http://www.intrafish.com/.

Biswell, Harold H. *Prescribed Burning in California Wildlands Vegetation Management*. Berkeley: University of California Press, 1989.

Bjørndal, Trond, Gunnar Knapp, and Audun Lem. *Salmon—A Study of Global Supply and Demand*. Globefish Research Programme 73. Report. Rome: FAO/GLOBEFISH, Fishery Industries Division, 2003.

*Boston Globe*. "Rescuing the Sea Pact," April 5, 2004.

Boyd, Robert. *Indians, Fire and the Land in the Pacific Northwest*. Corvallis: Oregon State University Press, 1999.

Brannon, Ernest L., Donald F. Amend, Matthew A. Cronin, James E. Lannan, Scott LaPatra, William J. McNeil, Richard E. Noble, et al. "The Controversy about Salmon Hatcheries," *Fisheries* 29, no. 9 (September 2004): 12–31.

British Columbia Environmental Assessment Office. "Impacts of Farmed Salmon Escaping from Net Pens," February 25, 1997. http://www.eao.gov.bc.ca/.

British Columbia Ministry of Agriculture, Food and Fisheries. *B.C. Salmon Aquaculture Policy*, January 31, 2002.

Brulle, Ramon Vanden, and Nick J. Gayeski. "Overwhelming Evidence: How Hatcheries Are Jeopardizing Salmon Recovery." *Washington Trout Report* 13, no. 1 (Spring 2003): 4–8.

Buell, Montgomery. "Waves of Change: Fishermen, Managers, and Ecology in the Bristol Bay Salmon Fishery, 1945–1980." PhD diss., Purdue University, 2002.

Burke, William T. "Anadromous Species and the New International Law of the Sea." *Ocean Development and International Law* 22 (1991): 95–131.

Burke, William T., Mark Freeberg, and Edward L. Miles. "United Nations Resolution on Driftnet Fishing: An Unsustainable Precedent for High Seas and Coastal Fisheries Management." *Ocean Development and International Law* 25 (1994): 127–86.

Buschmann, Alejandro H. *Impacto Ambiental de la Acuicultura: El Estado de la Investigacion en Chile y el Mundo* [The Environment of Aquaculture: The State of Research in Chile and the World]. Santiago, Chile: Terram Publicaciones: 2001.

Cameron, Fiona. "Of Marketing and Psychology—Selling Salmon." *IntraFish*, October 8, 2001. http://www.intrafish.com/.

Charron, Bertrand. "Salmon Farming and the Environment: An Overview of Some of the Attitudes Encountered." http://www.*intrafish.com*, February 24, 2000. http://www.intrafish.co.uk/intrafish-analysis/AO_2000_eng/index.php3?thepage=0.

Chilcote, Mark W., Steven A. Leider, and John L. Loch. "Differential Reproductive Success of Hatchery and Wild Summer-run Steelhead under Natural Conditions. *Transactions of the American Fisheries Society* 115 (1986): 726–35.

CNN. "Study: Farmed Salmon More Contaminated than Wild." *CNN.com*, January 9, 2004. http://www.cnn.com/2004/health/01/08/salmon.pollution.ap/index.html.

Coastal Alliance for Aquaculture Reform. "Farmed and Dangerous Action Center: Solutions and Alternatives." http://farmedanddangerous.org/solutions.htm.

———. "New Study Links Sea Lice to Wild Salmon Decline," March 2, 2003. http://farmedanddangerous.org/fad/website/files/Alex/Morton/release.pdf.

*Columbia Basin Bulletin*. "10. ODFW Developing State's First Hatchery Research Center," April 2, 2004. http://www.cbbulletin.com/.

Columbia River Inter-Tribal Fish Commission, ed. "Proceedings of the Tribal Fisheries Co-management Symposium." In *Tribal Fisheries Co-management Symposium*. Portland, Oregon, January 30, 2003.

Columbia River Inter-Tribal Fish Commission. "A Short Chronology of Treaty Fishing on the Columbia River." http://www.critfc.org/text/timeline.html.

Cone, Joseph, and Sandy Ridlington, eds. *The Northwest Salmon Crisis: A Documented History*. Corvallis: Oregon State University Press, 1996.

Connolly, Thomas J. "Cultural Stability and Change in the Prehistory of Southwest Oregon and Northern California." PhD diss., University of Oregon, 1986.

Cowan, Steve (Director, Producer, Habitat Media, Portland, Oregon). Personal communication, 2004.

David Suzuki Foundation. "New Analysis Links Salmon Farms, Sea Lice, and Broughton Pink Salmon Crash." News Release, July 2004. http://www.davidsuzuki.org/campaigns_and_programs/salmon_aquaculture/news_releases/newsaquaculture07200401.asp.

DeBano, Leonard F., Daniel G. Neary, and Peter F. Ffolliott. *Fire's Effects on Ecosystems.* New York: John Wiley and Sons, Inc., 1998.

DellaSala, Dominick A., Stewart B. Reid, Terrence J. Frest, James R. Strittholt, and David M. Olson. "A Global Perspective of Biodiversity of the Klamath-Siskiyou Ecoregion." *Natural Areas Journal* 19, no. 4 (1999): 300–319.

Duncan, Emma. "Fish Food for Thought." *WWF Newsroom*, February 18, 2003. http://www.panda.org/news_facts/newsroom/features/news.cfm?uNewsID=5921.

Eagle, Josh, Rosamond L. Naylor, and Whitney L. Smith. "Why Farm Salmon Outcompete Fishery Salmon." *Marine Policy* 28, no. 3 (May 2004): 259–70.

Elsner, Alan. "Fish Farms Are Ruining Alaska's Salmon Fishing Industry." Reuters News Source, September 26, 2001.

Environmental Assessment Office. "Salmon Aquaculture Review: Chapter 2. The Salmon Aquaculture Industry," Government of Canada: Environmental Assessment Office Report, August 1997. http://www.intrafish.com/laws-and-regulations/report_bc.

Environmental News Service. "Fish Farmers' Noise Blasts Whales from B.C. Waters," November 7, 2001.

Fladmark, Knut R. "Routes: Alternate Migration Corridors for Early Man in North America." *American Antiquity* 44, no. 1 (1979): 55–69.

Fleming, Ian A., Kjetil Hindar, Ingrid B. Mjolnerod, Bror Jonsson, Torveig Balstad, and Anders Lamberg. "Lifetime Success and Interactions of Farm Salmon Invading a Native Population." *Proceedings of the Royal Society of London* 267 (2000).

Gaudet, Dave. *Atlantic Salmon: A White Paper.* Juneau: Alaska Department of Fish and Game, March 5, 2002.

Goldburg, Rebecca J., and Tracy Triplet. *Murky Waters: Environmental Effects of Aquaculture in the U.S.* Environmental Defense Fund Report. New York: 1997.

Gray, Dennis J. "The Takelma and Their Athapascan Neighbors: A New Ethnographic Synthesis for the Upper Rogue River Area of Southwestern Oregon." Paper 37, Anthropology Dept., University of Oregon, 1987.

Gresh, Ted, Jim Lichatowich, and Peter Schoonmaker. "An Estimation of Historic and Current Levels of Salmon Production in the Northeast Pacific Ecosystem: Evidence of a Nutrient Deficit in the Freshwater Systems of the Pacific Northwest." *Fisheries* 25, no. 1 (January 2000): 15–21.

Gross, Mart, and Cory T. Robertson. "Aquaculture and the Future of Salmon." IUCN Commercial Captive Propagation Workshop, December 7–9, 2001.

Gunther, Erna. "Analysis of First Salmon Ceremony." *American Anthropologist* 28 (1926): 606–17.

Harden, Blaine. "Hatchery Salmon to Count as Wildlife." *Washington Post* (April 29, 2004): A1.

Harvard University, John F. Kennedy School of Government. "The Harvard Project on American Indian Economic Development: Honoring Nations: 2002 Honoree." http://www.ksg.harvard.edu/hpaied/hn/hn_2002_fish.htm.

Heizer, Robert. F. *Languages, Territories, and Names of California Indian Tribes.* Berkeley: University of California Press, 1966.

Hey, Ellen. "Global Fisheries Regulations in the First Half of the 1990s." *International Journal of Marine and Coastal Law* 11, no. 4 (1996): 459–90.

Hilborn, Ray, and Doug Eggers. "A Review of the Hatchery Programs for Pink Salmon in Prince William Sound and Kodiak Island, Alaska." *Transactions of the American Fisheries Society* 129 (2000): 333–50.

House, Freeman L. "Co-evolution." In *The New Settler Interviews.* Vol. 1, *Boogie at the Brink*, edited by Beth Robinson Bosk, 24–47. White River Junction, VT: Chelsea Green Publishing Company, 2000.

Hunter, David., James. Salzman, and Durwood Zaelke. *"International Environmental Law and Policy."* New York: Foundation Press, 1998.

Hyatt, Kim D. "Stewardship for Biomass or Biodiversity: A Perennial Issue for Salmon Management in Canada's Pacific Region." *Fisheries* 21, no. 10 (1996): 4–5.

"Independent Multidisciplinary Science Team Scientific Workshop Summary." In *Independent Multidisciplinary Science Team Scientific Workshop.* Corvallis, OR, October 21–22, 2003.

*Independent* (UK). "Russians Want Part of Alaska Returned," July 8, 1999.

Internet Guide To International Fisheries Law, http://www.oceanlaw.net/texts/summaries/driftnet.html, http://www.oceanlaw.net/texts/ga46_215.htm.

Johnston, Douglas M. "The Driftnetting Problem in the Pacific Ocean: Legal Considerations and Diplomatic Options." *Ocean Development and International Law* 21 (1990): 5–39.

Jorgensen, Joseph. *Western Indians.* San Francisco: W. H. Freeman and Co., 1980.

Joyner, Christopher C. "The United States and the New Law of the Sea." *Ocean Development and International Law* 27 (1996): 41–58.

*Juneau Empire.* "Between Worlds: How the Alaska Native Claims Settlement Act Reshaped the Destinies of Alaska's Native People." Special Report. Juneau: *Juneau Empire*, January 31, 1999.

Kaeriyama, Masahide. "Dynamics of Chum Salmon, *Oncorhynchus keta*, Populations Released from Hokkaido, Japan." *North Pacific Anadromous Fish Commission Bulletin* 1 (1998): 90–102.

———. "Hatchery Programmes and Stock Management of Salmonid Populations in Japan." Chap. 10 in *Stock Enhancement and Sea Ranching*, edited by Bari R. Howell, Erlend Moksness, and Terje Svasand. Oxford, England: Blackwell Science, Ltd., 1999.

Kaeriyama, Masahide, and Rizalita R. Edpalina. "Evaluation of the Biological Interaction between Wild and Hatchery Population for Sustainable Fisheries Management of Pacific Salmon." http://www.htokai.ac.jp/DM/~salmon/2ndSE.htm.

Kaeriyama, Masahide, and Hiroshi Mayama. "Rehabilitation of Wild Chum Salmon Population in Japan." *Technical Rep. Hokkaido Salmon Hatchery* 165 (1996): 41–52.

Kitada, Shuichi. "Effectiveness of Japan's Stock Enhancement Programmes: Current Perspectives." Chap. 8 in *Stock Enhancement and Sea Ranching*, edited by Bari R. Howell, Erlend Moksness, and Terje Svasand. Oxford, England: Blackwell Science, Ltd., 1999. Knapp, Gunnar P. "Alaska Salmon Ranching: An Economic Review of the Alaska Salmon Hatchery Programme." Chap. 37 in *Stock Enhancement and Sea Ranching*, edited by Bari R. Howell, Erlend Moksness, and Terje Svasand. Oxford, England: Blackwell Science, Ltd. 1999.

———. *Challenges and Strategies for the Alaska Salmon Industry*. Anchorage: Institute of Social and Economic Research, University of Alaska, December 20, 2001.

———. *Estimates of United States Production and Consumption of Salmon*. Anchorage: Institute of Social and Economic Research, University of Alaska, March 2000.

———. "Implications of Aquaculture for Wild Fisheries: The Case of Alaska Wild Salmon." *International Institute of Fisheries Economics and Trade*. (August 22, 2002).

———. "The Wild Salmon Industry: Five Predictions for the Future." *Fisheries Economics Newsletter*, no. 51 (May 2001): 1.

Knapp, Gunnar, and Fran Ulmer. "Alaska's Salmon Industry Must Find Way to Manage Fisheries Economically." *Seafood.com News*. (Originally published in *Anchorage Daily News*, July 8, 2004, B6.) http://news.seafoodnet.com.

Kroeber, Alfred L. "California Culture Provinces." *University of California Publications in American Archaeology and Ethnology* 17, no. 2 (1920): 151–69.

———. *Cultural and Natural Areas of Native North America*. Berkeley: University of California Press, 1963.

———. *Handbook of the Indians of California*. Bureau of American Ethnology Bulletin 78. Washington, DC: Smithsonian Institution, 1925.

Kroeber, Alfred L., and Samuel A. Barrett. "Fishing Among the Indians of Northwestern California." *Anthropological Records* 22 (1960): 1–156.

Lackey, Robert T. (Senior Fisheries Biologist Western Ecology Division, National Health and Environmental Effects Research Laboratory, U.S. Environmental Protection Agency). Personal communication, 2004.

Levin, Phillip S., and Michael S. Schiewe. "Preserving Salmon Biodiversity." *American Scientist* 89, no. 3 (2001): 220–27.

Lewis, Henry T. "Patterns of Indian Burning in California: Ecology and Ethnohistory." Chap. 2 in *Before the Wilderness*, edited by Thomas C. Blackburn and Kat Anderson. Menlo Park, CA: Ballena Press, 1993.

Lichatowich, James A. *Salmon Without Rivers: A History of the Pacific Salmon Crisis*. Washington, DC: Island Press, 1999.

Loy, Wesley. "On the Rocks: Historic Alaska Fishing Industry Faces Unprecedented Threat." *Alaska Daily News*, January 13, 2002.

Mahnken, Conrad, Gregory Ruggerone, William Waknitz, and Thomas Flagg. "A Historical Perspective on Salmonid Production from Pacific Rim Hatcheries." *North Pacific Anadromous Fish Commission Bulletin* 1 (1998): 38–53.

Mandryk, Carole A. S., Heiner Josenhans, Rolf W. Mathews, and Daryl W. Fedje. "Late Quaternary Paleoenvironments in Northwest North America: Implications for Inland vs. Coastal Migration Routes." *Quaternary Science Reviews* 20, nos. 1–3 (2001): 301–14.

Matson, Richard Ghia, and Gary Coupland. *The Prehistory of the Northwest Coast*. San Diego: Academic Press, 1995.

McMahon, Thomas E., and David S. deCalesta. "Effects of Fire on Fish and Wildlife." Chap. 18 in *Natural and Prescribed Fire in Pacific Northwest Forest*, edited by John D. Walstad, Steven R. Radosevich, and David V. Sandberg. Corvallis: Oregon State University Press, 1990.

McNair, Marianne. *Alaska Salmon Enhancement Program 2000 Annual Report*. Regional Information Report 5J01-01. Juneau: Division of Commercial Fisheries, Alaska Department of Fish and Game, 2001.

Meffe, Gary K. "Techno-arrogance and Halfway Technologies: Salmon Hatcheries on the Pacific Coast of North America." *Conservation Biology* 6 (1992): 350–54.

Mikhno, Igor. "Commercial Driftnet Fishing and Its Influence on Marine Ecosystems." Presentation at the First Annual Working Meeting of the International Bering Sea Forum, Petropavlovsk-Kamchatsky, Russia, May 28–31, 2004.

Murray, Alexander G., Ronald J. Smith, and Ronald M. Stagg. "Shipping and the Spread of Infectious Salmon Anemia in Scottish Aquaculture." *Emerging Infectious Diseases* 8, no. 1 (2002): 1–5.

Myers, Ransom A., Simon A. Levin, Russell Lande, Frances C. James, William W. Murdoch, and Robert T. Paine. "Hatcheries and Endangered Salmon." *Science* 303 (March 26, 2004): 1.

Nagasawa, Kazuya, and Soto-o Ito. "Distribution, Migration and Growth in the North Pacific Ocean of Sockeye Salmon (*Oncorhynchus nerka*) Produced from Laucustrine Form." Chap. 11 in *Stock Enhancement and Sea Ranching*, edited by Bari R. Howell, Erlend Moksness, and Terje Svasand. Oxford, England: Blackwell Science, Ltd., 1999.

National Marine Fisheries Service. *See* NMFS.

Naylor, Rosamond L., Josh Eagle, and Whitney L. Smith. "Salmon Aquaculture in the Pacific Northwest: A Global Industry with Local Impacts." *Environment* 45, no. 8 (October 2003): 18–39.

Naylor, Rosamund, Rebecca J. Goldburg, Jurgenne H. Primavera, Nils Kautsky, Malcolm C. M. Beveridge, Jason Clay, Carl Folke, et al. "Effect of Aquaculture on World Fish Supplies." *Nature* 405 (June 29, 2000): 1019.

Neilson, Ronald P., B. Smith, and I. C. Prentice. "Simulated Changes in Vegetation Distribution under Global Warming." In *The Regional Impacts of Climate Change: An Assessment of Vulnerability*, edited by R. T. Watson, M. C. Zinyowera, R. H. Moss, and David J. Dokken, 439–56. Cambridge, England: Cambridge University Press, 1998.

NMFS. *Status Review of Chinook Salmon from Washington, Idaho, Oregon, and California*. U.S. Department of Commerce, National Oceanic and Atmospheric Administration (NOAA). NOAA Technical Memorandum. NMFS-NWFSC-35, 1998.

———. *Status Review of Coho Salmon from Washington, Oregon, and California*. U.S. Department of Commerce, NOAA. NOAA Technical Memorandum. NMFS-NWFSC-24, 1995.

NMFS, International Fisheries Division. "2003 Report of the Secretary of Commerce to the Congress of the United States Concerning U.S. Actions Taken on Foreign Large-Scale High Seas Driftnet Fishing Pursuant to Section 206(e) of the Magnuson-Stevens Fishery and Conservation and Management Act, as

Amended by Public Law 104-297, the Sustainable Fisheries Act of 1996." NMFS International Fisheries Division, NOAA Fisheries, Office of Sustainable Fisheries. http://www.nmfs.noaa.gov/sfa/international/Congress/Reports/DriftnetRptCongress.pdf. NMFS, Northwest Region. "News in Northwest Fisheries: Ban on High Seas Drift Nets Being Enforced." Report. NOAA, June 1999. http://www.nwr.noaa.gov/1salmon/salmesa/pubs/fsdrift.htm.

O'Harra, Doug. "Pen-reared Atlantic Salmon Show Up in Alaskans' Nets." *Anchorage Daily News*, October 15, 2001.

Pacific Fisheries Resource Conservation Council. *Communique: Pink Salmon in Broughton Archipelago in Crisis: Report.* Vancouver, BC: 2002. http://www.fish.bc.ca/html/fish3011.htm.

Paone, Sergio. *Farmed and Dangerous: Human Health Risks Associated with Salmon Farming.* Tofino: Friends of Clayoquot Sound, November 2000.

———. *Industrial Disease: The Risk of Disease Transfer from Farmed Salmon to Wild Salmon.* Tofino: Friends of Clayoquot Sound, April 2000.

*People's Daily Online.* "China Sets up Marine Functional Zones," September 10, 2002. http://english.peopledaily.com.cn/200209/10/eng20020910_102974.shtml.

Pianin, Eric. "Toxins Cited in Farmed Salmon: Cancer Risk Is Lower in Wild Fish, Study Reports." *Washington Post*, January 9, 2004, A1. http://www.washingtonpost.com/ac2/wp-dyn/A733-2004Jan8?language=printer.

PricewaterhouseCoopers LLP, Canada. "A Competitiveness Survey of the British Columbia Salmon Farming Industry." Report for Aquaculture Development Branch, Ministry of Agriculture, Food and Fisheries, May 2003.

"The Promise of a Blue Revolution." *Economist* 368 (August 7, 2003): 19–21.

Quinault Indian Nation. "History of the Quinaults." http://209.206.175.157/history.htm.

Randall, Jeff. "Conflict in the Bering Sea: The Continuing Dispute over the Bering Sea Maritime Boundary Line." REECAS Newsletter (Fall 2003/Winter 2004): 11–12.

Redman, Charles L. *Human Impact on Ancient Environments.* Tucson: University of Arizona Press, 1999.

Reifenberg, Anne. "Taste Test: Wild vs. Farmed Salmon." *Wall Street Journal*, January 5, 2000, sec. NW3.

Rentschler, Kay. "Sushi Rice, California's New Gold Rush." *New York Times*, October 8, 2003.

Richardson, John. "Salmon Farming Fights for Its Life." *Portland (ME) Press Herald*, January 20, 2002.

Riddell, B. E. "Spatial Organization of Pacific Salmon: What to Conserve?" In *Genetic Conservation of Salmonid Fishes*, edited by Joseph G. Cloud and Gary H. Thorgaard, 23–41. New York: Plenum Press, 1993.

Riddle, George W. "Early Days in Oregon: A History of Riddle Valley." Riddle, OR: Riddle Parent-Teacher's Association, 1953. Roberts, Ronald J. "Positive Outlook for Salmon Industry." *Aquaculture Magazine* (September/October 2000): 56.

"Salmon Farms, Sea Lice, and Wild Salmon: A Watershed Watch Commentary on Risk, Responsibility, and the Public Interest." Coquitlam, BC: Watershed Watch Salmon Society, December 2001.

Schalk, Randall F. "Estimating Salmon and Steelhead Usage in the Columbia Basin Before 1850: The Anthropological Perspective." *Northwest Environmental Journal* 2, no. 2 (1986): 1–29.

Singer, Paul. "U.S. Delays Ratifying International Sea Pact." *Miami Herald*, April 13, 2004. http://www.miami.com/mld/miamiherald/business/national/8421603.htm?1c.

Stevens, Lee R. "Handbook for International Operations of U.S. Scientific Research Vessels: Part 2. Zones of Jurisdiction." University-National Oceanographic Laboratory System, January 1986. http://www.gso.uri.edu/unols/for_cln/for_cln.html.

Stockner, J. G., E. Rydin, and P. Hyenstrand. "Cultural Oligotrophication." *Fisheries* 25, no. 5 (May 2000): 7–14.

Sullivan, Kathleen, Thomas E. Lisle, C. Andrew Dolloff, Gordon Grant, and Leslie M. Reid. "Stream Channels: The Link Between Forest and Fishes." Contribution No. 57 in *Streamside Management: Forestry and Fisheries Interactions*, edited by Ernest O. Salo and Terrance W. Cundy, 39–97. Seattle: College of Forest Resources, University of Washington, 1987.

Swezey, Sean L., and Robert F. Heizer. "Ritual Management of Salmonid Fish Resources in California." Chap. 10 in *Before the Wilderness*, edited by Thomas C. Blackburn and Kat Anderson. Menlo Park, CA: Ballena Press, 1993.

Tonnesson, Stein. "Two Scenarios of Conflict Management." In *Workshop on the South China Sea Conflict*, 125–29. Oslo, 1999.

Tyedmers, Peter. "Salmon and Sustainability: The Biophysical Cost of Producing Salmon through the Commercial Salmon Fishery and the Intensive Salmon Culture Industry." PhD diss., University of British Columbia, Vancouver, 2000.

U.S. Census Bureau. "Annual Estimates of the Population by Race Alone or in Combination and Hispanic or Latino Origin for the United States and States." July 1, 2003. http://www.census.gov/popest/states/asrh/tables/SC-EST2003-05.pdf.

———. "2003 Population Estimates, Geographic Area: United States—County by State, and for Puerto Rico." U.S. Census Bureau, American FactFinder. http://factfinder.census.gov/servlet/GCTTable?_bm=y&-geo_id=&-ds_name=PEP_2003_EST&-_lang=en&-mt_name=PEP_2003_EST_GCTT1_US25&-format=US-25|US-25S&-CONTEXT=gct.

United Nations. "Status of the United Nations Convention on the Law of the Sea, of the Agreement Relating to the Implementation of Part XI of the Convention and of the Agreement for the Implementation of the Provisions of the Convention Relating to the Conservation and Management of Straddling Fish Stocks and Highly Migratory Fish Stocks." United Nations, Division for Ocean Affairs and the Law of the Sea, 2004. http://www.un.org/Depts/los/reference_files/status2003.pdf.

———. "United Nations Convention on the Law of the Sea of 10 December 1982: Overview and Full Text." United Nations, Division for Ocean Affairs and the Law of the Sea, 2004. http://www.un.org/Depts/los/convention_agreements/convention_overview_convention.htm.

———. "The United Nations Convention on the Law of the Sea: A Historical Perspective." United Nations, Division for Ocean Affairs and the Law of the Sea, 1998. http://www.un.org/Depts/los/convention_agreements/convention_historical_perspective.htm.

Vanden, Ramon, and Nick Gayeski. "Overwhelming Evidence: How Hatcheries Are Jeopardizing Salmon Recovery." *Washington Trout Report* 13, no. 1 (Spring 2003).

Volpe, John. "P-I Focus: Farming Is a Net-loss Proposition Ecologically, Socially, and Economically: A Salmon Scare." *Seattle Post-Intelligencer*, January 25, 2004. http://seattlepi.nwsource.com/opinion/157539_focus25.html.

———. *Super Un-natural: Atlantic Salmon in B.C. Waters.* Vancouver, BC: David Suzuki Foundation, October 2001.

Waples, Robin S. "Evolutionarily Significant Units and the Conservation of Biological Diversity under the Endangered Species Act." *American Fisheries Society Symposium* 17 (1995): 8–27.

———. "Pacific Salmon, *Oncorhynchus spp.*, and the Definition of 'Species' under the Endangered Species Act." *Marine Science Review* 53, no. 3 (1991): 11–22.

Washington Department of Fish and Wildlife. "Summary of Columbia River Fall Salmon and Summer Steelhead Returns for 2002." http://wdfw.wa.gov/fish/salmon_columbia02.htm.

Waterman, Thomas T., and Alfred L. Kroeber. "The Kepal Fish Dam." *University of California Publications in American Archaeology and Ethnology* 35, no. 6 (1938).

Weber, Michael L., ed. "Summary of Papers Presented at a Panel: Perspectives on Sustainability in Aquaculture." In *Aquaculture America 2000.* New Orleans: February 5, 2000.

———. "What Price Farmed Fish: A Review of the Environmental and Social Costs of Farming Carnivorous Fish." Washington, DC: SeaWeb Aquaculture Clearinghouse, 2003.

Woody, Elizabeth, Jim Lichatowich, Richard Manning, Freeman L. House, and Seth Zuckerman. *Salmon Nation: People, Fish, and Our Common Home.* Edited by Edward C. Wolf and Seth Zuckerman. Portland, OR: Ecotrust, 2003.

World Wildlife Fund (WWF). *The Status of Wild Atlantic Salmon: A River by River Assessment.* WWF Norway, May 2001.

Young, Gordon. "Wild Pink Salmon Crash Blamed on BC Fish Farm Lice." *Ichthyology at the Florida Museum of Natural History: In the News*, November 25, 2002. http://www.flmnh.ufl.edu/fish/InNews/salmoncrash2002.htm.

## CHAPTER 2 MAPS

### Indigenous Peoples of the North Pacific, c. 1880

This map was prepared as a collaborative effort by two leading senior linguists from the Department of Anthropology, National Museum of Natural History, Smithsonian Institution. Dr. Igor Krupnik conducted original research and design for the Asian side, and Dr. Ives Goddard, foremost linguist of North American native languages, spearheaded research for the North American side. Dr. Goddard and his associates researched and designed the definitive maps of native languages for the *Handbook of North American Indians* (edited by William C. Sturtevant and published by the Smithsonian Institution, Washington, DC) in the following volumes:

Damas, David, ed. *Arctic.* Handbook of North American Indians 5. Washington, DC: Smithsonian Institution, 1984.

Goddard, Ives, ed. *Languages.* Handbook of North American Indians 17. Washington, DC: Smithsonian Institution, 1996.

Heizer, Robert F., ed. *California.* Handbook of North American Indians 8. Washington, DC: Smithsonian Institution, 1978.

Helm, June, ed. *Subarctic.* Handbook of North American Indians 6. Washington, DC: Smithsonian Institution, 1981.

Suttles, Wayne, ed. *Northwest Coast.* Handbook of North American Indians 7. Washington, DC: Smithsonian Institution, 1990.

### Native Americans and Salmon Coevolution

*Peer reviewer:*
*Frank Lake, USDA Forest Service/Oregon State University, Corvallis, Oregon, United States*

DellaSala, Dominick A., Stewart B. Reid, Terrence J. Frest, James R. Strittholt, and David M. Olson. "A Global Perspective of Biodiversity of the Klamath-Siskiyou Ecoregion." *Natural Areas Journal* 19, no. 4 (1999): 300–319.

Kroeber, Alfred L. "California Culture Provinces, Map 1" in "Provinces and Sub-provinces of Native Civilization on the Pacific Coast of the United States." *University of California Publications in American Archaeology and Ethnology* 17, no. 2 (1920): 151–69.

National Oceanic and Atmospheric Administration (NOAA), National Marine Fisheries Service (NMFS), Coho Evolutionary Significant Units, Protected Resources Division. Portland, Oregon, 1998.

Olson, David M., Eric Dinerstein, Eric D. Wikramanayake, Neil D. Burgess, George V. N. Powell, Emma C. Underwood, Jennifer A. D'Amico, et al. "Terrestrial Ecoregions of the World: A New Map of Life on Earth." *BioScience* 51, no. 11 (2001): 933–38.

### Catch Composition

Hare, Steven R., Nathan J. Mantua, and Robert C. Francis. "Inverse Production Regimes: Alaska and West Coast Pacific Salmon." *Fisheries* 24 (1999): 6–14.

Ianovskaia, N.V., N. N. Sergeeva, E. A. Bogdan, A.V. Kudriavtseva, I. K. Kalashnikova, and E. A. Romanova. *Ulovy tikhookeanskikh lososei, 1900–1986 g.g.* [Harvest of Pacific Salmonids, 1900–1986]. Moscow, Russia: VNIRO, 1989.

North Pacific Anadromous Fish Commission. *Statistical Yearbook.* http://www.npafc.org/.

Rogers, Don E. *Estimates of Annual Salmon Runs from the North Pacific, 1951–1998.* Seattle: University of Washington Fisheries Research Institute, 1999.

Shepard, M. P. C., C. D. Shepard, and A. W. Argue. "Historic Statistics of Salmon Production around the Pacific Rim." *Canadian Manuscript Report of Fisheries and Aquatic Sciences* 1819 (1985): 1–297.

### Hatcheries

*Peer reviewers:*
*James Brady, Owner, North Cape Fisheries Consulting, Anchorage, Alaska, United States*

*David Johnson, Biologist, Washington Department of Fish and Wildlife, Olympia, Washington, United States*

*Sue Lehmann, Biologist, Enhancement Support and Assessment, Unit Habitat and Enhancement Branch, Fisheries and Oceans Canada, Vancouver, British Columbia, Canada*

*Alice Low, Senior Fisheries Biologist, Native Anadromous Fish and Watershed Branch, California Department of Fish and Game, Sacramento, California, United States*

*Bryan Ludwig, Director, Science and Evaluation, Hatchery Manager, Freshwater Fisheries Society of BC, Victoria, British Columbia, Canada*

*Sergei Makeev, Director, Sakhalin Wild Nature Fund, Aniva, Russia*

*George Nandor, Hatchery Production Coordinator, Oregon Department of Fish and Wildlife, Eastern Oregon University, LaGrande, Oregon, United States*

*Ki-baek Seong, Yangyang Inland Fisheries Research Institute, Yangyang, South Korea*

*Mikhail Skopets, Fish Biologist, Russian Academy of Sciences, Magadan, Russia*

*Judy Urrutia, Fish Planting/Hatchery Support, California Department of Fish and Game, Lands and Facilities Branch, Sacramento, California, United States*

*Sergei F. Zolotukhin, Director, Salmon Laboratory, KhabarovskTINRO, Khabarovsk, Russia*

Eronova, Elena (Junior Researcher, Khabarovsk Branch of TINRO Center, Russia). Personal communication, 2000.

Farrington, Craig. Alaska Salmon Enhancement Program. Regional Information Report No. 5Jo3-05. Juneau: Alaska Department of Fish and Game, Division of Commercial Fisheries, 2003.

Kaeriyama, Masahide (Professor of Marine Ecology, Department of Marine Science and Technology, School of Engineering, Hokkaido Tokai University, Minamisawa, Minami-ku, Sapporo, Japan). Personal communication, 2003.

Korolev, Mikhail R. (Assistant Director, Sevvostrybvod, Kamchatrybvod, Petropavlovsk-Kamchatsky, Russia). Personal communication, 2001.

Lehmann, Sue. *Canadian Enhanced Salmonid Production During 1978–2002 (1977–2001 brood years).* NPAFC Report, Doc. 710, Rev 1. 10p. +i. Department of Fisheries and Oceans, Vancouver, BC, 2003.

Lehmann, Sue (Assessment Biologist, Habitat and Enhancement Branch, Department of Fisheries and Oceans, Vancouver, BC). Personal communication, 2004.

Low, Alice (Senior Biologist, California Department of Fish and Game Sacramento, CA). Personal communication, 2004.

McGee, Steve. (Fishery Biologist, Coordinator, Private Nonprofit Hatcheries Program, Alaska Department of Fish and Game, Division of Commercial Fisheries, Juneau, AK). Personal communication, 2004.

Nagata, Mitsuhiro (Research Scientist, Hokkaido Fish Hatchery, Hokkaido Department of Fisheries and Forestry, Hokkaido Government). Personal communication, 2003.

Nandor, George (Hatchery Production Coordinator, Oregon Department of Fish and Wildlife, Salem, OR). Personal communication, 2004.

National Salmon Resources Center. Salmon Database 1999, *Oncorhynchus* spp (Salmon Database 9(1)). (Database on biological assessment of Pacific salmon populations in Japan.) Fisheries Agency of Japan, 2000.

Okamoto, Yutaka (Director, Okamoto International Affairs Research Institute, Tokyo, Japan). Personal communication, 2003.

Oregon Department of Fish and Wildlife. "Hatchery Facilities GIS database." Oregon Department of Fish and Wildlife, Natural Resources Information Program. Dataset: ODFW Hatchery Facilities. http://rainbow.dfw.state.or.us/nrimp/information/odfwgisdata.htm.

Putivkin, Sergei (Senior Researcher, Magadan Branch of TINRO Center, Russia). Personal communication, 2001.

Semenchenko, Anatolii (Senior Research Scientist, Pacific Fisheries Research Center [TINRO], Primorsky Krai, Vladivostok, Russia). Personal communication, 2000.

Zhul'kov, Aleksandr (Senior Researcher, SakhNIRO, Yuzhno-Sakhalinsk, Russia). Personal communication, 2001.

**Fish Farming**

*Peer reviewers:*

*Andy Appleby, Aquaculture Coordinator, Washington Department of Fish and Game, Olympia, Washington, United States*

*Rick Deegan, Information Officer, Planning and Information Branch, British Columbia Ministry of Fisheries, Victoria, British Columbia, Canada*

*Arve Mogster, Board Member, Pacific Aquaculture Caucus, Manchester, Washington, United States*

Appleby, Andrew (Washington Department of Fish and Wildlife Aquaculture Coordinator). Personal communication, 2001.

Ministry of Agriculture, Food and Fisheries, Resource and Community Planning Unit, Dataset marine finfish-110.shp. Provided by Rick Deegan, Resource Analyst, Marine Salmon Net Pen Operations of British Columbia, Agriculture Development Branch, Victoria, British Columbia, December 2000.

**Salmon Trade**

Anderson, James L. *The International Seafood Trade.* Boca Raton: CRC Press, 2003.

Bjørndal, Trond, Gunnar Knapp, and Audun Lem. *Salmon—A Study of Global Supply and Demand.* Globefish Research Programme 73. Rome: FAO/GLOBEFISH, Fishery Industries Division, 2003.

Fisheries Commodities Production and Trade, 1976–2001: *FAO Yearbook. Fishery Statistics, Commodities* 93. 2001.

**Marine Jurisdictions**

*Peer reviewers:*

*Paul E. Niemeier, Office of Sustainable Fisheries, NOAA Fisheries (National Marine Fisheries Service), Silver Spring, Maryland, United States*

*Stetson Tinkham, Deputy Director, Office of Marine Conservation, U.S. Department of State, Washington, DC, United States*

*Dorothy Zbicz, Officer, Bureau of Oceans and International Environmental and Scientific Affairs, U.S. Department of State, Washington, DC, United States*

Environmental Systems Research Institute, Inc. *Digital Chart of the World Data Dictionary*. Redlands, CA: ESRI, 1993.

Ross, D. A., and T. A. Landry. *Marine Scientific Research Boundaries*. Woods Hole, MA: International Marine Science Cooperation Program, Woods Hole Oceanographic Institution, 1986.

United Nations. *China Maritime Claims Map*. United Nations, Division for Ocean Affairs and the Law of the Sea, Office of Legal Affairs, 1999.

**Shared Stocks**

*Peer reviewers:*

*Sandy Johnston, Canadian Chair, Yukon/Transboundary Rivers Resource Management Panel, Fisheries and Ocean Canada, Whitehorse, Yukon, Canada*

*Scott Kelley, Regional Management Biologist, Alaska Department of Fish and Game, Alaska, United States*

*Don Kowal, Executive Secretary, Pacific Salmon Commission, Vancouver, British Columbia, Canada*

*Gary S. Morishima, Technical Advisor, Natural Resources, Quinault Management Center, Mercer Island, Washington, United States*

*Glen Oliver, Pacific Salmon Treaty Research Supervisor, Alaska Department of Fish and Game, United States*

*Leon Shaul, Coho Salmon Project Leader, Commercial Fisheries Division, Alaska Department of Fish and Game, United States*

*Melanie Sullivan, Resource Management Biologist, Fisheries and Oceans Canada, Pacific Region, Nanaimo, British Columbia, Canada*

*Laurie Weitkamp, Conservation Biology Division, Northwest Fisheries Science Center, NOAA Fisheries (National Marine Fisheries Service), NMFS/NWFSC, Seattle, Washington, United States*

Alaska Department of Fish and Game, Commercial Fisheries Division, Region 2, Bering River Management District. Juneau, 2004.

Miles, Edward, Stephen Gibbs, David Fluharty, Christine Dawson, and David Teeter. *The Management of Marine Regions: The North Pacific*. Berkeley: University of California Press, 1982.

Pacific Salmon Commission/Joint Interceptions Committee (PSC/JIC). *Third Report on the Parties' Estimate of Salmon Interceptions: 1980–1991*. Report JIC (93)-1. Prepared for the Research and Statistics Committee, Pacific Salmon Commission, 1993.

Ross, D. A., and T. A. Landry. *Marine Scientific Research Boundaries*. Woods Hole, MA: International Marine Science Cooperation Program, Woods Hole Oceanographic Institution, 1986.

**Protected Areas**

*Peer reviewers:*

*Maxim Dubinin, Coordinator, Protected Areas of Russia Project, Biodiversity Conservation Center, Moscow, Russia*

*Bruce Hill, Northern British Columbia Coordinator, Canadian Parks and Wilderness Society, British Columbia, Canada*

*Valery Neronov, Institute of Ecology and Evolution, Russian Academy of Sciences, Moscow, Russia*

*Konstantin Zgurovsky, Biologist, WWF Russia, Moscow, Russia*

IUCN and UNEP. 1:50,000–1:2,000,000. "World Database on Protected Areas, 2003." *UNEP World Conservation Monitoring Centre*. http://sea.unep-wcmc.org/. (The WDPA 2004 was based on the 2003 version launched at the World Congress on Protected Areas. It includes substantial updates to the previous version and new records. Currently, the WDPA 2004 is the best global database on protected areas.)

Newell, Josh. *The Russian Far East: A Reference Guide for Conservation and Development. McKinleyville, CA: Daniel and Daniel, 2004.*

**CHAPTER 3**

Augerot, Xanthippe. "An Environmental History of the Salmon Management Philosophies of the North Pacific: Japan, Russia, Canada, Alaska and the Pacific Northwest United States." PhD diss., Oregon State University, 2000.

Beamish, Richard J., and Daniel R. Bouillon. "Pacific Salmon Production Trends in Relation to Climate." *Canadian Journal of Fisheries and Aquatic Science* 50 (1993): 1002–16.

Bering Sea Task Force. Report to Governor Tony Knowles. Alaska: 1999.

Bilby, Robert E., Brian R. Fransen, Peter A. Bisson, and Jason K. Walker. "Response of Juvenile Coho Salmon (*Oncorhynchus kisutch*) and Steelhead (*Oncorhynchus mykiss*) to the Addition of Salmon Carcasses to Two Streams in Southwestern Washington, U.S.A." *Canadian Journal of Fisheries and Aquatic Science* 55 (1998): 1909–18.

Bjørndal, Trond, Gunnar Knapp, and Audun Lem. *Salmon—A Study of Global Supply and Demand*. Globefish Research Programme 73. Rome: FAO/GLOBEFISH, Fishery Industries Division, 2003.

British Columbia Adventure Network. "Drainage." Physiogeography of British Columbia. B.C. Adventure. 2004. http://www.bcadventure.com/adventure/frontier/physio/drain.htm.

Bryant, Dirk, Daniel Nielsen, and Laura Tangley. *Last Frontier Forests: Ecosystems and Economies on the Edge*. Report. Washington, DC: World Resources Institute, 1997.

Cederholm, C. Jeff, Matt D. Kunze, Takeshi Murota, and Atuhiro Sibatani. "Pacific Salmon Carcasses: Essential Contributions of Nutrients and Energy for Aquatic and Terrestrial Ecosystems." *Fisheries* 24, no. 10 (October 1999): 6–15.

Chang, Kenneth. "Scientist Links Man to Climate over the Ages." *New York Times*, December 10, 2003, late edition, sec. A, 28.

Chigirinskii, A. I. "Promysel' Tikhookeanskikh lososei v Beringovom more." [Harvest of Pacific Salmon in the Bering Sea.] *Izvestiia Tikhookeanskogo nauchno-issledovatel'skogo instituta rybnogo khoziaistvo i okeanografii (TINRO)* [Proceedings of the Pacific

Scientific Research Institute for Fisheries and Oceanography (TINRO)] 116 (1994): 142–51.

Dynesius, Mats, and Christer Nilsson. "Fragmentation and Flow Regulation of River Systems in the Northern Third of the World." *Science* 266 (1994): 753–62.

*Encyclopædia Britannica Online*, s.v. "Kamchatka River." http://www.britannica.com/eb/article?eu=45498 (accessed July 12, 2004).

Fairbanks, Richard G. "A 17,000-year Glacio-eustatic Sea Level Record: Influence of Glacial Melting Rates on the Younger Dryas Event and Deep-ocean Circulation." *Nature* 342, no. 6250 (1989), 637–42.

Forman, Richard T. T., and Michel Godron. "Landscape Dynamics." Part 3 in *Landscape Ecology*. New York: John Wiley and Sons, Inc., 1986.

Francis, Robert C., and Steven R. Hare. "Decadal Scale Regime Shifts in the Large Marine Ecosystems of the Northeast Pacific: A Case for Historical Science." *Fisheries Oceanography* 3, no. 4 (1994): 279–91. http://www.iphc.washington.edu/Staff/hare/html/papers/francis-hare/abst_f-h.html.

Gilbertsen, Neal. "The Global Salmon Industry and Its Impacts in Alaska." *Alaska Economic Trends* 23, no. 10 (October 2003): 3–11. http://labor.state.ak.us/trends/oct03.pdf.

Global River Discharge (RivDIS). "Ishikari Summary and PlotsOhashi RivDIS Summary." Global River Discharge (RivDIS). Oak Ridge National Laboratory Distributed Active Archive Center. http://daac.ornl.gov/rivdis/stations/text/japan/173/summary.html.

Hunt, George L., Jr., and Phyllis J. Stabeno. "Climate Change and the Control of Energy Flow in the Southeastern Bering Sea." *Progress in Oceanography* 55 (2002): 5–22.

The Independent Institute. "New Report on What the EPA Isn't Telling Us." *Independent* 13, no. 3 (2003): 3.

Japan River Association. "Ishikari River." Japan River Association. No Date. http://www.japanriver.or.jp/river_law/kasenzu/kasenzu_gaiyou/hokkaido_r/007ishikari.htm.

Kammerer, J. C. "Water Fact Sheet: Largest Rivers in the United States." Open-File Report 87-242. U.S. Department of the Interior, United States. Geological Society Survey, 1987. Revised Department of the Interior. May 1990. http://water.usgs.gov/pubs/of/ofr87-242/.

Khomenko, Z. N., editor. *Spravochnik po fizicheskoi geografii Sakhalinskoi oblasti* (Guide to the Physical Geography of Sakhalin Ooblast). Iuzhno-Sakhalinsk, Russia: Sakhalin Book Publishing, 2003.

Kliashtorin, L. B., and F. N. Rukhlov. "Long-term Climate Change and Pink Salmon Stock Fluctuations." *North Pacific Anadromous Fish Commission Bulletin* 1 (1998): 464–79.

Lee, Peter, Dmitry Aksenov, Lars Laestadius, Ruth Nogueron, and Wynet Smith. *Canada's Large Intact Forest Landscapes*. Edmonton, AB: Global Forest Watch Canada, 2003.

Lindberg, G. U. "Principles of the Distribution of Fisheries and the Geological History of the Far-Eastern Seas." Translated by W. E. Ricker. *Fisheries Research Board of Canada Translation Series* 141 (1958): 47–51.

Longhurst, Alan R. "Biogeographic Provinces of the North Pacific Ocean." In *Ecological Geography of the Sea*. Philadelphia: Academic Press, 1998.

Loughlin, Thomas R., and Kiyotaka Ohtani, eds. *Dynamics of the Bering Sea*. Fairbanks, AK: University of Alaska, 1999.

Mantua, Nathan. "The Pacific Decadal Oscillation." In *Encyclopedia of Global Environmental Change*, edited by R. E. Munn, Michael C. MacCracken, and John S. Perry. Vol. 1. New York: John Wiley and Sons, Inc., 2002.

Mantua, Nathan J., Steven R. Hare, Yuan Zhang, John M. Wallace, and Robert C. Francis. "A Pacific Interdecadal Climate Oscillation with Impacts on Salmon Production." *Bulletin of the American Meteorological Society* 78 (June 1997): 1069–79.

McMillan, John, and James Starr. *Hoh River Basin: Refugia Status Report*. Portland, OR: Wild Salmon Center, 2000.

McNutt, Lyn. "How Does Ice Cover Vary in the Bering Sea from Year to Year?" *Bering Climate*. http://www.beringclimate.noaa.gov/essays_mcnutt.html.

Meehan, William R., ed. *Influences of Forest and Rangeland Management on Salmonid Fishes and Their Habitats*. American Fisheries Society Special Publication 19. Bethesda, MD: American Fisheries Society, 1991.

Morrison, John, Michael C. Quick, and Michael G. G. Foreman. "Climate Change in the Fraser River Watershed: Flow and Temperature Projections." *Journal of Hydrology* 263, nos. 1–4 (June 10, 2002): 230–44.

Naiman, Robert J., and Robert E. Bilby, eds. *River Ecology and Management: Lessons from the Pacific Coastal Ecoregion*. New York: Springer-Verlag, 1998.

National Assessment Synthesis Team, U.S. Global Change Research Program. *Climate Change Impacts on the United States: The Potential Consequences of Climate Variability and Change, Overview: Pacific Northwest*. http://www.usgcrp.gov/usgcrp/Library/nationalassessment/overviewpnw.htm.

National Oceanic and Atmospheric Administration (NOAA). "North Pacific Ocean Theme Page." *NOAA Pacific Marine Environmental Laboratory*. http://www.pmel.noaa.gov/np/pages/seas/npmap4.html.

Natural Resources Canada. "Rivers." *The Atlas of Canada*. Natural Resources Canada. 2004. http://atlas.gc.ca/site/english/learningresources/facts/rivers.html.

North Pacific Anadromous Fish Commission. "Catch and Releases of Salmon." http://www.npafc.org/statistics/general/statistics.htm.

Olson, David M., Eric Dinerstein, Eric D. Wikramanayake, Neil D. Burgess, George V. N. Powell, Emma C. Underwood, Jennifer A. D'Amico, et al. "Terrestrial Ecoregions of the World: A New Map of Life on Earth." *BioScience* 51 (November 2001): 933–38.

Pearcy, William G. *Ocean Ecology of North Pacific Salmonids*. Seattle: University of Washington Press, 1992.

Peterson, William T., and Franklin B. Schwing. "A New Climate Regime in Northeast Pacific Ecosystems." *Geophysical Research Letters* 30, no. 17 (September 9, 2003): 1896–99.

Pidwirny, Michael. "Introduction to the Hydrosphere." Chapter 8 in *Fundamentals of Physical Geography*. Kelowna, BC: Okanagan University, Department of Geography, 2004. http://www.physicalgeography.net/weblinks_ch8.html.

Pielou, E. C. *After the Ice Age: The Return of Life to Glaciated North America*. Chicago: University of Chicago Press, 1991.

———. *Fresh Water*. Chicago: University of Chicago Press, 1998.

Ritter, Dale F. *Process Geomorphology*. Dubuque, IA: William C. Brown Publishers, 1978.

River Bureau, Ministry of Land, Infrastructure, and Transport. "Table: Major Rivers in Japan." River Bureau, Ministry of Land, Infrastructure, and Transport. 1995. http://www.mlit.go.jp/river/english/table.html.

Rojas-Burke, Joe. "Uncertain Currents." *Oregonian*, February 4, 2004, p. B9–10.

Shugart, Herman, Roger Sedio, and Brent Sohngen. "Forests and Climate Change: Potential Impacts on U.S. Forest Resources." Report for the Pew Center of Global Climate Change. Arlington, VA: February 2003.

Sugimoto, Takashige, and Kazuaki Tadokoro. "Interannual-interdecadal Variations in Zooplankton Biomass, Chlorophyll Concentration and Physical Environment in the Subarctic Pacific and Bering Sea." *Fisheries Oceanography* 6, no. 2 (1997): 74–93.

Tomczak, Matthias, and J. Stuart Godfrey. "The Pacific Ocean." Chap. 8 in *Regional Oceanography: An Introduction*. 2nd ed. Delhi, India: Daya Publishing House, 2003.

Union of Concerned Scientists. "Global Environment: Forests and Climate Change: Recognizing Forests' Role in Climate Change." http://www.ucsusa.org/global_environment/biodiversity/page.cfm?pageID=526.

U.S. Environmental Protection Agency. "The Water Cycle." http://www.epa.gov/Region7/kids/wtrcycle.htm.

Van Andel, Tjeerd H. *New Views on an Old Planet: Continental Drift and the History of Earth*. Cambridge, England: Cambridge University Press, 1985.

Welch, David W. "New Developments in Ocean Salmon Research." *EEZ Technology* 4 (1999): 203–10.

———. "Thermal Limits and Ocean Migrations of Sockeye Salmon (*Oncorhynchus nerka*): Long-term Consequences of Global Warming." *Canadian Journal of Fisheries and Aquatic Science* 55 (1998): 937–48.

Welch, David W., A. I. Chigirinskii, and Yukimasa Ishida. "Upper Thermal Limits on the Oceanic Distribution of Pacific Salmon (*Oncorhynchus spp.*) in the Spring." *Canadian Journal of Fisheries and Aquatic Science* 52 (1995): 489–503.

Wild Salmon Center: Wild Salmon Center. *Anadyr River Watershed: Rapid Assessment Report*. Portland, OR: Wild Salmon Center, June 2003.

World Resources Institute. "First Scientific Assessment of Condition of World's Forests Shows Much More than Tropical Forests at Risk." News release, March 4, 1997. http://newsroom.wri.org/newsrelease_text.cfm?NewsReleaseID=107.

———. "Reports Conclude Much of World's Remaining Intact Forests at Risk." News release, April 3, 2002. http://pubs.wri.org/pubs_description.cfm?PubID=3717.

World Wildlife Fund. "Conservation Science: Delineation of Ecoregions." http://www.worldwildlife.org/science/ecoregions/delineation.cfm.

Zolotukhin, Sergei. (Head of Laboratory, Khabarovsk Branch of TINRO Center, Russia). Personal communication, 2001.

## CHAPTER 3 MAPS

### Extent of Glaciation

*Peer reviewer:*
*Dr. David Welch, Department of Fisheries and Oceans, Sciences Branch, Pacific Region, Pacific Biological Station, Nanaimo, British Columbia, Canada*

Ray, N., and J. M. Adams. "A GIS-based Vegetation Map of the World at the Last Glacial Maximum (25,000–15,000 BP)." *Internet Archaeology* 11 (2001). http://lgb.unige.ch/~ray/lgmveg/.

### Terrestrial Ecoregions

Olson, David M., Eric Dinerstein, Eric D. Wikramanayake, Neil D. Burgess, George V. N. Powell, Emma C. Underwood, Jennifer A. D'Amico, et al. "Terrestrial Ecoregions of the World: A New Map of Life on Earth." *BioScience* 51, no. 11 (2001): 933–38.

### Sea Ice

*Peer reviewers:*
*AGC(SW) Greg Rose, USN, Liaison/Public Affairs, Ice Reconnaissance LCPO, National Ice Center, Naval Ice Center, NOAA, Washington, DC, United States*

Arctic Sea Ice Charts Dataset. Provided by Chief Greg Rose, National Ice Center/Naval Ice Center. March 14, 1996, 2001.

### Primary Production

*Peer reviewers:*
*Lisa Eisner, Auke Bay Laboratory, Alaska Fisheries Science Center, NOAA, Juneau, Alaska, United States*

*Dave Foley, CoastWatch Coordinator, Pacific Fisheries Environmental Laboratory, NOAA, Pacific Grove, California, United States*

*Paul Harrison, Department of Earth and Ocean Sciences, University of British Columbia, Vancouver, British Columbia, Canada*

Seawifs satellite imagery, May and July 1998 and 2002. Courtesy of NASA Goddard Space Flight Center DAAC and Orbimage, Inc. Provided by Dave Foley, NOAA CoastWatch, 2003.

### Sea Surface Temperature

*Peer reviewers:*
*Lisa Eisner, Auke Bay Laboratory, Alaska Fisheries Science Center, NOAA, Juneau, Alaska, United States*

*Dave Foley, CoastWatch Coordinator, Pacific Fisheries Environmental Laboratory, NOAA, Pacific Grove, California, United States*

Pathfinder AVHRR, May and July 1998 and 2002. Courtesy of NASA Jet Propulsion Laboratory PODAAC and the California Institute of Technology. Provided by Dave Foley, NOAA CoastWatch, 2003.

**CHAPTER 4**

Augerot, Xanthippe. "An Environmental History of the Salmon Management Philosophies of the North Pacific: Japan, Russia, Canada, Alaska and the Pacific Northwest United States." PhD diss., Oregon State University, 2000.

Baker, Timothy T., Alex C. Wertheimer, Robert D. Burkett, Ronald Dunlap, Douglas M. Eggers, Ellen I. Fritts, Anthony J. Gharrett, Rolland A. Holmes, and Richard L. Wilmot. "Status of Pacific Salmon and Steelhead Escapements in Southeast Alaska." *Fisheries* 21, no. 10 (1996): 6–18.

Behnke, Robert J. *Native Trout of Western North America*. Bethesda, MD: American Fisheries Society, 1992.

Groot, Cornelius, Leo Margolis, and W. C. Clarke, eds. *Physiological Ecology of Pacific Salmon*. Vancouver: University of British Columbia Press, 1995.

Gustafson, R. G., Robin S. Waples, J. M. Myers, G. J. Bryant, O. W. Johnson, and L. A. Weitkamp. "Pacific Salmon Extinctions: Lost Diversity, Populations, and ESUs." NMFS draft paper, 2003.

Huntington, Charles, Willa Nehlsen, and Jon Bowers. "A Survey of Healthy Native Stocks of Anadromous Salmonids in the Pacific Northwest and California." *Fisheries* 21, no. 3 (1996): 6–14.

Lichatowich, James A. *Salmon Without Rivers: A History of the Pacific Salmon Crisis*. Washington, DC: Island Press, 1999.

Nehlsen, Willa., Jack. E. Williams, and James A. Lichatowich. "Pacific Salmon at the Crossroads: Stocks at Risk from California, Oregon, Idaho, and Washington." Fisheries 16, no. 2 (1991): 4–21.

Shepard, M. P., C. D. Shepard, and A. W. Argue. "Historic Statistics of Salmon Production around the Pacific Rim." *Canadian Manuscript Report of Fisheries and Aquatic Sciences* 1819 (September 1985): 1–297.

Slaney, Tim L., Kim D. Hyatt, Tom G. Northcote, and Robert J. Fielden. "Status of Anadromous Salmon and Trout in British Columbia and Yukon." *Fisheries* 23, no. 10 (1996): 20–35.

Waples, Robin S., R. G. Gustafson, L. A. Weitkamp, J. M. Myers, O. W. Johnson, P. J. Busby, J. J. Hard, et al. "Characterizing Diversity in Salmon from the Pacific Northwest." *Journal of Fish Biology* 59, suppl. A (2001): 1–41.

**CHAPTER 4 MAPS**

**All Distribution**
Alaska Department of Fish and Game. Anadromous Waters Catalog and Atlas. Anchorage, AK: Habitat and Restoration Division, 2003. http://www.habitat.adfg.state.ak.us/geninfo/anadcat/anadcat.shtml.

Fisheries and Oceans Canada. "Fisheries Information Summary System Salmon Distribution Zones." Vancouver, BC: Habitat and Enhancement Branch, 2001. http://www-heb.pac.dfo-mpo.gc.ca/maps/themesdata_e.htm.

Gritsenko, O.F. , ed. *Atlas of marine distribution of Pacific salmons during the spring-summer feeding and pre-spawning migrations.* A.E. Babyrev, A.K. Gruzevich, E.M. Klovatch, V.I. Karpenko, N.V. Klovatch, and S.S. Kozlov. Moscow: VNIRO Publishing, 2002.

HYDRO1k. "HYDRO1k Elevation Derivative Database." Sioux Falls, SD: U.S. Geological Survey, Resource Observation System Data Center, 1998. http://edcwww.cr.usgs.gov/landdaac/gtopo30/hydro/.

Pacific States Marine Fisheries Commission. "Anadromous Fish Distributions." Gladstone, OR: StreamNet, 2003. http://www.streamnet.org/online-data/GISData.html.

**Chum Distribution**
*Peer reviewers:*
*Robert Behnke, Professor Emeritus, Department of Fishery and Wildlife Biology, Colorado State University, Fort Collins, Colorado, United States*

*Jinping Chen, Doctoral Student, Institute of Zoology, Chinese Academy of Sciences, Beijing, China*

*Jim Irvine, Scientist, Department of Fisheries and Oceans, Nanaimo, British Columbia, Canada*

*J. Johnson, FDD Project Biologist, Alaska Department of Fish and Game, Anchorage, Alaska, United States*

*Peter Moyle, Professor, Department of Wildlife, Fish, and Conservation Biology, University of California, Davis, California, United States*

*Mikhail Skopets, Fish Biologist, Russian Academy of Sciences, Magadan, Russia*

*Alex Wertheimer, NOAA National Marine Fisheries Service, Auke Bay Lab, Juneau, Alaska, United States*

Chilcote, Mark, Chip Dale, Kathryn Kostow, Howard Schaller, and Hal Weeks. *Wild Fish Management Policy*. Biennial Progress Report. Oregon Department of Fish and Wildlife, 1992.

Department of Fisheries and Oceans. "Fraser River Chum Salmon." Vancouver, BC: Fraser River Action Plan, Fishery Management Group, 1996.

Hallock, Robert J., and Donald H. Fry. "Five Species of Salmon, *Oncorhynchus*, in the Sacramento River, California." *California Fish and Game* 53, no. 1 (1967): 5–22.

Johnson, Orlay W., W. Stewart Grant, Robert G. Kope, Kathleen Neely, F. William Waknitz, and Robin S. Waples. Status Review of Chum Salmon from Washington, Oregon, and California. U.S. Department of Commerce, National Oceanic and Atmospheric Administration (NOAA). NOAA Technical Memorandum. NMFS-NWFSC-32, 1997.

Kostow, Kathryn, ed. *Biennial Report of the Status of Wild Fish in Oregon*. Oregon Department of Fish and Wildlife, 1995. http://www.dfw.state.or.us/ODFWhtml/Research&Reports/WildFishRead.html. (Web page now unavailable).

Moyle, Peter B., Ronald M. Yoshiyama, Eric D. Wikramanayake, and Jack E. Williams. *Fish Species of Special Concern in California*. Report, 2nd ed. Rancho Cordova, CA: California Department of Fish and Game, Inland Fisheries Division, 1995.

Nehlsen, Willa., Jack. E. Williams, and James A. Lichatowich. "Pacific Salmon at the Crossroads: Stocks at Risk from California, Oregon, Idaho, and Washington." Fisheries 16, no. 2 (1991): 4–21.

Scofield, N. B. "The Humpback and Dog Salmon Taken in San Lorenzo River." *California Fish and Game* 2, no. 1 (1916): 41.

**Pink Distribution**

*Peer reviewers:*

*Robert Behnke, Professor Emeritus, Department of Fishery and Wildlife Biology, Colorado State University, Fort Collins, Colorado, United States*

*Jinping Chen, Doctoral Student, Institute of Zoology, Chinese Academy of Sciences, Beijing, China*

*Jim Irvine, Scientist, Department of Fisheries and Oceans, Nanaimo, British Columbia, Canada*

*J. Johnson, FDD Project Biologist, Alaska Department of Fish and Game, Anchorage, Alaska, United States*

*Peter Moyle, Professor, Department of Wildlife, Fish, and Conservation Biology, University of California, Davis, California, United States*

*Mikhail Skopets, Fish Biologist, Russian Academy of Sciences, Magadan, Russia*

*Alex Wertheimer, NOAA National Marine Fisheries Service, Auke Bay Lab, Juneau, Alaska, United States*


Ayers, R. J. "Pink Salmon Caught in Necanicum River." *Oregon Fish Commission Resource Briefs* 6, no. 2 (1955): 20.

Basham, Larry R., and Lyle G. Gilbreath. "Unusual Occurrence of Pink Salmon (*Oncorhynchus gorbuscha*) in the Snake River of Southeastern Washington." *Northwest Science* 52 (1978): 32–34.

Evermann, Barton W., and H. Walton Clark. "A Distributional List of the Species of Freshwater Fishes Known to Occur in California." *California Fish and Game* 35 (1931).

Fiscus, Hugh (Fish Biologist, Washington Department of Fish and Wildlife). Personal communication, January 1995.

Hallock, Robert J., and Donald H. Fry. "Five Species of Salmon, *Oncorhynchus*, in the Sacramento River, California." *California Fish and Game* 53, no. 1 (1967): 5–22.

Hard, Jeffrey J., Robert G. Kope, W. Stewart Grant, F. William Waknitz, L. Ted Parker, and Robin S. Waples. *Status Review of Pink Salmon from Washington, Oregon, and California*. U.S. Department of Commerce, NOAA. NOAA Technical Memorandum. NMFS-NWFSC-25, 1996.

Heard, William R. "Life History of Pink Salmon (*Oncorhynchus gorbuscha*)." In *Pacific Salmon Life Histories*, edited by Cornelius Groot and Leo Margolis, 121–230. Vancouver: University of British Columbia Press, 1991.

Herrmann, R. B. "Occurrence of Juvenile Pink Salmon in a Coastal Stream South of the Columbia River." *Oregon Fish Commission Resource Briefs* 7, no. 1 (1959): 81.

Mathisen, Ole A. "Spawning Characteristics of Pink Salmon (*Oncorhynchus gorbuscha*) in the Eastern North Pacific Ocean." *Aquaculture Fisheries Management* 25, no. 2 (1994): S147–S156.

Moyle, Peter B. (Professor, Department of Wildlife, Fish and Conservation Biology, University of California, Davis). Personal communication, September 1994.

Moyle, Peter B., Ronald M. Yoshiyama, Eric D. Wikramanayake, and Jack E. Williams. *Fish Species of Special Concern in California*. Report, 2nd ed. Rancho Cordova, CA: California Department of Fish and Game, Inland Fisheries Division, 1995.

Nehlsen, Willa, Jack. E. Williams, and James A. Lichatowich. "Pacific Salmon at the Crossroads: Stocks at Risk from California, Oregon, Idaho, and Washington." Fisheries 16, no. 2 (1991): 4–21.

Scofield, N. B. "The Humpback and Dog Salmon Taken in San Lorenzo River." *California Fish and Game* 2, no. 1 (1916): 41.

Smedley, Stephen C. "Pink Salmon in Prairie Creek, California." *California Fish and Game* 38, no. 2 (1952): 275.

Snyder, John O. "Salmon of the Klamath River, California." *California Fish and Game* 10, no. 4 (1931).

Taft, Alan C. "Pink Salmon in California." *California Fish and Game* 24, no. 2 (1938): 197–98.

U.S. Army Corps of Engineers. *1994 Annual Fish Passage Report: Columbia and Snake Rivers for Salmon, Steelhead and Shad*. North Pacific Division, Portland, OR, and Walla Walla, WA, 1994.

———. *2002 Annual Fish Passage Report, Columbia and Snake Rivers for Salmon, Steelhead, Shad, and Lamprey*. North Pacific Division, Portland, OR, and Walla Walla, WA, 2002.

Weeks, Hal (Marine Ecologist, Oregon Department of Fish and Wildlife). Personal communication, September 1994.

Williams, R. Walter, Richard M. Laramie, and J. J. Ames. *A Catalog of Washington Streams and Salmon Utilization*. Vol. 1, *Puget Sound*. Washington Department of Fisheries, Olympia, WA, 1975.


**Sockeye Distribution**

*Peer reviewers:*

*Robert Behnke, Professor Emeritus, Department of Fishery and Wildlife Biology, Colorado State University, Fort Collins, Colorado, United States*

*Mikhail Skopets, Fish Biologist, Russian Academy of Sciences, Magadan, Russia*

*Alex Wertheimer, NOAA National Marine Fisheries Service, Auke Bay Lab, Juneau, Alaska, United States*


Bartlett, Grace. *The Story of Wallowa Lake*. N.p., 1967.

Bendire, Charles. "Notes on Salmonidae of the Upper Columbia." Proceedings of the U.S. National Museum 6 (1881): 81–87.

Bjornn, Ted C., Donovan R. Craddock, and Donald R. Corley. "Migration and Survival of Redfish Lake, Idaho, Sockeye Salmon, *Oncorhynchus nerka*." *Transactions of the American Fisheries Society* 97, no. 4 (1968): 360–73.

Bryant, Floyd G., and Zell E. Parkhurst. "Part 4: Washington Streams from the Klickitat and Snake Rivers to Grand Coulee Dam, with Notes on the Columbia and Its Tributaries above Grand Coulee Dam (Area III)." Special Scientific Report: Fisheries, no. 37. In *Survey of the Columbia River and Its Tributaries with Special Reference to the Management of Its Fishery Resources*. Washington, DC: U.S. Department of the Interior, U.S. Fish and Wildlife Service, 1950.

Chapman, Don, Charles M. Peven, Tracy Hillman, Albert Giorgi, and Fred Utter. *Status of Sockeye Salmon in the Mid-Columbia Region*. Report. Don Chapman Consultants, Inc., Boise, ID, 1995.

Chapman, Don W., William S. Platts, D. Park, and M. Hill. *Status of Snake River Sockeye Salmon*. Final Report for Pacific Northwest Utilities Conference Committee. Don Chapman Consultants, Inc., Boise, ID, 1990.

Chapman, Wilbert M. "Observations on Migration of Salmonid Fishes in the Upper Columbia River." *Copeia* 1941, no. 4 (1941): 240–42.

Cobb, John N. *The Salmon Fisheries of the Pacific Coast*. Report of the Commissioner of Fisheries for the Fiscal Year 1910 and Special Papers. Washington, DC: Department of Commerce and Labor, Government Printing Office, 1911. Columbia Basin Fish and Wildlife Authority (CBFWA). *Integrated System Plan for Salmon and Steelhead Production in the Columbia River Basin*. Portland, OR: Columbia Basin System Planning, Northwest Power Planning and Conservation Council, 1990.

Cramer, Steven P. *The Feasibility for Reintroducing Sockeye and Coho Salmon in the Grande Ronde River and Coho and Chum Salmon in the Walla Walla River*. Progress Report prepared for Nez Perce Tribe, Umatilla Confederate Tribes, Warm Springs Confederated Tribes, and Oregon Department of Fish and Wildlife. Gresham, OR: S. P. Cramer and Associates, 1990.

Davidson, Frederick A. *Historical Notes on Development of Yakima River Basin*. Mimeo Report. Toppenish, WA: Yakama Indian Nation, 1953.

Evermann, Barton W. "A Preliminary Report on Salmon Investigation in Idaho in 1894." *Bulletin of the United States Fish Commission* 15 (1896): 253–84.

———. "A Report upon Salmon Investigations in the Headwaters of the Columbia River, in the State of Idaho, in 1895, Together with Notes upon the Fishes Observed in that State in 1894 and 1895." *Bulletin of the United States Fish Commission* 16 (1896): 151–202.

Evermann, Barton W. and S. E. Meek. "A Report upon Salmon Investigations in the Columbia River Basin and Elsewhere on the Pacific Coast in 1896." *Bulletin of the United States Fish Commission* 17 (1898): 15–84.

Fish, Frederic F., and Mitchell G. Hanavan. *A Report upon the Grand Coulee Fish-maintenance Project 1939–1947*. Special Scientific Report: Fisheries, no. 55. Washington, DC: U.S. Fish and Wildlife Service, 1948.

French, Robert R., and Roy J. Wahle. *Salmon Escapements above Rock Island Dam, 1954–60*. Special Scientific Report: Fisheries, no. 493. Washington, DC: U.S. Department of the Interior, U.S. Fish and Wildlife Service, Bureau of Commercial Fisheries, 1965.

———. *Salmon Runs—Upper Columbia River, 1956–57*. Special Scientific Report: Fisheries, no. 364. Washington, DC: U.S. Department of the Interior, U.S. Fish and Wildlife Service, Bureau of Commercial Fisheries: 1960.

Gustafson, Richard G, Thomas C. Wainwright, Gary A. Winans, F. William Waknitz, L. Ted Parker, and Robin S. Waples. *Status Review of Sockeye Salmon from Washington and Oregon*. U.S. Department of Commerce, NOAA. NOAA Technical Memorandum. NMFS-NWFSC-33, 1997.

Jordan, D. Starr, and Barton W. Evermann. *The Fishes of North and Middle America: A Descriptive Catalogue of the Species of Fish-like Vertebrates Found in the Waters of North America, North of the Isthmus of Panama. Part I.* Bulletin 47. Washington, DC: Government Printing Office, U.S. National Museum, 1896.

Kostow, Kathryn, ed. *Biennial Report of the Status of Wild Fish in Oregon*. Oregon Department of Fish and Wildlife, 1995. http://www.dfw.state.or.us/ODFWhtml/Research&Reports/WildFishRead.html. (Web page now unavailable).

Mullan, James W. *Determinants of Sockeye Salmon Abundance in the Columbia River, 1880s–1982: A Review and Synthesis*. Biological Report 86, no. 3. U.S. Fish and Wildlife Service, 1986.

Nehlsen, Willa. *Historical Salmon and Steelhead Runs of the Upper Deschutes River Basin and Their Environments*. Report to Portland General Electric Company, Portland, OR, 1995.

Nehlsen, Willa, Jack. E. Williams, and James A. Lichatowich. "Pacific Salmon at the Crossroads: Stocks at Risk from California, Oregon, Idaho, and Washington." Fisheries 16, no. 2 (1991): 4–21.

Parkhurst, Zell E. "Part 6: Snake River System from the Mouth through the Grande Ronde River (Area V)." Special Scientific Report: Fisheries, no. 39. In *Survey of the Columbia River and Its Tributaries with Special Reference to the Management of Its Fishery Resources*. Washington, DC: U.S. Department of the Interior, U.S. Fish and Wildlife Service, 1950.

Rutter, C. "Natural History of the Quinnat Salmon: A Report of Investigations in the Sacramento River, 1896–1901." *Bulletin of the United States Fish Commission* 22 (1904): 65–141.

Scofield, N. B. "The Humpback and Dog Salmon Taken in San Lorenzo River." *California Fish and Game* 2, no. 1 (1916): 41.

Toner, Richard C. "A Study of Some of the Factors Associated with the Reestablishment of Blueback Salmon into the Upper Wallowa River System." Appendix A in *Environmental Survey Report Pertaining to Salmon and Steelhead in Certain Rivers of Eastern Oregon and the Willamette River and Its Tributaries. Part 1, Survey Reports of Eastern Oregon Rivers*, edited by Robert N. Thompson and James B. Haas. Clackamas, OR: Oregon Fish Commission, Resource Division, 1960.

Waples, Robin S., Orlay W. Johnson, and Robert P. Jones, Jr. *Status Review for Snake River Sockeye Salmon*. U.S. Department of Commerce, NOAA. NOAA Technical Memorandum. NMFS-NWFSC-195, 1991.

**Chinook Distribution**

*Peer reviewers:*

*Robert Behnke, Professor Emeritus, Department of Fishery and Wildlife Biology, Colorado State University, Fort Collins, Colorado, United States*

*Jim Irvine, Scientist, Department of Fisheries and Oceans, Nanaimo, British Columbia, Canada*

*Peter Moyle, Professor, Department of Wildlife, Fish, and Conservation Biology, University of California, Davis, California, United States*
*Brian Riddell, Science Advisor/Research Scientist, Pacific Biological Station, Department of Fisheries and Oceans, Nanaimo, British Columbia, Canada*

*Mikhail Skopets, Fish Biologist, Russian Academy of Sciences, Magadan, Russia*

*Alex Wertheimer, NOAA National Marine Fisheries Service, Auke Bay Lab, Juneau, Alaska, United States*

Bengeyfield, W., et al. *Evaluation of Restoring Historic Passage of Anadromous Fish at BC Hydro Facilities*. Burnaby, BC: BC Hydro, Power Supply Environment, 2001.

Bryant, Floyd G. "Part 2: Washington Streams from the Mouth of the Columbia River to and Including the Klickitat River

(Area I)." Special Scientific Report: Fisheries, no. 62. In *Survey of the Columbia River and Its Tributaries with Special Reference to the Management of Its Fishery Resources.* Washington, DC: U.S. Department of the Interior, U.S. Fish and Wildlife Service, 1949.

Bryant, Floyd G., and Zell E. Parkhurst. "Part 4: Washington Streams from the Klickitat and Snake Rivers to Grand Coulee Dam, with Notes on the Columbia and Its Tributaries above Grand Coulee Dam (Area III)." Special Scientific Report: Fisheries, no. 37. In *Survey of the Columbia River and Its Tributaries with Special Reference to the Management of Its Fishery Resources.* Washington, DC: U.S. Department of the Interior, U.S. Fish and Wildlife Service, 1950.

Chorneau, Tom. "Investment May Not Coax Fishery Back." *Santa Rosa Press Democrat*, May 12, 1998. http://www.pressdemo.com/water/5_12a.html.

Duncan, John. "North Island Fisheries Award." Vancouver Island North, BC News Releases, April 22, 2002. http://www.duncanmp.com/news/2002/award.html.

Evermann, Barton W. "A Preliminary Report on Salmon Investigation in Idaho in 1894." *Bulletin of the United States Fish Commission* 15 (1896): 253–84.

Fish, Frederic F., and Mitchell G. Hanavan. *A Report upon the Grand Coulee Fish-maintenance Project 1939–1947.* Special Scientific Report: Fisheries, no. 55. Washington, DC: U.S. Fish and Wildlife Service, 1948.

French, Robert R., and Roy J. Wahle. *Salmon Escapements above Rock Island Dam, 1954–60.* Special Scientific Report: Fisheries, no. 493. Washington, DC: U.S. Department of the Interior, U.S. Fish and Wildlife Service, Bureau of Commercial Fisheries, 1965.

———. *Salmon Runs—Upper Columbia River, 1956–57.* Special Scientific Report: Fisheries, no. 364. Washington, DC: U.S. Department of the Interior, U.S. Fish and Wildlife Service, Bureau of Commercial Fisheries: 1960.

Fulton, Leonard A. *Spawning Areas and Abundance of Chinook Salmon,* Oncorhynchus tshawytscha, *in the Columbia River Basin: Past and Present.* Special Scientific Report: Fisheries, no. 571. Washington, DC: U.S. Department of the Interior, U.S. Fish and Wildlife Service, Bureau of Commercial Fisheries, 1968.

———. *Spawning Areas and Abundance of Steelhead Trout and Coho, Sockeye and Chum Salmon in the Columbia River Basin—Past and Present.* Special Scientific Report: Fisheries, no. 618. Washington, DC: U.S. Department of Commerce, NOAA, National Marine Fisheries Service (NMFS), 1970.

Haas, James B. *Fishery Problems Associated with Brownlee, Oxbow, and Hells Canyon Dams on the Middle Snake River.* Investigational Report 4. Portland: Fish Commission of Oregon, 1965.

Higgins, Patrick. T., Soyka Dobush, and David Fuller. "Factors in Northern California Threatening Stocks with Extinction." Arcata, CA: Humboldt Chapter of the American Fisheries Society, 1992.

Kostow, Kathryn, ed. *Biennial Report of the Status of Wild Fish in Oregon.* Oregon Department of Fish and Wildlife, 1995. http://www.dfw.state.or.us/ODFWhtml/Research&Reports/WildFishRead.html. (Web page now unavailable).

Myers, James M., Robert G. Kope, Gregory J. Bryant, David Teel, Lisa J. Lierheimer, Thomas C. Wainwright, W. Stewart Grant, et al. *Status Review of Chinook Salmon from Washington, Idaho, Oregon, and California.* U.S. Department of Commerce, NOAA. NOAA Technical Memorandum NMFS-NWFSC-35, 1998.

Nehlsen, Willa, Jack. E. Williams, and James A. Lichatowich. "Pacific Salmon at the Crossroads: Stocks at Risk from California, Oregon, Idaho, and Washington." *Fisheries* 16, no. 2 (1991): 4–21.

Nickelson, Thomas E., Jay W. Nicholas, Alan M. McGie, Robert B. Lindsay, Daniel L. Bottom, Rod J. Kaiser, and Steven E. Jacobs. "Status of Anadromous Salmonids in Oregon Coastal Basins." Oregon Department of Fish and Wildlife, Research and Development Section, Corvallis, OR, and Ocean Salmon Management, Newport, OR, 1992.

Nielson, Reed S. "Part 5: Oregon Streams from the Deschutes River to the Walla Walla River (Area IV)." Specific Scientific Report: Fisheries, no. 38. In *Survey of the Columbia River and Its Tributaries with Special Reference to the Management of Its Fishery Resources.* Washington, D.C: U.S. Department of the Interior, U.S. Fish and Wildlife Service, 1950.

Northwest Power Planning Council. "Compilation of Information on Salmon and Steelhead Losses in the Columbia River Basin." Appendix D in *The 1987 Columbia River Basin Fish and Wildlife Program.* Portland, OR: Northwest Power Planning Council, 1996.

Parkhurst, Zell E. "Part 6: Snake River System from the Mouth through the Grande Ronde River (Area V)." Special Scientific Report: Fisheries, no. 39. In *Survey of the Columbia River and Its Tributaries with Special Reference to the Management of Its Fishery Resources.* Washington, DC: U.S. Department of the Interior, U.S. Fish and Wildlife Service, 1950.

———. "Part 7: Snake River from above the Grande Ronde River through the Payette River (Area VI)." Special Scientific Report: Fisheries, no. 40. In *Survey of the Columbia River and Its Tributaries with Special Reference to the Management of Its Fishery Resources.* Washington, DC: U.S. Department of the Interior, U.S. Fish and Wildlife Service, 1950.

———. "Part 8: Snake River above Payette River to Upper Salmon Falls (Area VII)." Special Scientific Report: Fisheries, no. 57. In *Survey of the Columbia River and Its Tributaries with Special Reference to the Management of Its Fishery Resources.* Washington, DC: U.S. Department of the Interior, U.S. Fish and Wildlife Service, 1950.

Parkhurst, Zell E., Floyd G. Bryant, and Reed S. Nielson. "Part 3." Special Scientific Report: Fisheries, no. 36. In *Survey of the Columbia River and Its Tributaries with Special Reference to the Management of Its Fishery Resources.* Washington, DC: U.S. Department of the Interior, U.S. Fish and Wildlife Service, 1950.

Troffe, Peter M. *Living Landscapes: Freshwater Fishes of the Columbia Basin in British Columbia.* Victoria, BC: Royal British Columbia Museum, 1999. http://livinglandscapes.bc.ca/cbasin/peter_myles/pdf/fish1e.pdf.

Washington Department of Fisheries, Washington Department of Wildlife, and Western Washington Treaty Indian Tribes. "Columbia River Salmon and Steelhead Stocks." Table 14 in *1992 Washington State Salmon and Steelhead Stock Inventory.* Olympia, WA: 1993. http://wdfw.wa.gov/fish/sassi/sassi92.pdf.

———. "Puget Sound Salmon and Steelhead Stocks." Table 12 in *1992 Washington State Salmon and Steelhead Stock Inventory.* Olympia, WA: 1993. http://wdfw.wa.gov/fish/sassi/sassi92.pdf.

———. "Washington Coastal Salmon and Steelhead Stocks." Table 13 in *1992 Washington State Salmon and Steelhead Stock Inventory*. Olympia, WA: 1993. http://wdfw.wa.gov/fish/sassi/sassi92.pdf.

Willis, Raymond A., Robert N. Thompson, James B. Haas, Melvin D. Collins, and Roy E. Sams. *Environmental Survey Report Pertaining to Salmon and Steelhead in Certain Rivers of Eastern Oregon and the Willamette River and Its Tributaries*. Clackamas, OR: Research Division, Fish Commission of Oregon, 1960.

Yoshiyama, Ronald M., Eric R. Gerstung, Frank W. Fisher, and Peter B. Moyle. "Historical and Present Distribution of Chinook Salmon in the Central Valley Drainage of California." In *Sierra Nevada Ecosystem Project: Final Report to Congress, Vol. III, Assessments, Commissioned Reports, and Background Information* 309–362. Davis, CA: University of California, Centers for Water and Wildland Resources, 1996.

**Coho Distribution**

*Peer reviewers:*

*Robert Behnke, Professor Emeritus, Department of Fishery and Wildlife Biology, Colorado State University, Fort Collins, Colorado, United States*

*Jim Irvine, Scientist, Department of Fisheries and Oceans, Nanaimo, British Columbia, Canada*

*Peter Moyle, Professor, Department of Wildlife, Fish, and Conservation Biology, University of California, Davis, California, United States*

*Mikhail Skopets, Fish Biologist, Russian Academy of Sciences, Magadan, Russia*

*Alex Wertheimer, NOAA National Marine Fisheries Service, Auke Bay Lab, Juneau, Alaska, United States*

Bengeyfield, W. et al. *Evaluation of Restoring Historic Passage of Anadromous Fish at BC Hydro Facilities*. Burnaby, BC: BC Hydro, Power Supply Environment, 2001.

Brown, Larry R., and Peter B. Moyle. Status of Coho Salmon in California. Report to the National Marine Fisheries Service. Davis: University of California, Davis, CA, Department of Wildlife and Fisheries Biology, 1991.

Bryant, Floyd G. "Part 2: Washington Streams from the Mouth of the Columbia River to and Including the Klickitat River (Area I)." Special Scientific Report: Fisheries, no. 62. In *Survey of the Columbia River and Its Tributaries with Special Reference to the Management of Its Fishery Resources*. Washington, DC: U.S. Department of the Interior, U.S. Fish and Wildlife Service, 1949.

Bryant, Floyd G., and Zell E. Parkhurst. "Part 4: Washington Streams from the Klickitat and Snake Rivers to Grand Coulee Dam, with Notes on the Columbia and Its Tributaries above Grand Coulee Dam (Area III)." Special Scientific Report: Fisheries, no. 37. In *Survey of the Columbia River and Its Tributaries with Special Reference to the Management of Its Fishery Resources*. Washington, DC: U.S. Department of the Interior, U.S. Fish and Wildlife Service, 1950.

Capital Regional District. *Bowker Creek Watershed Management Plan*. 2003. http://www.crd.bc.ca/es/bowker/bcwmp.pdf.

Crocker, Liz. *Mill Hill, Thetis Lake, and Francis/King: A Cultural History of Three Regional Parks*. Report. Victoria, BC: Capital Regional District Parks, 1999.

Fisheries and Oceans Canada. *Salmon Update: Review of 2000 Salmon Season*. NR-PR-01-008E. January 24, 2001. http://www-comm.pac.dfo-mpo.gc.ca/pages/release/p-releas/2001/nr008_e.htm.

French, Robert R., and Roy J. Wahle. *Salmon Escapements above Rock Island Dam, 1954–60*. Special Scientific Report: Fisheries, no. 493. Washington, DC: U.S. Department of the Interior, U.S. Fish and Wildlife Service, Bureau of Commercial Fisheries, 1965.

———. *Salmon Runs—Upper Columbia River, 1956–57*. Special Scientific Report: Fisheries, no. 364. Washington, DC: U.S. Department of the Interior, U.S. Fish and Wildlife Service, Bureau of Commercial Fisheries: 1960.

Fulton, Leonard A. *Spawning Areas and Abundance of Steelhead Trout and Coho, Sockeye and Chum Salmon in the Columbia River Basin—Past and Present*. Special Scientific Report: Fisheries, no. 618. Washington, DC: U.S. Department of Commerce, National Oceanic and Atmospheric Administration, National Marine Fisheries Service, 1970.

Kostow, Kathryn, ed. *Biennial Report of the Status of Wild Fish in Oregon*. Oregon Department of Fish and Wildlife, 1995. http://www.dfw.state.or.us/ODFWhtml/Research&Reports/WildFishRead.html. (Web page now unavailable).

Nehlsen, Willa, Jack. E. Williams, and James A. Lichatowich. "Pacific Salmon at the Crossroads: Stocks at Risk from California, Oregon, Idaho, and Washington." *Fisheries* 16, no. 2 (1991): 4–21.

Northwest Power Planning Council. "Compilation of Information on Salmon and Steelhead Losses in the Columbia River Basin." Appendix D in *The 1987 Columbia River Basin Fish and Wildlife Program*. Portland, OR: Northwest Power Planning Council, 1986.

Parkhurst, Zell E. "Part 6: Snake River System from the Mouth through the Grande Ronde River (Area V)." Special Scientific Report: Fisheries, no. 39. In *Survey of the Columbia River and Its Tributaries with Special Reference to the Management of Its Fishery Resources*. Washington, DC: U.S. Department of the Interior, U.S. Fish and Wildlife Service, 1950.

———. "Part 7: Snake River from above the Grande Ronde River through the Payette River (Area VI)." Special Scientific Report: Fisheries, no. 40. In *Survey of the Columbia River and Its Tributaries with Special Reference to the Management of Its Fishery Resources*. Washington, DC: U.S. Department of the Interior, U.S. Fish and Wildlife Service, 1950.

———. "Part 8: Snake River above Payette River to Upper Salmon Falls (Area VII)." Special Scientific Report: Fisheries, no. 57. In *Survey of the Columbia River and Its Tributaries with Special Reference to the Management of Its Fishery Resources*. Washington, DC: U.S. Department of the Interior, U.S. Fish and Wildlife Service, 1950.

Parkhurst, Zell E., Floyd G. Bryant, and Reed S. Nielson. "Part 3." Special Scientific Report: Fisheries, no. 36. In *Survey of the Columbia River and Its Tributaries with Special Reference to the Management of Its Fishery Resources*. Washington, DC: U.S. Department of the Interior, U.S. Fish and Wildlife Service, 1950.

Weitkamp, Laurie A., Thomas C. Wainwright, Gregory J. Bryant, George B. Miller, David J. Teel, Robert G. Kope, and Robin S. Wapels. *Status Review of Coho Salmon from Washington, Oregon, and California*. U.S. Department of Commerce, NOAA. NOAA Technical Memorandum. NMFS-NWFSC-24, 1995.

**Masu Distribution**

*Peer reviewers:*

*Jinping Chen, Doctoral Student, Institute of Zoology, Chinese Academy of Sciences, Beijing, China*

*Mikhail Skopets, Fish Biologist, Russian Academy of Sciences, Magadan, Russia*

*Mitsuhiro Nagata, Research Scientist, Hokkaido Fish Hatchery, Hokkaido, Japan*

Kato, Fumihiko. "Life Histories of Masu and Amago Salmon (*Oncorhynchus masou* and *Oncorhynchus rhodurus*)." In *Pacific Salmon Life Histories*, edited by Cornelius Groot and Leo Margolis, 447–522. Vancouver: University of British Columbia Press, 1991.

Machidori, S., and F. Kato. "Spawning Populations and Marine Life of Masu Salmon." *International North Pacific Fisheries Commission*. Bulletin no. 43, 1984.

**Steelhead Distribution**

*Peer reviewers:*
*Robert Behnke, Professor Emeritus, Department of Fishery and Wildlife Biology, Colorado State University, Fort Collins, Colorado, United States*

*Jim Irvine, Scientist, Department of Fisheries and Oceans, Nanaimo, British Columbia, Canada*

*Kirill Kuzishchin, Associate Professor, Department of Ichthyology, Moscow State University, Moscow, Russia*

*Peter Moyle, Professor, Department of Wildlife, Fish, and Conservation Biology, University of California, Davis, California, United States*

*Dmitrii S. Pavlov, Member of the Russian Academy of Sciences, Moscow State University, Moscow, Russia*

*Ksenia Savvaitova, Biology Faculty, Moscow State University, Moscow, Russia*

*Mikhail Skopets, Fish Biologist, Russian Academy of Sciences, Magadan, Russia*

*Alex Wertheimer, NOAA National Marine Fisheries Service, Auke Bay Lab, Juneau, Alaska, United States*

Bryant, Floyd G. "Part 2: Washington Streams from the Mouth of the Columbia River to and Including the Klickitat River (Area I)." Special Scientific Report: Fisheries, no. 62. In *Survey of the Columbia River and Its Tributaries with Special Reference to the Management of Its Fishery Resources.* Washington, DC: U.S. Department of the Interior, U.S. Fish and Wildlife Service, 1949.

Bryant, Floyd G., and Zell E. Parkhurst. "Part 4: Washington Streams from the Klickitat and Snake Rivers to Grand Coulee Dam, with Notes on the Columbia and Its Tributaries above Grand Coulee Dam (Area III)." Special Scientific Report: Fisheries, no. 37. In *Survey of the Columbia River and Its Tributaries with Special Reference to the Management of Its Fishery Resources.* Washington, DC: U.S. Department of the Interior, U.S. Fish and Wildlife Service, 1950.

Evermann, Barton W. "A Preliminary Report on Salmon Investigation in Idaho in 1894." *Bulletin of the United States Fish Commission* 15 (1896): 253–84.

French, Robert R., and Roy J. Wahle. *Salmon Escapements above Rock Island Dam, 1954–60.* Special Scientific Report: Fisheries, no. 493. Washington, DC: U.S. Department of the Interior, U.S. Fish and Wildlife Service, Bureau of Commercial Fisheries, 1965.

———. *Salmon Runs—Upper Columbia River, 1956–57.* Special Scientific Report: Fisheries, no. 364. Washington, DC: U.S. Department of the Interior, U.S. Fish and Wildlife Service, Bureau of Commercial Fisheries: 1960.

Fulton, Leonard A. *Spawning Areas and Abundance of Steelhead Trout and Coho, Sockeye and Chum Salmon in the Columbia River Basin—Past and Present.* Special Scientific Report: Fisheries, no. 618. Washington, DC: U.S. Department of Commerce, NOAA, NMFS, 1970.

Glavin, Terry. *Protecting the Public Interest in the Conservation of Wild Salmon in British Columbia: A Strategy for the Conservation of Pacific Salmon.* Report for the Sierra Club of British Columbia, January 2003. http://www.sierraclub.ca/national/postings/wild-salmon-report.pdf.

Haas, James B. *Fishery Problems Associated with Brownlee, Oxbow, and Hells Canyon Dams on the Middle Snake River.* Investigational Report 4. Portland, OR: Fish Commission of Oregon, 1965.

Kostow, Kathryn, ed. *Biennial Report of the Status of Wild Fish in Oregon.* Oregon Department of Fish and Wildlife, 1995. http://www.dfw.state.or.us/ODFWhtml/Research&Reports/WildFishRead.html (Web page now unavailable).

Nehlsen, Willa, Jack E. Williams, and James A. Lichatowich. "Pacific Salmon at the Crossroads: Stocks at Risk from California, Oregon, Idaho, and Washington." *Fisheries* 16, no. 2 (1991): 4–21.

Nielson, Reed S. "Part 5: Oregon Streams from the Deschutes River to the Walla Walla River (Area IV)." Specific Scientific Report: Fisheries, no. 38. In *Survey of the Columbia River and Its Tributaries with Special Reference to the Management of Its Fishery Resources.* Washington, DC: U.S. Department of the Interior, U.S. Fish and Wildlife Service, 1950.

Northwest Power Planning Council. "Compilation of Information on Salmon and Steelhead Losses in the Columbia River Basin." Appendix D in *The 1987 Columbia River Basin Fish and Wildlife Program.* Portland, OR: Northwest Power Planning Council, 1996.

Parkhurst, Zell E. "Part 7: Snake River from above the Grande Ronde River through the Payette River (Area VI)." Special Scientific Report: Fisheries, no. 40. In *Survey of the Columbia River and Its Tributaries with Special Reference to the Management of Its Fishery Resources.* Washington, DC: U.S. Department of the Interior, U.S. Fish and Wildlife Service, 1950.

———. "Part 8: Snake River above Payette River to Upper Salmon Falls (Area VII)." Special Scientific Report: Fisheries, no. 57. In *Survey of the Columbia River and Its Tributaries with Special Reference to the Management of Its Fishery Resources.* Washington, DC: U.S. Department of the Interior, U.S. Fish and Wildlife Service, 1950.

Parkhurst, Zell E., Floyd G. Bryant, and Reed S. Nielson. "Part 3." Special Scientific Report: Fisheries, no. 36. In *Survey of the Columbia River and Its Tributaries with Special Reference to the Management of Its Fishery Resources.* Washington, DC: U.S. Department of the Interior, U.S. Fish and Wildlife Service, 1950.

Willis, Raymond A., Robert N. Thompson, James B. Haas, Melvin D. Collins, and Roy E. Sams. *Environmental Survey Report Pertaining to Salmon and Steelhead in Certain Rivers of Eastern Oregon and the Willamette River and Its Tributaries.* Clackamas, OR: Research Division, Fish Commission of Oregon, 1960.

**Risk of Extinction: All Species**

*Peer reviewers:*
*Dan Bottom, Research Fishery Biologist, NOAA Fisheries (National Marine Fisheries Service), Newport, Oregon, United States*

*Jeff Rodgers, Oregon Plan Monitoring Coordinator, Oregon Department of Fish and Wildlife, Corvallis, Oregon, United States*

*Mikhail Skopets, Fish Biologist, Russian Academy of Sciences, Magadan, Russia*

*Alex Wertheimer, NOAA National Marine Fisheries Service, Auke Bay Lab, Juneau, Alaska, United States*


Baker, Timothy T., Alex C. Wertheimer, Robert D. Burkett, Ronald Dunlap, Douglas M. Eggers, Ellen I. Fritts, Anthony J. Gharrett, Rolland A. Holmes, and Richard L. Wilmot. "Status of Pacific Salmon and Steelhead Escapements in Southeast Alaska." *Fisheries* 21, no. 10 (1996): 6–18.

Huntington, Charles, Willa Nehlsen, and Jon Bowers. "A Survey of Healthy Native Stocks of Anadromous Salmonids in the Pacific Northwest and California." *Fisheries* 21, no. 3 (1996): 6–14.

Nehlsen, Willa, Jack E. Williams, and James A. Lichatowich. "Pacific Salmon at a Crossroads: Stocks at Risk from California, Oregon, Idaho, and Washington." *Fisheries* 16, no. 2 (1991): 4–21.

Slaney, Tim L., Kim D. Hyatt, Tom G. Northcote, and Robert J. Fielden. "Status of Anadromous Salmon and Trout in British Columbia and Yukon." *Fisheries* 23, no. 10 (1996): 20–35.


*The following colleagues completed and submitted data forms and/or synthesized expert judgment and literature for the purpose of developing a comprehensive assessment of risk across the Pacific Rim for this atlas.*

Greg Bryant, Fisheries Biologist, NOAA Fisheries, United States

Elena Eronova, Junior Research Scientist, Khabarovsk Branch of TINRO Center, Russia

Kim Hyatt, Stock Assessment Scientist, Department of Fisheries and Oceans, Nanaimo, British Columbia, Canada

Leon Khorevin, Senior Research Scientist, SakhNIRO, Russia

Mikhail Korolev, Deputy Director, Sevvostrybvod, Russia
Kirill Kuzishchin, Biology Faculty, Moscow State University, Russia

Evgenii Muzurov, Chief Federal Inspector, Russian Federation North-West Basin Department for Protection, Conservation and Control of Fishery Resources, Russia

Sergei Putivkin, Researcher, Magadan Branch of TINRO Center, Russia

Dr. Vladimir Radchenko, Director, SakhNIRO, Russia

Jeff Rodgers, Oregon Plan Monitoring Coordinator, Oregon Department of Fish and Wildlife, United States

Aleksandr Rogatnykh, Director (former), Magadan Branch of TINRO Center, Russia

Dorie Roth, GIS Analyst, Ecotrust, Portland, Oregon, United States

Dmitrii S. Pavlov, Member of the Russian Academy of Sciences, Moscow State University, Moscow, Russia

Ksenia Savvaitova, Biology Faculty, Moscow State University, Russia

Anatolii Semenchenko, Senior Research Scientist, TINRO–Center. Russia

Mikhail Skopets, Fish Biologist, Russian Academy of Sciences, Magadan, Russia

Valentina Urnysheva, Lead Ichthyologist (retired), Sevvostrybvod, Russia

Vladimir Volobuev, Senior Research Scientist (now Director), Magadan Branch of TINRO Center

Dr. Aleksandr Zhul'kov (deceased), Senior Research Scientist, SakhNIRO, Russia

Dr. Sergei Zolotukhin, Director, Salmon Laboratory, Khabarovsk Branch of TINRO Center, Russia

**Risk Peer Reviewers and Data Contributors**

*The following individuals were consulted for peer review and contributed supplemental data in establishing risk for species across the Pacific Rim.*

Dr. Robert Behnke, Professor Emeritus, Colorado State University, United States

Greg Bryant, Fisheries Biologist, NOAA Fisheries, United States

Jinping Chen, Doctoral Student, Institute of Zoology, Chinese Academy of Sciences, China

Igor' A. Chereshnev, Institute of Biological Problems of the North, Russian Academy of Science, Far East Branch, Russia

Tim Haverland, GIS Analyst/Programmer, Commercial Fish Division, Alaska Department of Fish and Game, United States

Kim Hyatt, Stock Assessment Scientist, Department of Fisheries and Oceans, Canada

Sandy Johnston, Head of Stock Assessment and Fisheries Management, Yukon/Transboundary Region, Department of Fisheries and Oceans, Canada

Masahide Kaeriyama, Professor, Department of Marine Sciences and Technology, Hokkaido Tokai University, Japan.

Robert Kope, Research Fishery Biologist, Northwest Fisheries Science Center, NOAA Fisheries, United States

Mikhail Korolev, Deputy Director, Sevvostrybvod, Russia

Kirill Kuzishchin, Biology Faculty, Moscow State University, Russia

Paul McElhaney, NOAA Fisheries, United States

Peter Moyle, Professor of Fish Biology, Department of Wildlife, Fish, and Conservation Biology, University of California, Davis, United States

Evgenii Muzurov, Chief Federal Inspector, Russian Federation North-West Basin Department for Protection, Conservation and Control of Fishery Resources, Russia

Mitsuhiro Nagata, Research Scientist, Hokkaido Fish Hatchery, Japan

Yutaka Okamoto, Director, Okamoto International Affairs Research Institute, Japan

Dmitrii S. Pavlov, Member of the Russian Academy of Sciences, Moscow State University, Russia

Aleksandr Rogatnykh, Director (former), Magadan Branch of TINRO Center, Russia

Ksenia Savvaitova, Biology Faculty, Moscow State University, Russia

Jim Seeb, Program Director/Scientist, Genetics Laboratory, Alaska Department of Fish and Game, United States

Anatolii Semenchenko, Senior Research Scientist, TINRO–Centre, Russia

Ki-Baek Seong, Yangyang Inland Fisheries Research Institute, National Fisheries Research and Development Institute, South Korea

Dr. Alex Wertheimer, NOAA National Marine Fisheries Service, Auke Bay Lab, United States

Dr. Aleksandr Zhul'kov (deceased), Senior Research Scientist, SakhNIRO, Russia

Dr. Sergei Zolotukhin, Director, Salmon Laboratory, Khabarovsk Branch of TINRO Center, Russia

**CHAPTER 5**


Amano, Reiko. "Construction Dams in Japan with a Special Focus on the Nagara River." Submission to the Fourth Regional Consultation of the World Commission on Dams in Hanoi, Vietnam, February 26–27, 2000. http://nagara.ktroad.ne.jp/english/wcdsubf.html.

American Fisheries Society. "AFS Policy Statement #3: Nonpoint Source Pollution (Full Text)." Bethesda, MD: American Fisheries Society, 2004. http://www.fisheries.org/html/Public_Affairs/Policy_Statements/ps_3.shtml.

Antosh, Nelson. "New Drilling Rig in Tundra Faces Chilling Challenges." *Houston Chronicle*, February 22, 2003.

Ashton, Linda. "Scientists Seek New Energy Sources." *Los Angeles Times*, November 18, 2001. http://www.latimes.com/news/science/wire/sns-ap-exp-frozen-gas1118nov18.story.

Atkinson, Clinton E. "Salmon Fisheries of the Soviet Far East." Master's thesis, University of Washington, 1964.

Augerot, Xanthippe. "An Environmental History of the Salmon Management Philosophies of the North Pacific: Japan, Russia, Canada, Alaska and the Pacific Northwest United States." PhD diss., Oregon State University, 2000.

Aurora Research Institute. Arctic Salmon Workshop Executive Summary. Inuvik, Canada: 2000.

Bahls, Peter. *How Healthy Are Healthy Stocks? Case Studies of Three Salmon and Steelhead Stocks in Oregon and Washington, Including Population Status, Threats, and Monitoring Recommendations*. Report for the Native Fish Society. Portland, OR: David Evans and Associates, Inc., 2000.

Baker, Timothy T., Alex C. Wertheimer, Robert D. Burkett, Ronald Dunlap, Douglas M. Eggers, Ellen I. Fritts, Anthony J. Gharrett, Rolland A. Holmes, and Richard L. Wilmot. "Status of Pacific Salmon and Steelhead Escapements in Southeastern Alaska. *Fisheries* 21 (1996): 6–18.

Behnke, Robert J. *Native Trout of Western North America*. Bethesda, MD: American Fisheries Society, 1992.

———. *Trout and Salmon of North America*. New York: The Free Press, a Chanticleer Press Edition, a Division of Simon and Schuster, Inc., 2002.

Bonneville Power Administration. "The Geologic Story of the Columbia Basin." *The Power of the Columbia*. http://www.bpa.gov/power/pl/columbia/4-geology.htm.

Bransten, Jeremy. "Russia: Virgin Forests Under Threat." *Radio-Free Europe/Radio Liberty*. April 4, 2002. http://www.rferl.org/features/2002/04/04042002085100.asp.

British Columbia Offshore Oil and Gas Team. "Offshore Oil and Gas in BC: A Chronology of Activity." http://www.offshoreoilandgas.gov.bc.ca/offshore-oil-and-gas-in-bc/chronology.htm.

———. "Offshore Oil and Gas in BC: Frequently Asked Questions." http://www.offshoreoilandgas.gov.bc.ca/faq/.

Brooke, James. "Enviro-PacRim: Let a Hundred Russian Kilowatts Bloom." *New York Times*, March 23, 2004, World Business.

Buell, Montgomery. "Waves of Change: Fishermen, Managers, and Ecology in the Bristol Bay Salmon Fishery, 1945–1980." PhD diss., Purdue University, 2002.

Buklis, Lawrence S. "A Description of Economic Changes in Commercial Salmon Fisheries in a Region of Mixed Subsistence and Market Economies." *Arctic* 52, no. 1 (March 1999): 40–48.

"Cascadia Scorecard: Seven Key Trends Shaping the Northwest." Seattle: Northwest Environment Watch, 2004. Chapter 6, "Forests," 57-61.

Cederholm, C. Jeff, Matt D. Kunze, Takeshi Murota, and Atuhiro Sibatani. "Pacific Salmon Carcasses: Essential Contributions of Nutrients and Energy for Aquatic and Terrestrial Ecosystems." *Fisheries* 24, no. 10 (October 1999): 6–15.

Center for Columbia River History: A Regional Partnership. "Dams of the Columbia Basin and Their Effects on the Native Fishery." http://www.ccrh.org/comm/river/dams.htm.

Coast Range Association. "Salmon and Forests: A Report on the Salmon River Watershed." http://www.coastrange.org/salmon.htm.

Cooke, Steven J., Scott G. Hinch, Anthony P. Farrell, Michael F. Lapointe, Simon R. M. Jones, J. Stevenson Macdonald, David A. Patterson, Michael C. Healey, and Glen van der Kraak. "Abnormal Migration Timing and High en Route Mortality of Sockeye Salmon in the Fraser River, British Columbia." *Fisheries* 29, no. 2 (2004): 22–33.

Earthjustice. "Fishermen's Lawsuit Seeks Water for Salmon in Klamath Basin." *Earthjustice: News Room*, April 24, 2002. http://earthjustice.org/news/display.html?ID=358.

Earthworks and Oxfam America. "Dirty Metals: Mining, Communities and the Environment." Report, 2004. http://www.nodirtygold.org/dirty_metals_report.cfm.

"Fingerprints of Global Warming on Wild Animals and Plants: A Globally Coherent Fingerprint of Climate Change Impacts Across Natural Systems." Nature 421:37–42; 57–60. January 2003.

Finney, Bruce. P., Irene Gregory-Eaves, Jon Sweetman, Marianne S.V. Douglas, and John P. Smol. "Impacts of Climate Change and Fishing on Pacific Salmon Abundance over the Past 300 Years." *Science* 290 (October 27, 2000): 795–99. http://www.sciencemag.org/.

"Flooding the Land, Warming the Earth: Greenhouse Gas Emissions from Dams." Berkeley, CA: International Rivers Network, June 2002.

Friends of the Earth Japan and International Rivers Network. "No More Dam Illusions: The Growing Success of Dam Opponents in Japan." In *Dammed River, Damned Lies: What the Water Establishment Doesn't Want You to Know*. Briefing kit created for Third World Water Forum, Kyoto, Japan, March 2003. http://www.irn.org/.

Fukuwaka, Masa-aki, and Toshiya Suzuki. "Density-dependence of Chum Salmon in Coastal Waters of the Japan Sea." *North Pacific Anadromous Fish Commission Bulletin* 2 (2000): 75–82.

Gall, G. A. E., and P. A. Crandell. "The Rainbow Trout." *Aquaculture* 100 (1992): 1–10. Quoted in R. Froese and D. Pauly, eds., "*Oncorhynchus mykiss*." *FishBase*. http://fishbase.org/Summary/SpeciesSummary.cfm?ID=239&genusname=Oncorhynchus&speciesname=mykiss.

Gilis, V. A. (Sovet narodnykh deputatov Ust-kamchatskii District, Kamchatka, Russia). Letter to Kamchatrybvod regarding proposed Elovka zakaznik, July 17, 2000.

Glumov, Ivan F., and R. R. Murzin. "The Strategy of the Mineral Resource Use on the Continental Shelf of Russia: The Near-term Outlook." Paper. Ministry of Natural Resources of Russia, June 2002.

Government of British Columbia, Ministry of Water, Land, and Air Protection. *Indicators of Climate Change for British Columbia, 2002*. http://wlapwww.gov.bc.ca/air/climate/indicat/index.html.

Gresh, Ted, Jim Lichatowich, and Peter Schoonmaker. "An Estimation of Historic and Current Levels of Salmon Production in the Northeast Pacific Ecosystem: Evidence of a Nutrient Deficit in the Freshwater Systems of the Pacific Northwest." *Fisheries*, 25, no. 1 (January 2000): 15–21.

Hartman, G. F., J. C. Scrivener, and M. J. Miles. "Impacts of Logging in Carnation Creek, a High-energy Coastal Stream in British Columbia, and Their Implication for Restoring Fish Habitat." *Canadian Journal of Fisheries and Aquatic Sciences* 53, suppl. no. 1 (1996): 237–51.

Hayhoe, Katharine, Daniel Cayan, Christopher B. Field, Peter C. Frumhoff, Edwin P. Maurer, Norman L. Miller, Susanne C. Moser, et al. "Emissions Pathways, Climate Change, and Impacts on California." *Proceedings of the National Academy of Sciences of the United States of America* (PNAS) 101, no. 34 (August 24, 2004): 12422–27. http://www.pnas.org/cgi/content/full/101/34/12422?maxtoshow=&HITS=10&hits=10&RESULTFORMAT=1&andorexacttitle=and&andorexacttitleabs=and&andorexactfulltext=and&searchid=1095366449615_6412&stored_search=&FIRSTINDEX=0&sortspec=relevance&volume=101&firstpage=12422.

Healey, Michael C. "Life History of Chinook Salmon (*Oncorhynchus tshawytscha*)." In *Pacific Salmon Life Histories*, edited by Cornelius Groot and Leo Margolis, 313–93. Vancouver: University of British Columbia Press, 1991.

Huntington, Charles, Willa Nehlsen, and Jon Bowers. "A Survey of Healthy Native Stocks of Anadromous Salmonids in the Pacific Northwest and California." *Fisheries* 21, no. 3 (1996): 6–14.

ICOLD (International Commission on Large Dams). *World Register of Dams 1998*. Paris: ICOLD, 1998.

Imhof, Aviva. "Dams Contribute Significantly to Climate Change, WCD Finds." Special issue: *World Commission on Dams, World Rivers Review* 15, no. 6 (Dec. 2000): 16.

Institute for Fisheries Resources. *Fishlink Sublegals* 4, no. 17 (2001).

International Rivers Network. "Crisis on the Klamath." "Warming the Earth: Hydropower Threatens Efforts to Curb Climate Change." http://www.irn.org/.

Kaeriyama, Masahide, and Hirhosi Mayama. "Rehabilitation of Wild Chum Salmon Population in Japan." *Hokkaido Salmon Hatchery Technical Report* 165 (1996): 41–52.

Kanto, Ichiro, and Osamu Ishikawa. "Trends of Japanese Markets." *Newsletter of the North Pacific Anadromous Fish Commission* 3, no. 2 (Summer 1999).

Knapp, Gunnar P. "Alaska Salmon Ranching: An Economic Review of the Alaska Salmon Hatchery Programme." Chap. 37 in *Stock Enhancement and Sea Ranching*, edited by Bari R. Howell, Erlend Moksness, and Terje Svasand. Oxford, England: Blackwell Science, Ltd., 1999.

———. "Implications of Aquaculture for Wild Fisheries: The Case of Alaska Wild Salmon." In *International Institute of Fisheries Economics and Trade*. Wellington, New Zealand: August 22, 2002.

———. "The Wild Salmon Industry: Five Predictions for the Future." *Fisheries Economics Newsletter*, no. 51 (May 2001): 1.

Knudsen, E. Eric. "Managing Pacific Salmon Escapements: The Gaps Between Theory and Reality." In *Sustainable Fisheries Management: Pacific Salmon*, edited by E. Eric Knudsen, Cleveland R. Steward, Donald D. MacDonald, Jack E. Williams, and Dudley W. Reiser, 237–61. Boca Raton: Lewis Publishers, 2000.

Kommersant Publishing House. "Amur Region: General Information." *Kommersant: Russia's First Independent Media Holding Online*. http://www.kommersant.com/tree.asp?rubric=5&node=377&doc_id=-34.

Korolev, Mikhail R. "Vybor i analiz proektnykh rechnykh system dlia proekta PROON-GEF 'Sokhranenie bioraznoobraziia lososevykh Kamchatki i ikh ustoichivoe ispol'zovanie'" [Selection and Analysis of the Project River Systems for UNDP/GEF Project: The Conservation and Sustainable Use of Wild Salmonid Diversity in Kamchatka]. In *Sokhranenie bioraznoobraziia Kamchatki i prilegaiushchikh morei, Materialy regional'noi konferentsii 11–12 aprelia 2000 g.g. Petropavlovsk-Kamchatskii*. [Conservation of Biodiversity of Kamchatka and Coastal Waters, Materials of Regional Scientific Conference Petropavlovsk-Kamchatsky, April 11–12, 2000], 70–72. Petropavlovsk-Kamchatsky, Kamchatksii Institut Ekologii i prirodopol'zovaniia DVO RAN and Kamchatskaia Liga Nezavisimykh Ekspertov, Russia: 2000.

Kostow, Kathryn. "The Status of Salmon and Steelhead in Oregon." In *Pacific Salmon and Their Ecosystems: Status and Future Options*, edited by Deanna J. Stouder, Peter A. Bisson, and Robert J. Naiman, 145–79. New York: Chapman and Hall, 1995.

Kuzishchin, Kirill V., Sergei V. Maksimov, V. Y. Upriamov, V. K. Larin, Natal'ia V. Varnaskaia, and Guido R. Rahr. "K probleme ustoichivogo ispol'zovaniia rybnykh resursov Zapadnoi Kamchatki: opredelenie rechnykh basseinov, prioritetnykh dlia sokhraneniia bioroznobraziia lososevykh ryb." [Regarding Sustainable Use of Western Kamchatka Fish Resource: Identification of Priority River Basins for Salmonid Biodiversity Conservation.] In *Doklady vtoroi Kamchatskoi Oblastnoi nauchno-prakticheskoi konferentsii, Problemy okhrany i ratsional'nogo ispol'zovaniia bioresursov Kamchatki* [Papers of the Second Regional Scientific-practical Conference, Issues of Protection and Rational Use of Kamchatka Bioresources], 35–40. Petropavlovsk-Kamchatsky, Kamchatrybvod and Kamchatsky Pechatny Dvor, Russia, October 3–6, 2000.

Lichatowich, James A. *Salmon Without Rivers: A History of the Pacific Salmon Crisis*. Washington, DC: Island Press, 1999.

Living Oceans Society. "Frequently Asked Questions." http://www.livingoceans.org/oilgas/faq.shtml; "About the Moratorium in BC." http://www.livingoceans.org/oilgas/index.shtml; "Environmental Impacts of Offshore Oil and Gas." http://www.livingoceans.org/oilgas/impacts.shtml.

Malik, Lila N., Nikolai I. Koronkevich, Irina S. Zaitseva, and Elena A. Barabanova. "Briefing Paper on Development of Dams in the Russian Federation and Other NIS Countries." Case study prepared as an input to the World Commission on Dams, Cape Town, 2000. http://www.dams.org/kbase/studies/ru/.

Mantua, Nathan. "The Pacific Decadal Oscillation." In *Encyclopedia of Global Environmental Change*, edited by R. E. Munn, Michael C. MacCracken, and John S. Perry. Vol. 1. New York: John Wiley and Sons, Inc., 2002.

Mantua, Nathan, and Robert C. Francis. "Natural Climate Insurance for Pacific Northwest Salmon and Salmon Fisheries: Finding Our Way through the Entangled Bank." In *Fish in Our Future? Perspectives on Fisheries Sustainability*, edited by Donald D. MacDonald and E. Eric Knudsen. Bethesda, MD: American Fisheries Society, forthcoming.

Mantua, Nathan, Robert C. Francis, David Fluharty, and Philip W. Mote. "Climate and the Pacific Northwest Salmon Crisis: A Case of Discordant Harmony." In *Rhythms of Change: Climate Impacts on the Pacific Northwest*, edited by Edward Miles and Amy K. Snover. Cambridge, MA: MIT Press, forthcoming.

Maksimov, Sergei V., and Xanthippe Augerot. "Problemy i perspektivy sokhraneniia vodnykh bioresursov Kamchatki" [Problems and Prospects for Kamchatka Aquatic Bio-resources Conservation.] In *Tezisy vtoroi nauchno-prakticheskoi konferentsii, Problemy okhrany i ratsional'nogo ispol'zovaniia bioresursov Kamchatki* [Abstracts of the Second Regional Scientific-practical Conference, Issues of Protection and Rational Use of Kamchatka Bioresources], 71–72. Petropavlovsk-Kamchatsky, Kamchatrybvod, Russia, October 3–6, 2000.

McCully, Patrick. "Large Dams Fail WCD 'Core Values.'" Special issue: *World Commission on Dams, World Rivers Review* 15, no. 6 (Dec. 2000): 3.

McGinn, Anne Platt. "Blue Revolution: The Promises and Pitfalls of Fish Farming." *World Watch* (March/April 1998): 10–19.

McKee, Jeffrey K., Paul W. Sciulli, C. David Fooce, and Thomas A. Waite. "Forecasting Global Biodiversity Threats Associated with Human Population Growth." *Biological Conservation* 115, no. 1 (2004): 161–64.

Morrison, John, Michael C. Quick, and Michael G. G. Foreman. "Climate Change in the Fraser River Watershed: Flow and Temperature Projections." *Journal of Hydrology* 263 (2002): 230–44.

Myers, K. W., R. V. Walker, et al. *Migrations and Abundance of Salmonids in the North Pacific, 2000*. Seattle: High Seas Salmon Research Program, University of Washington, School of Aquatic and Fishery Sciences, 2000.

Natural Resources Defense Council and Defenders of Wildlife. *Effects of Global Warming on Trout and Salmon in U.S. Streams*. May 2002.

Nehlsen, Willa, Jack E. Williams, and James A. Lichatowich. "Pacific Salmon at the Crossroads: Stocks at Risk from California, Oregon, Idaho, and Washington." *Fisheries* 16, no. 2 (1991): 4–21.

Neilson, Ronald P., B. Smith, and I. C. Prentice. "Simulated Changes in Vegetation Distribution under Global Warming." In *The Regional Impacts of Climate Change: An Assessment of Vulnerability*, edited by R. T. Watson, M. C. Zinyowera, R. H. Moss, and David J. Dokken, 439–56. Cambridge, England: Cambridge University Press, 1998.

"New UNEP Study on Forests: Russia and Mexico Provide Least Protection for Their Forests." *Gallon Environment Letter* 6, no. 8 (April 16, 2002).

Northwestern Division Pacific Salmon Coordination Office. "Columbia River Basin—Dams and Salmon." U.S. Army Corps of Engineers, Northwest Division. http://www.nwd.usace.army.mil/ps/colrvbsn.htm.

"Official Warns of Poor Condition of Russian Dams." *Radio-Free Europe/Radio Liberty Newsline* 8, no. 142 (July 28, 2004). http://www.rferl.org/newsline/2004/07/280704.asp.

Offshore-Environment. "Interesting Facts about Oil, Gas and Ocean Environment." http://www.offshore-environment.com/facts.html.

Organization of the Petroleum Exporting Countries. "FAQs: Questions about the Petroleum Industry." http://www.opec.org/.

Oregon Natural Resources Council. "The Klamath Basin-Rogue Basin Water Transfer." http://www.onrc.org/programs/klamath/watertransf.html.

Ortolano, Len, Katherine Kao Cushing, et al. "Grand Coulee Dam and the Columbia Basin Project, USA." Case study report prepared as an input to the World Commission on Dams, Cape Town, 2000. http://www.dams.org/kbase/studies/us/.

Page, L. M., and B. M. Burr. *A Field Guide to Freshwater Fishes of North America North of Mexico*. Boston: Houghton Mifflin Company, 1991. Quoted in R. Froese and D. Pauly, eds., "*Oncorhynchus gorbuscha*." *FishBase*. http://fishbase.org/Summary/SpeciesSummary.cfm?ID=240&genusname=Oncorhynchus&speciesname=gorbuscha; "*Oncorhynchus keta*." http://fishbase.org/Summary/SpeciesSummary.cfm?ID=241&genusname=Oncorhynchus&speciesname=keta; "*Oncorhynchus kisutch*." http://fishbase.org/Summary/SpeciesSummary.cfm?ID=245&genusname=Oncorhynchus&speciesname=kisutch; "*Oncorhynchus nerka*." http://fishbase.org/Summary/SpeciesSummary.cfm?ID=243&genusname=Oncorhynchus&speciesname=nerka; "*Oncorhynchus tshawytscha*." http://fishbase.org/Summary/SpeciesSummary.cfm?ID=244&genusname=Oncorhynchus&speciesname=tshawytscha.

Population Reference Bureau. *2003 World Population Data Sheet of the Population Reference Bureau*. http://www.prb.org/.

———. "Using Global Population Projections." In *2003 World Population Data Sheet Highlights*. http://www.prb.org/.

Pottinger, Lori. "WCD Report Confirms Social, Economic, Environmental Harms from Dams." Special issue: *World Commission on Dams, World Rivers Review* 15, no. 6 (Dec. 2000): 1.

Province of British Columbia. "2001 Census Profile: British Columbia." *BC Stats*. http://www.bcstats.gov.bc.ca/data/cen01/profiles/csd_txt.htm.

Rogers, Paul. "Salmon Kill Blamed on Water Sent to Farmers." *San Jose Mercury News*, January 5, 2003. http://www.mercurynews.com/mld/mercurynews/news/4878385.htm?1c.

*Russia Journal Daily*. "First Power Unit of Bureya Plant Launched." Russian News Ticker, July 1, 2003. http://www.russiajournal.com/.

———. "UES to Invest 7.8 Bln Rbl in Bureya Plant." Russian News Ticker, June 12, 2003. http://www.russiajournal.com/.

Salo, Ernest O. "Life History of Chum Salmon (*Oncorhynchus keta*)." In *Pacific Salmon Life Histories*, edited by Cornelius Groot and Leo Margolis, 231–309. Vancouver: University of British Columbia Press, 1991.

Seong, Ki-Baek (Yangyang Inland Fisheries Research Institute, National Fisheries Research and Development Institute). Personal communication, 2000.

Shashkov, Alexander, and Olga Fronina. "Poaching Puts Russia into Annual Loss of 15 Bln Dlrs." *Russian Far East News*, 2004.

Slaney, Tim L., Kim D. Hyatt, Thomas G. Northcote, and Rob J. Fielden. "Status of Anadromous Salmon and Trout in British Columbia and Yukon." *Fisheries* 21 (1996): 20–35.

Speer, Lisa (Natural Resources Defense Council). Personal communication, December 2004.

State of Alaska, Department of Military and Veterans Affairs, Division of Emergency Services, Office of Public Affairs. "Alaska Honors National Dam Safety Day, May 31." News release, May 26, 2000. http://www.ak-prepared.com/pr24.htm.

Stockner, J. G., E. Rydin, and P. Hyenstrand. "Cultural Oligotrophication." *Fisheries* 25, no. 5 (May 2000): 7–14.

Takeuchi, Kuniyoshi, and Joji Harada. "Operation, Monitoring and Rehabilitation of Dams/Reservoirs in Japan: Institutional Framework and Empirical Studies." Contributing paper to the World Commission on Dams, 1998. In *Dams and Development: A New Framework for Decision-making: The Report of the World Commission on Dams*. London: Earthscan, 2000.

Taylor, J. E., III. "Making Salmon: The Political Economy of Fishery Science and the Road Not Taken." *Journal of the History of Biology* 31 (1998): 33–59.

Titus, Robert G., Don C. Erman, and William M. Snider. *History and Status of Steelhead in California Coastal Drainages South of San Francisco Bay*. Berkeley: California Department of Fish and Game, University of California, Berkeley, 2000.

United Nations Environment Programme, Division of Technology, Industry, and Economics. International Environmental Technology Centre. "Planning and Management of Lakes and Reservoirs: An Integrated Approach to Eutrophication." *Newsletter and Technical Publications*, December 1999. http://www.unep.or.jp/ietc/publications/techpublications/TechPub-11/index.asp.

U.S. Army Corps of Engineers, U.S. Department of the Interior, U.S. Department of Agriculture, National Oceanic and Atmospheric Administration. "Fiscal Year 2001–2004 Funding for Columbia and Snake River and Coastal Salmon Recovery," August 22, 2003. http://www.salmonrecovery.gov/CEQ_08-22-2003/revisedsalmonbudgetfactsheetFINAL.pdf.

U.S. Census Bureau. "California QuickFacts." *State and County QuickFacts*. http://quickfacts.census.gov/qfd/states/06000.html.

U.S. Department of the Interior, Bureau of Reclamation. "The History of Hydropower Development in the United States." "The Role of Hydropower Development in the U.S. Energy Equation." *Reclamation: Managing Water in the West*. http://www.usbr.gov/.

U.S. Department of the Interior, Mineral Management Service, 2000. Gulf of Mexico OCS Oil and Gas Lease Sale 181, Draft Environmental Impact Statement (DEIS), p. IV-50.

U.S. Department of Labor Occupational Safety and Health Administration. "Oil and Gas Well Drilling and Servicing eTool." http://www.osha.gov/SLTC/etools/oilandgas/index.html.

U.S. Environmental Protection Agency. "EPA Global Warming: Impacts-Coastal Fisheries." April 6, 2001. http://yosemite.epa.gov/oar/globalwarming.nsf/content/ImpactsFisheries.html.

U.S. Environmental Protection Agency, Toxics Release Inventory (TRI) Program. *TRI On-site and Off-site Disposal or Other Releases, by State, 2002*. Data table. http://www.epa.gov/tri/tridata/tri02/press/rel-state-2002.pdf.

———. *TRI On-site and Off-site Disposal or Other Releases, 2002: Metal Mining (SIC code 10)*. Data table. http://www.epa.gov/tri/tridata/tri02/press/Rel-stateSIC10-2002.

U.S. Fish and Wildlife Service. "Potential Impacts of Proposed Oil and Gas Development on the Arctic Refuge's Coastal Plain: Historical Overview and Issues of Concern." *Arctic National Wildlife Refuge*. http://arctic.fws.gov/issues1.htm#section4.

U.S. National Archives and Records Administration. "Research Room: Records of the Bonneville Power Administration [BPA]." http://www.archives.gov/research_room/federal_records_guide/bonneville_power_administration_rg305.html.

Vendiola, Shelly (Indigenous Environmental Network, Indigenous Women's Network). Personal communication, November 25, 2003.

Watson, R. T., M. C. Zinyowera, R. H. Moss, R. E. Basher, M. Beniston, O. F. Canziani, et al. Summary for Policymakers—The Regional Impacts of Climate Change: An Assessment of Vulnerability. Intergovernmental Panel on Climate Change, 1998.

Woody, Elizabeth, Jim Lichatowich, Richard Manning, Freeman L. House, and Seth Zuckerman. *Salmon Nation: People, Fish, and Our Common Home*. Edited by Edward C. Wolf and Seth Zuckerman. Portland, OR: Ecotrust, 2003.

World Bank Group. "Industry Sector Guidelines: Oil and Gas Development (Onshore)." In *Pollution Prevention and Abatement Handbook 1998: Toward Cleaner Production*, 359–62. World Bank Group, 1999. http://lnweb18.worldbank.org/ESSD/envext.nsf/51ByDocName/PollutionPreventionandAbatementHandbook.

World Commission on Dams. "Briefing Paper: Russian Federation and Other NIS Countries: Final Paper—Executive Summary." Case Studies, November 2000. http://www.dams.org/kbase/studies/ru/ru_exec.htm.

———. "Case Study: USA: Grand Coulee Dam and Columbia River Basin: Final Paper—Executive Summary. November 2000. http://www.dams.org/kbase/studies/us/us_exec.htm.

———. "Dams and Development: A New Framework for Decision-making." "Fact Sheet for Part I: The Global Review." "Fact Sheet for Part II: The Way Forward, Frequently Asked Questions." In *The Report of the World Commission on Dams*. CD-ROM. London: Earthscan Publications, Ltd., November 16, 2000.

World Wildlife Fund. "Farewell to Bureya." *WWF Russia Bulletin*, Spring 2003.

Yoshiyama, Ronald M., Eric R. Gerstung, Frank W. Fisher, and Peter B. Moyle. "Chinook Salmon in the California Central Valley: An Assessment." *Fisheries* 25, no. 2 (February 2000): 6–20.

Zinke Environment Consulting for Central and Eastern Europe. "Case Study: The Nagara River." http://www.zinke.at/index.html.

## CHAPTER 5 MAPS

### Human Population Density

*Peer reviewers:*
*Barry Edmonston, Director, Population Research Center, Portland State University, Portland, Oregon, United States*

*Irina Sharkova, Research Assistant Professor, Population Research Center and School of Urban Studies and Planning, Portland State University, Portland, Oregon, United States*

LandScan Global Population Database. Oak Ridge, TN: Oak Ridge National Laboratory. http://www.ornl.gov/gist/2002. The LandScan data set is a worldwide population database compiled on a 30" x 30" latitude/longitude grid. Census counts (at subnational level) were apportioned to each grid cell based on likelihood coefficients of proximity to roads, slope, land cover, nighttime lights, and other data sets. LandScan has been developed as part of the Oak Ridge National Laboratory (ORNL).

Landsat Satellite Image (Russia and China). Landsat 7 ETM+ Imagery, August 12, 1999, WRS Path/Row 113/27. Provided by Tom Stone, Senior Research Associate, Woods Hole Research Center. Woods Hole, MA, 2003.

### Logging in Frontier Forests

Bryant, Dirk, Daniel Nielsen, and Laura Tangley. *Last Frontier Forests: Ecosystems and Economies on the Edge*. Washington, DC: World Resources Institute, 1997.

Olson, David M., Eric Dinerstein, Eric D. Wikramanayake, Neil D. Burgess, George V. N. Powell, Emma C. Underwood, Jennifer A. D'Amico, et al. "Terrestrial Ecoregions of the World: A New Map of Life on Earth." *BioScience* 51, no. 11 (2001): 933–38.

### Mineral Development

*Peer reviewers:*
*Josh Newell, PhD Candidate, Department of Geography, University of Washington, Seattle, Washington, United States*

*Paul Robinson, Research Director, Southwest Research and Information Center, Albuquerque, New Mexico, United States*

*Alan Septoff, Mineral Policy Center's Reform Campaign Director, Washington, DC, United States*

Environmental Mining Council of British Columbia. Dataset entitled, Producing and Recently Closed Mines and Advanced Exploration Projects. www.miningwatch.org/emcbc/mapping/nw.htm#Download.

Nokleberg, Warren J. , Timothy D. West, Kenneth M. Dawson, Vladimir I. Shpikerman, Thomas K. Bundtzen, Leonid M. Parfenov, James W.H. Monger, Vladimir V. Ratkin, Boris V. Baranov, Stanislauv G. Byalobzhesky, Michael F. Diggles, Roman A. Eremin, Kazuya Fujita, Steven P. Gordey, Mary E. Gorodinskiy, Nikolai A. Goryachev, Tracey D. Feeney, Yuri F. Frolov, Arthur Grantz, Alexander I. Khanchuck, Richard D. Koch, Boris A. Natalin, Lev M. Natapov, Ian O. Norton, William W. Patton, Jr., George Plafker, Anany I. Pozdeev, Ilya S. Rozenblum, David W. Scholl, Sergei D. Sokolov, Gleb M. Sosunov, David B. Stone, Rowland W. Tabor, Nickolai V. Tsukanov, Tracy L. Vallier, 1998. Summary terrane, mineral deposit, and metallogenic belt maps of the Russian Far East, Alaska, and the Canadian Cordillera: U.S. Geological Survey Open-File Report 98-136, 1998. http://wrgis.wr.usgs.gov/open-file/of98-136/

### Sakhalin Pipeline

*Peer reviewers:*
*Richard A. Fineberg, President and Principal Investigator, Research Associates, Ester, Alaska, United States*
*Sergei Makeev, Director, Sakhalin Wild Nature Fund, Aniva, Russia*

Oil field and pipeline data maps. Courtesy of Sakhalin Environment Watch. Provided by Dmitrii Lisitsyn, Director, and Elena Mezhennaia, GIS. Yuzhno Sakhalinsk, Russia, 2003. http://www.sakhalin.environment.ru.

### Dams

Bonneville Power Administration. BPA Hydro Site Database, MGL-911, July 31, 1995. Provided by Steven Bellcoff, Bonneville Power Administration.

Sockeye Salmon Distribution, 1:100,000. StreamNet Project, May 2003. Provided by Travis Butcher, Pacific States Marine Fisheries Commission. ftp://ftp.streamnet.org/pub/streamnet/ASCII_Data/GeneralizedFishDistribution.zip.

"Update on Power Plant Development at Existing Dams on the Columbia River in Canada." Handout from the 4th World Fisheries Congress, Vancouver, BC, May 2–6, 2004.

### Climate Change

*Peer reviewer:*
*Nathan Mantua, University of Washington, Department of Atmospheric Sciences, JISAO: the Joint Institute for the Study of the Atmosphere and Oceans, Seattle, Washington, United States*

Neilson, Ronald P., and D. Marks. "A Global Perspective of Regional Vegetation and Hydrologic Sensitivities from Climatic Change." *Journal of Vegetation Science* 5 (1994): 715–30.

## CHAPTER 6

Augerot, Xanthippe. "An Environmental History of the Salmon Management Philosophies of the North Pacific: Japan, Russia, Canada, Alaska and the Pacific Northwest United States." PhD diss., Oregon State University, 2000.

Brulle, Ramon Vanden, and Nick J. Gayeski. "Overwhelming Evidence: How Hatcheries Are Jeopardizing Salmon Recovery." *Washington Trout Report* 13, no. 1 (Spring 2003): 4–8.

Church, Michael. "The Landscape of the Pacific Northwest." In *Carnation Creek and Queen Charlotte Islands Fish/Forestry Workshop: Applying 20 Years of Coast Research to Management Solutions*, edited by Dan L. Hogan, Peter J. Tschaplinski, and Stephen Chatwin, 13–22. Report. Victoria, BC: British Columbia Ministry of Forests Research Program, 1998.

"Climate Change 2001: Impacts, Adaptation and Vulnerability." In *Contribution of Working Group II to the Third Assessment Report of the Intergovernmental Panel on Climate Change*, edited by James J. McCarthy, Osvaldo F. Canziani, Neil A. Leary, David J. Dokken, and Kasey S. White. New York: Cambridge University Press, 2001.

*Columbia Basin Bulletin*. "10. ODFW Developing State's First Hatchery Research Center," April 2, 2004. http://www.cbbulletin.com/.

Frissell, Christopher A. *A New Strategy for Watershed Restoration and Recovery of Pacific Salmon in the Pacific Northwest*. Eugene, OR: The Pacific Rivers Council, 1993.

Frissell, Christopher A., and Richard K. Nawa. "Incidence and Causes of Physical Failure of Artificial Habitat Structures in Streams of Western Oregon and Washington." *North American Journal of Fisheries Management* 10 (1992): 199–214.

Frissell, Christopher A., William J. Liss, Charles E. Warren, and M. D. Jurly. "A Hierarchical Framework for Stream Habitat Classification: Viewing Streams in a Watershed Context." *Environmental Management* 10 (1986): 199–214.

Glavin, Terry. *A Strategy for the Conservation of Pacific Salmon*. Sierra Club of British Columbia, January, 2003.

Gustafson, R. G., Robin S. Waples, J. M. Myers, G. J. Bryant, O. W. Johnson, and L. A. Weitkamp. "Pacific Salmon Extinctions: Lost Diversity, Populations, and ESUs." NMFS draft paper, 2003.

Huntington, Charles W., Willa Nehlsen, and Jon Bowers. *Healthy Native Stocks of Anadromous Salmonids in the Pacific Northwest and California*. Portland, OR: Oregon Trout, 1994.

Hyatt, Kim D. "Stewardship for Biomass or Biodiversity: A Perennial Issue for Salmon Management in Canada's Pacific Region." *Fisheries* 21, no. 10 (1996): 4–5.

Lichatowich, James A. *Salmon Without Rivers: A History of the Pacific Salmon Crisis*. Washington, DC: Island Press, 1999.

Meffe, Gary K. "Techno-arrogance and Halfway Technologies: Salmon Hatcheries on the Pacific Coast of North America." *Conservation Biology* 6 (1992): 350–54.

National Research Council. *Upstream: Salmon and Society in the Pacific Northwest*. Washington, DC: National Academy Press, 1996.

Nehlsen, Willa, Jack E. Williams, and James Lichatowich. "Pacific Salmon at the Crossroads: Stocks at Risk from California, Oregon, Idaho, and Washington." *Fisheries* 16 (1991): 4–21.

Northwest Power Planning Council. *Inaugural Annual Report of the Columbia Basin Fish and Wildlife Program, 1978–1999*. Council document 2001–2. Portland, OR: February, 2001. http://www.nwcouncil.org/library/2001/2001-2/default.htm.

Phelan, Sean. "A Pacific Rim Approach to Salmon Management: Redefining the Role of Pacific Salmon International Consensus." *Environmental Law* 33 (2003): 247–89.

Platts, W. S., and R. L. Nelson. "Stream Habitat and Fisheries Response to Livestock Grazing and Instream Improvement Structures, Big Creek, Utah." *Journal of Soil and Water Conservation* 40 (1985): 374–79.

Rahr, Guido R., James A. Lichatowich, Raymond Hubley, and Shauna M. Whidden. "Sanctuaries for Native Salmon: A Conservation Strategy for the 21st Century." *Fisheries* 23, no. 4 (1998): 6–7, 36.

Reisenbichler, Reginald R., and Steve P. Rubin. "Genetic Changes from Artificial Propagation of Pacific Salmon Affect the Productivity and Viability of Supplemented Populations." *ICES Journal of Marine Science* 56 (1999): 459–66.

Riddell, B. E. "Spatial Organization of Pacific Salmon: What to Conserve?" In *Genetic Conservation of Salmonid Fishes*, edited by Joseph G. Cloud and Gary H. Thorgaard, 23–41. New York: Plenum Press, 1993.

Sedell, James R., Gordon H. Reeves, F. Richard Hauer, Jack A. Stanford, and C. P. Hawkins. "Role of Refugia in Recovery from Disturbances: Modern Fragmented and Disconnected River Systems. *Environmental Management* 14 (1990): 711–24.

Stanford, Jack A., J. V. Ward, William J. Liss, Christopher Frissell, Richard N. Williams, James Lichatowich, and Charles C. Coutant. "A General Protocol for Restoration of Regulated Rivers." *Regulated Rivers: Research and Management* 12 (1996): 391–413.

Waples, Robin S., R. G. Gustafson, L. A. Weitkamp, J. M. Myers, O. W. Johnson, P. J. Busby, J. J. Hard, et al. "Characterizing Diversity in Salmon from the Pacific Northwest." *Journal of Fish Biology* 59 suppl. A (2001): 1–41.

Woody, Elizabeth, Jim Lichatowich, Richard Manning, Freeman L. House, and Seth Zuckerman. *Salmon Nation: People, Fish, and Our Common Home*. Edited by Edward C. Wolf and Seth Zuckerman. Portland, OR: Ecotrust, 2003.

# Acknowledgments

**The "we" involved in this atlas includes** more than one hundred scientists and other salmonphiles from a half-dozen countries whose expertise and generosity are represented in the preceding pages. Highlights of their involvement are sprinkled on a timeline spanning a decade, and while I can't begin to thank the countless people involved (many are listed here), I hope I can pay tribute to the scope and vision of this international collaborative effort.

The seeds of this book were planted in 1994 with the three-year North Pacific Rim Salmon Project, launched at Oregon State University (OSU) with $250,000 in congressional funding arranged by then-U.S. Senator Mark Hatfield (Oregon). Managed by the USEPA through a cooperative agreement with OSU's Center for Analysis of Environmental Change, the Pacific Rim Project was led by Dan Bottom and Jeff Rodgers, staffed by researchers Cathy Dey, Cathy Baldwin, and me (then a graduate student).

After I completed my coursework at OSU, I transferred my interest in the Pan-Pacific salmon stock status report—which had then evolved into research throughout the Pacific Rim to fill in datasets on this vast land/seascape—to the Wild Salmon Center in Portland, Oregon. To continue the trajectory of our work, it was necessary to build a more diverse, trans-Pacific collective. Teaming up with Oregon Sea Grant, we hosted a workshop in 1998 for our far-flung researchers throughout the Pacific Rim to begin what would become an international collaborative effort to create our ecoregions. With the support of Kim Hyatt, Jeff Rodgers, and Sergey Zolotukhin, I moved ahead with follow-up trips, phone calls, correspondence to gather all the risk of extinction data (represented in chapter 4) during the course of the next three years. Thanks to Guido Rahr at Wild Salmon Center for giving a home to this project out of OSU and to former Ecotrust GIS Analyst Dorie Roth for partnering with me in prior drafts of this manuscript.

That's part one of the story. Part two transpired in March 2003 with the launch of the State of the Salmon Consortium. With generous support from the Gordon and Betty Moore Foundation, State of the Salmon was formed to create a comprehensive, integrated source of information and knowledge on North Pacific salmon. My fellow co-director, Ecotrust vice president of fisheries Edward Backus, is the founder of the visionary GIS team Interrain Pacific and the engine behind the geographic information systems mapping in this book. Wild Salmon Center president Guido Rahr and Ecotrust president Spencer Beebe have been central to the project's development—captains, navigators, helmsmen, and crew all at once. Our State of the Salmon team—Brian Caouette, Ben Donaldson, Dana Foley, Andrew Fuller, Ray Hollander, Cathy Pearson, Pete Rand, Greg Robillard, and Charles Steinback—have each had an essential role in this effort. I'd like to single out Ben, for his analytical assistance and technical support, and Pete, for the ample time and expertise he provided this project.

Three members of our team have devoted themselves to this project for more than a year. Special thanks to Dana, for her tireless enthusiasm and skills for holding the entire project together as it grew and for help in crafting its voice; to Charles, for his masterful technical knowledge of GIS, matched only by his professionalism; and to Andrew, for his fresh and creative design which shaped this science and countless details into a cohesive book.

Jade Chan, our copy editor, manuscript stylist, and paragon of patience, toiled with us for months, and her attention to detail and dedication were admired and appreciated. Editorial dexterity was gracefully provided by Kacy Curtis, who organized map peer review; Sarah Reich, contributor and researcher extraordinaire; and Grace M and Anna Suessbrick, copy editors and proofreaders. Our colleagues at Wild Salmon Center and Ecotrust also deserve thanks for their patience in enduring any number of demands for on-the-spot consultations, facts, opinions, and support—particularly Ian Gill, RJ Kopchak, Craig Jacobson, Dave Martin, John McMillan, Howard Silverman, and Liz Woody. Seth Zuckerman and Ted Wolf, friends and former Ecotrust colleagues, provided invaluable wordsmithing and expertise.

This book was truly a collective venture by all of the salmonphiles who read and reread various iterations of prose, studied evolving forms of our maps, and generously contributed their accumulated knowledge and wisdom. Many of them are listed as peer reviewers in our sourcing section under specific maps, but I'd like to extend particularly special thanks to Mark W. Chilcote, Frank K. Lake, Jan Konigsberg, Peter Tyedmers, and Steve Wise for their contributions to earlier versions of this atlas.

We are also indebted to esteemed linguists Drs. William Goddard and Igor Krupnik of the Smithsonian Institution, who undertook a complex project with spontaneity, passion, and brilliance. The inspiring Carl Safina, President of the Blue Ocean Institute, generously provided us with commentary and encouragement.

Lastly, our dozens of map and text reviewers are among the foremost leaders in our fields, and we owe a debt of gratitude to them all. Each sentence of this book has the momentum of several or many behind it. Among the luminaries cited, I single out the following people—the premier North Pacific salmon and trout specialists in our field—who contributed momentum, content, and guidance: Robert Behnke, Dick Beamish, Stan Gregory, Masahide Kaeriyama, Jim Lichatowich, Peter Moyle, Phil Mundy, Dimitri Pavlov, Bill Pearcy, Evelyn Pinkerton, Ksenia Savvaitova, Jack Stanford, and Robin Waples.

On a personal note, I would like to take this opportunity to thank my family for their support and patience as this second "dissertation" came to life.

Truly a collaborative effort including ten years and scores of committed people, the spirit of this atlas and of Pacific salmon study is crystallized for me in one moment during that first international conference we held back in March 1998 for our Pacific Rim researchers. So much planning had gone into the logistics of getting folks to Portland, and I was awash in minutiae. But when I walked into the conference room and saw perhaps thirty specialists who had been working on our stock status report to date, I was overwhelmed. I scanned the room and saw familiar and unfamiliar faces, from Russia, Japan, and Canada, and from Alaska, Washington, Oregon, and California. Up until that point, we had been working together via e-mail, by fax, and on the phone, usually one-on-one; but now, for the first time, we were together in one room under the umbrella of our common goal. And as we met to define our salmon ecoregions and discuss threats to salmon throughout their range, we all recognized that the moment was magical. We understood that it takes a collective to make progress on behalf of North Pacific salmon. I hope that epiphany has been conveyed in the pages of this book.

Xanthippe Augerot
Wild Salmon Center
Portland, January 2005

## CONTRIBUTORS TO ECOREGIONS

In order to establish salmon ecoregions Levels 1–4, Xanthippe Augerot, Dan Bottom, and Jeff Rodgers convened the North Pacific Salmon Workshop at Oregon State University, Corvallis, Oregon, United States, held May 4–6, 1999. The following colleagues were in attendance, and their input was invaluable:

CANADA: Kim Hyatt, Science Branch, Pacific Biological Station, Canada Department of Fisheries and Oceans; David Welch, Science Branch, Pacific Biological Station, Canada Department of Fisheries and Oceans

JAPAN: Yutaka Okamoto, Okamoto International Affairs Research Institute; Takemi Ichimura, Tokyo Life Science Laboratory

RUSSIA: Igor Chereshnev, Institute of Biological Problems of the North, Far East Branch, Russian Academy of Sciences; Elena Eronova, Khabarovsk Branch TINRO; Mikhail Korolyov, Sevvostrybvod; Evgeny Muzurov, Sevvostrybvod; Ksenia Savvaitova, Biology Faculty, Moscow State University; Anatoly Semenchenko, TINRO-Centre; Alexander Zhulkov, SakhNIRO; Sergey Zolotukhin, Khabarovsk TINRO

ALASKA: Tim Haverland, Commercial Fish Division, Alaska Department of Fish and Game; Alan Springer, University of Alaska–Fairbanks; Ed Weiss, Habitat Division, Alaska Department of Fish and Game

WASHINGTON, OREGON, CALIFORNIA, AND IDAHO: Greg Bryant, NOAA Fisheries; David Gordon, Pacific Environment; Mike Beltz, The Ecology Center; Dorie Roth, Ecotrust; Stanley Gregory, Oregon State University; David Heller, US Forest Service Region 6; David Hulse, University of Oregon; Paul McElhany, NOAA Fisheries; Rich Lincoln, Washington Department of Fish and Wildlife; Willa Nehlsen, US Fish and Wildlife Service; Tom Nickelson, Oregon Department of Fish and Wildlife; Ron Nielsen, Pacific Northwest Research Station, US Forest Service; William Pearcy, Oregon State University; Guido Rahr, Wild Salmon Center.

Additional data contributors not attending the first workshop included: Sergei Putivkin, Magadan TINRO; Valentina Urnysheva, Sevvostrybvod; Alexander Rogatnykh, Magadan TINRO; Vladimir Volobuev, Magadan TINRO; Leon Khorevin, SakhNIRO; Vladimir Radchenko, SakhNIRO. ■

# Index

## P



## Q



## R



## S

## T



## U



## V



## W